Ecological Studies, Vol. 199

Analysis and Synthesis

Edited by

I.T. Baldwin, Jena, Germany
M.M. Caldwell, Logan, USA
G. Heldmaier, Marburg, Germany
R.B. Jackson, Durham, USA
O.L. Lange, Würzburg, Germany
H.A. Mooney, Stanford, USA
E.-D. Schulze, Jena, Germany
U. Sommer, Kiel, Germany

Ecological Studies

Volumes published since 2001 are listed at the end of this book.

John D. Stednick
Editor

Hydrological and Biological Responses to Forest Practices

The Alsea Watershed Study

Foreword by George W. Brown and James T. Krygier

Springer

John D. Stednick
Department of Forest, Rangeland,
and Watershed Stewardship
Colorado State University
Fort Collins, CO 80523
USA
jds@cnr.colostate.edu

ISBN: 978-0-387-94385-5   e-ISBN: 978-0-387-69036-0

Library of Congress Control Number: 2007938054

Printed on acid-free paper.

9 8 7 6 5 4 3 2 1

springer.com

You cannot stay on the summit forever;
you have to come down again. So why
bother in the first place? Just this: What
is above knows what is below, but what
is below does not know what is above.
In climbing, take careful note of the
difficulties along your way; for as you go
up, you can observe them. Coming
down, you will no longer see them, but
you will know they are there if you have
observed them well. There is an art
of finding one's direction in the lower
regions by the memory of what one saw
higher up. When one can no longer see,
one can at least still know.

Rene Dumal, *Mount Analogue*

Foreword

The Alsea Logging and Aquatic Resources Study, commissioned by the Oregon Legislature in 1959, marked the beginning of four decades of research in the Pacific Northwest devoted to understanding the impacts of forest practices on water quality, water quantity, aquatic habitat, and aquatic organism populations. While earlier watershed research examined changes in runoff and erosion from various land uses, this study was the first watershed experiment to focus so heavily on aquatic habitat and organism response to forest practices.

The Alsea Watershed Study, as it came to be known, extended over 15 years with seven years of pretreatment calibration measurements, a year of treatment, and seven years of post-treatment monitoring. The research was a cooperative effort with scientists from Oregon State University, Oregon Department of Fish and Wildlife, the U.S. Geological Survey, and the U.S. Environmental Protection Agency. Cooperating landowners included the Georgia-Pacific Corporation, the U.S. Forest Service, and a local rancher. It was a remarkable 15-year partnership marked by excellent cooperation among the participants and outstanding coordination among the scientists, many of whom participated actively for the entire period.

The Alsea Study was an important landmark in changing forest practices in Oregon and the Pacific Northwest. The study, among other things, demonstrated the importance of maintaining riparian vegetation in protecting water quality and fish habitat during timber harvest operations. This conclusion led directly to regulations in the Oregon Forest Practices Act of 1971 that required leaving riparian vegetation in harvest units. Other such findings and their application will be discussed in the chapters that follow.

The "new" Alsea Watershed Study has likewise made important contributions to our understanding of long-term effects of forest practices. The original study was limited by funding to 15 years and, while a very long commitment for most organizations and legislatures today, still represents a very short period in the life of a forest. Being able to revisit the study watersheds 20 years later provides additional perspectives about watershed response to treatments. It also emphasizes the importance of long-term monitoring when such critical environmental issues as forest practices and aquatic resources are concerned, both to identify problems and to understand where no problems exist. Long-term monitoring also provides an opportunity to better understand and evaluate the range of natural variability in climate, population dynamics, and other factors. For example, the 1960s were in a wet cycle that included major floods while the 1970s and 1980s were much drier and included some significant El niño activity in the Pacific, which affected salmon populations.

This book contains information gleaned from the original Alsea Watershed Study and the "new" study of 20 years later. It also includes analyses and syntheses of other relevant work, much of which originated from questions generated by the original study. Like the original study, this book represents the cooperative effort of many scientists from several disciplines. It comes at a time when mangers and policy makers are searching for ways to restore the runs of salmon and steelhead to rivers and streams of the Pacific Northwest. We hope it will be a valuable contribution in that important effort.

Corvallis, OR   George W. Brown and James T. Krygier [1]
March 1999

---

[1] George W. Brown is Dean Emeritus and the late James T. Krygier was Professor Emeritus, College of Forestry, Oregon State University. Krygier was one of the scientists who began the Alsea Study, served as a co-leader of the research team, helped finish the study in 1973, and is now deceased. Brown began work on the study as a graduate student of Krygier's in 1964 and continued on the scientist team as a faculty member from 1966 to 1973.

Preface

My original idea to reactivate the Alsea streamflow monitoring program was part of a sabbatical leave supported by the National Council for Air and Stream Improvement, Inc. (NCASI) and the Department of Forest Engineering, Oregon State University (OSU). The camaraderie of George Ice and the late Ben Stout (NCASI) in particular made for an enjoyable leave and I want to acknowledge the late Hank Froehlich and Jim Kiser (OSU) for our discussions on what does it all mean. My plan was to finish this book at OSU, but that opportunity was not presented. Nonetheless, the book took longer to complete than I originally thought. The interest by other investigators and new support prompted me to not let the perfect be the enemy of the good. It was my final decisions that went to press, thus, any errors or omissions are my responsibility not those of the authors or reviewers.

This book does not completely cover all the lessons learned. Even with the conveniences of modern science, data loggers fail, water level floats leak, and water quality samples disappear between the field and the lab. Those are the lessons that have to be experienced to be learned, or shared over a beer. This compilation may not recognize all the people who contributed to efforts in the Alsea watersheds but any oversight on my part was not intended.

A note of appreciation goes to James D. Hall at OSU for his willingness to check the details and find the nits, especially in the references. It was special to have one of the original investigators be part of this effort. Thanks to C.A. Troendle for our continued discussions on forest hydrology. Thanks to my family, especially Susan.

Gould, CO   John D. Stednick
October 2005

Contents

Contributors

Paul W. Adams
Department of Forest Engineering
Oregon State University
Corvallis, OR 97331

Charles W. Andrus
Adolfson Associates
Portland, OR 97204

Robert L. Beschta
Department of Forest Engineering
Oregon State University
Corvallis, OR 97331

Peter A. Bisson
USDA Forest Service
Pacific Northwest Research Station
Olympia, WA 98512-9193

Sarah Greene
USDA Forest Service
Forest Science Laboratory
Corvallis, OR 97331

Stanley V. Gregory
Department of Fisheries
and Wildlife
Oregon State University
Corvallis, OR 97331

James D. Hall
Department of Fisheries and
Wildlife
Oregon State University
Corvallis, OR 97331

Anne Hairston-Strang
Maryland Department of Natural
Resources
Annapolis, MD 21401

George G. Ice
National Council for Air and Stream
Improvement, Inc.
Corvallis, OR 97339

William L. Jackson
USDI National Park Service
Water Resources Division
Fort Collins, CO 80525

Timothy J. Kern
ASRC Management Services
Fort Collins, CO 80524

Arthur McKee
Flathead Lake Biological Station
University of Montana
Polson, MT 59860

Thomas Nickelson
Oregon Department of Fish
and Wildlife
Corvallis, OR 97333

John S. Schwartz
Department of Civil and
Environmental Engineering
University of Tennessee
Knoxville, TN 37996

John D. Stednick
Department of Forest, Rangeland,
and Watershed Stewardship
Colorado State University
Fort Collins, CO 80523

Randall C. Wildman
Department of Fisheries and Wildlife
Oregon State University
Corvallis, OR 97331

Chapter 1
The Alsea Watershed Study

James D. Hall and John D. Stednick

After World War II, increasing demand for natural resources in the Pacific Northwest led to potential conflicts concerning their use. In particular, the demand for lumber in housing construction caused an upsurge in the rate of logging. This led to concerns that salmonid resources were being adversely affected by logging practices then in place (McKernan et al. 1950). In a farsighted move, the governor of Oregon established a Natural Resources Committee in state government in the early 1950s, made up of the executives of all state agencies concerned with natural resources. This committee was charged with coordinating the management of resources and resolving conflicts where possible. To that end, committee members began planning for an integrated study of the natural resources in a river basin. In 1954, the governor's committee held a well-attended public meeting in the Alsea River Basin to gather comments and recommendations from concerned residents who had established their own committee structure (Anonymous 1954). Based on that meeting, and on recommendations from the governor's committee, the Oregon legislature passed a bill providing an appropriation of $50,000 to establish a basin study, with an effective date of 1 July 1957. The study was to be administered by the Natural Resources Committee, which selected the Alsea River Basin for analysis. The overall goal of the Alsea Basin Study of Integrated

Several researchers contributed useful information to this chapter, including John Corliss, Bob Phillips, Don Chapman, and the late Jim Krygier. We particularly acknowledge the contributions to the study made by Don Chapman, the original coordinator, who, along with Jim Krygier, developed the conceptual framework of the study. Without their sound beginnings, and Dr. Krygier's careful work throughout, the study could not have been successful. Much of the historical account in this chapter comes from unpublished material in files retained by James Hall, who also holds a substantial portion of the original data. The College of Forestry at Oregon State University has archived most of the hydrologic data.

James D. Hall
Department of Fisheries and Wildlife, Oregon State University, Corvallis, OR 97331
james.hall@oregonstate.edu

Land–Water Management was to learn how to obtain maximum productivity of a river basin for the greatest public good (Chapman et al. 1961).

This was the first long-term study to assess the effects of land-use activities on water and salmonid resources in the temperate coniferous forests of the western United States. Specific objectives of the investigation, as initially conceived were:

- to determine the effects of forestry, mining, industrial, agricultural, munici-pal, fishery, game, recreational, and other practices occurring in the watershed on quantity and quality of water for all uses
- to determine management practices that would avoid or minimize damaging effects of current land and water practices on aquatic resources to determine whether land and water management practices could be modified to increase water resources (biological and physical) in the watershed and to evaluate various modifications of such practices
- to investigate conflicts among various resource interests involved in the basin
- to improve public understanding of the interrelationships among all natural resources.

These were ambitious goals. Given the limited initial appropriation, the committee settled on three primary components for the first two years of work: a water survey, a soil–vegetation survey, and a logging–aquatic resources study. The logging–aquatic resources study was a major emphasis, intended to run for up to 15 years. Later it was referred to as the Alsea Watershed Study (AWS) and it is the focus of this volume. Before describing that study in detail, we provide a brief description of the two other components.

Water Survey

The objective of the water survey was to determine water yield, water quality, and rainfall patterns in the Alsea River Basin. Recording stream gauges were installed on each major tributary of the Alsea River: North Fork Alsea River, South Fork Alsea River, Five Rivers, Fall Creek, and Drift Creek. The main Alsea River was gauged using the existing U.S. Geological Survey (USGS) station at Tidewater. The USGS participated in the stream-gauging program on a matching-funds basis with Oregon State College (now Oregon State University). Streamflow and water temperature records were begun in August 1958. Five to seven years of streamflow data were deemed necessary to define baseline conditions.

Water quality samples were collected at eight stations throughout the river basin several times a year for chemical analysis by the Oregon State Sanitary Authority (now Department of Environmental Quality). These data provided an estimate of seasonal changes in water quality (Chapman et al. 1961).

Soil and Vegetation Survey

The soil and vegetation survey was a cooperative venture financed jointly by the U.S. Department of Agriculture Forest Service (USFS), Oregon State University, U.S. Department of Interior Bureau of Land Management (BLM), and U.S. Department of Agriculture Soil Conservation Service (SCS) (now Natural Resources Conservation Service). The soil and vegetation survey was considered a prime source of information to aid land managers in making resource management decisions. The survey was headed by John Corliss, Department of Soils, Oregon Agricultural Experiment Station, at Oregon State University, who was assisted by C.T. Dyrness. The objectives of the study were to provide:

- a basic inventory of soil and vegetation in the Alsea River Basin
- characterization of soil and vegetation units
- interpretations for use and management of forest soils and vegetation
- new techniques and procedures for making forest soil–vegetation surveys
- baseline data for additional soil and vegetation research in the Alsea River Basin.

The soil–vegetation survey for the Alsea River Basin began in 1958 with a review of the previous efforts of forestland soil survey in western Oregon and Washington and of methods used in California to characterize forest soils and vegetation. The effect of landscape forms on forest management and relation to soil and geologic features was also studied. From this preliminary review, a method was devised to simultaneously delineate broad soil–landscape–vegetation areas on aerial photos and acetate overlays. A contractor provided a new flight of high-resolution aerial photos needed for the complex interpretation. These delineations were then field-checked and revised as necessary. First-season work required more revisions as mappers gained experience with the soil, landscape form, vegetation patterns, and their interrelationships. Combined information on soils, landscape form, and vegetation provided a closer agreement with an index of forest growth productivity than the three factors used alone (Chapman et al. 1961).

This effort was one of the first to combine soil, landscape form, and vegetation in the same survey. It showed the feasibility and desirability of a combined survey, and did so at a cost comparable to that of conventional surveys of agricultural soils alone (Chapman et al. 1961). However, this new approach to understanding the relations among landforms, soils, and vegetation encountered problems when it came time to publish the results. Owing to an interagency agreement, the soil information had to be published by the SCS, but this agency was not able to publish the vegetation data. As a consequence, a comprehensive data report that included both soil and vegetation maps (Corliss and Dyrness 1964) was prepared and distributed in limited numbers to local cooperators but never formally published. However, the complete soil survey data were published (Corliss 1973), and a summary of the soil–vegetation survey was published in a symposium proceedings (Corliss and Dyrness 1965).

Logging–Aquatic Resources Study

The general objective of the largest component of the basin study, now referred to as the Alsea Watershed Study (AWS), was to determine the effects of different logging methods on physical and biotic characteristics of small Oregon coastal streams. Many agencies were to be involved, and principal emphasis was on effects on streamflow, sediment production, fish habitat, and salmonid populations. The study was designed for a 15-year period, which was to include 7 years of pretreatment, 1 year of logging, and 7 years of posttreatment.

Several times during this period, funding became a significant issue and the study continuity was threatened. One of the most significant threats came in 1959. The governor's committee submitted a budget of $226,000 for the second biennium of the program (1959–1961), envisioning significant expansion of existing projects and the addition of others. The Oregon legislature, however, had reservations about the study. They reduced the budget request to $50,000 and transferred administration of the work to the Agricultural Experiment Station at Oregon State College effective 1 July 1959, and directed that the money "be used for the purpose of *completing* [emphasis ours] the Alsea Basin study and for no other purpose." There was general agreement in the governor's committee that the views of the legislature should not be ignored, but the committee was reluctant to abandon the logging study. Fortunately, there were aspects of the basin study that could be completed within the biennium, and other agencies stepped in to support the logging study and allow it to continue to completion.

In particular, the Research Division of the Oregon State Game Commission (now Oregon Department of Fish and Wildlife), already involved in the logging study, made a major commitment to support the field operations. In addition to employing one or two full-time biologists on the project, they maintained a field laboratory on site throughout the study and provided one or two full-time technicians who did routine sampling, kept fish screens running during storm events, and assisted in all aspects of the work. Support of the Game Commission was key to success of the program. The Agricultural Experiment Station allocated money to continue support of the coordinator position and to provide funds to match the USGS contribution to streamflow and sediment analysis. Faculty from the School of Forestry at Oregon State University provided solid leadership throughout the study. Other agencies also maintained or increased their commitment to the study.

The 15-year study cost just over $1 million. In addition to the Governor's Natural Resources Committee, major cooperators in the initial phases of

the logging–aquatic resources study, and some of the key personnel, were:
(1) Oregon State University, School of Forestry (James Krygier, Coordinator;
George Brown) and Agricultural Experiment Station and Department of Fisheries and Wildlife (Donald Chapman, Coordinator, 1957–1963; James Hall,
Coordinator 1963–1973); (2) Oregon State Game Commission, Research Division (Homer Campbell, Research Supervisor; Robert Phillips, Project Leader
1959–1965; Richard Lantz, Project Leader 1966–1973); and (3) U.S. Geological
Survey (Robert Williams, David Harris). Two other groups deserve mention.
The USFS (Siuslaw Forest Supervisors Rex Wakefield and Spencer Moore,
along with a number of district rangers) was immensely helpful, particularly
during the timber sale and harvest, handling details usually reserved for experimental forests. The Georgia-Pacific Corporation made numerous contribu-
tions, financial as well as the loan of equipment, and further helped by
coordinating their harvest schedule on Needle Branch and lower Deer Creek.
Other cooperators included the Departments of Entomology and Botany at
Oregon State University; the U.S. Public Health Service (later Federal Water
Quality Administration, then Environmental Protection Agency); Oregon State
Sanitary Authority; U.S. Bureau of Commercial Fisheries through the Fish
Commission of Oregon; and Fred Williamson, a private landowner on Needle
Branch.

Three small watersheds in the Alsea River Basin were selected in which to
investigate the effects of logging (Fig. 1.1). The design involved comparison of
complete clearcutting of a small watershed with patchcutting of a larger
watershed. The third watershed remained undisturbed as a control. The paired
watershed study approach compares pre- and posttreatment relations for
control and treatment watersheds. Differences in the posttreatment relation
are attributed to the treatment.

Study Area

The three streams chosen for study, Deer Creek, Flynn Creek, and Needle
Branch, are tributaries of Drift Creek, which flows into Alsea Bay near Waldport. The watersheds lie about 16 km from the Pacific Ocean, at elevations
between 140 and 490 m. The climate is maritime, with mean annual precipitation approximately 250 cm, almost all of which falls as rain from October
through March. Snow is rare and summers are dry. Air temperatures generally
range from about $-7°C$ to $32°C$. The geology is typical of the Tyee sandstone
formation of the northern Oregon Coast Range (Corliss and Dyrness 1965).
Streamflow varies greatly over the seasons. Freshets occur from November
through February, the principal time of salmon spawning, and streamflows
go very low during the dry summers (Hall and Lantz 1969).

Two salmonid fish species, coho salmon (*Oncorhynchus kisutch*) and coastal
cutthroat trout (*O. clarkii*), are found in all three of the study streams. A small

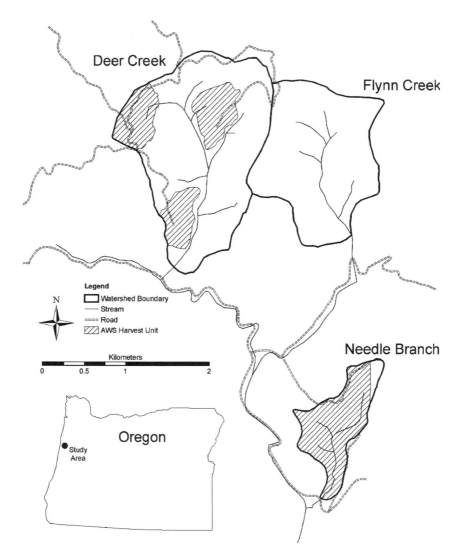

Fig. 1.1 Location of the Alsea watersheds in the Oregon Coast Range

number of steelhead (*O. gairdneri*) spawned in Deer Creek. Adult Chinook salmon (*O. tshawytscha*) rarely entered Deer Creek, and never spawned there. In general, adult coho salmon enter the streams to spawn from November through February. Fry emerge February through May and, after one year of residence in the streams, most juvenile salmon migrate to sea the following spring. Many of the cutthroat trout remain in the watersheds for most or all of their life, though there is some downstream movement of juveniles in the spring, and a small number of sea-run adults enter the watersheds in winter. In addition to these salmonids, there is a population of the reticulate sculpin *(Cottus perplexus)*. Its biomass is about equal to that of the coho salmon or cutthroat trout (Hall and Lantz 1969).

Deer Creek

Most of the Deer Creek watershed (303 ha) (Table 1.1) is on USFS land; a very small portion of the lower watershed, below the stream gauge, is owned by the Georgia-Pacific Corporation (now Plum Creek Timber Company). This watershed was cut in three patches. The overstory vegetation was Douglas-fir *(Pseudotsuga menziesii)*, representing two age classes (50–70 and 70–110 years old), along with red alder *(Alnus rubra)* 40–60 years old, with a few 20- to 40-year-old alder present in the lower watershed. Primary understory species were salmonberry *(Rubus spectabilis)*, vine maple *(Acer circinatum)*, and sword fern *(Polystichum munitum)* (Moring and Lantz 1975).

   Mean summer minimum streamflow during the prelogging period was 8.5 liters per second $(L \cdot s^{-1})$, or 0.30 cubic feet per second (cfs). Peak winter flow was 5.7 cubic meters per second $(m^3 \cdot s^{-1})$ or 201 cfs. Annual mean water temperature was 9.6°C; daily temperatures ranged from a minimum of 1.1°C to a maximum of 16.1°C. The diel range was 0.5°C to 2.2°C (Hall and Lantz 1969).

Flynn Creek

The Flynn Creek watershed (202 ha) is on USFS land (Table 1.1). The watershed was undisturbed and served as the control. Vegetation consisted of mixed hardwood and conifer stands. In 1959, Douglas-fir stands represented

Table 1.1 Characteristics of study watersheds as of 1990

	Flynn Creek	Deer Creek	Needle Branch
Treatment	control	patchcut	clearcut
Watershed area (ha)	202	303	71
Stream length accessible to anadromous fish (m)	1433	2324	966
Median watershed slope (%)	27	39	30
Vegetation (%)			
Conifer	36	64	80
Hardwood	64	36	20
Area in roads (%)	0	4	5
Area harvested (%)	0	41[1]	82[2]

[1] 26% of the watershed was harvested in 1966 in three clearcuts; an additional 15% was harvested in three clearcuts, in 1978, 1987, and 1988.
[2] 13 ha in the headwaters of Needle Branch had been logged in 1956.

two age classes: 30–50 and 70–110 years old. The older stand had regenerated after the great Alsea fire of about 1850. Red alder stands were 30–70 years old. Understory vegetation species included salal (*Gaultheria shallon*), sword fern, vine maple, salmonberry, and isolated groups of bracken fern (*Pteridium aquilinum*) (Moring and Lantz 1975).

During the prelogging period, mean summer minimum streamflow in Flynn Creek was 4.5 $L \cdot s^{-1}$ (0.16 cfs), and peak winter flow reached 3.9 $m^3 \cdot s^{-1}$ (137 cfs). Annual mean water temperature was 9.7°C; daily temperatures ranged from a minimum of 2.2°C to a maximum of 16.6°C. Diel ranges were 0.5° to 2.2°C (Hall and Lantz 1969).

Needle Branch

Needle Branch is the smallest of the three study watersheds, encompassing 71 ha (Table 1.1). This watershed was clearcut. Most of the watershed is owned by Georgia-Pacific Corporation, but about 16 ha of the lower watershed are owned by a private individual. Timber species present included Douglas-fir, western redcedar (*Thuja plicata*), and red alder. Douglas-fir stands were 70–110 years old, and cedar stands were 30–50 years old. A small stand of 30- to 50-year-old Oregon white oak (*Quercus garryana*) was also present. Understory vegetation was primarily vine maple and sword fern, although salal, bracken fern, salmonberry, thimbleberry (*Rubus parviflorus*), and dewberry (*R. vitifolius*) were also present.

The summer streamflow in Needle Branch is extraordinarily low for a stream that supports a significant run of salmon in the winter. During the prelogging period, the mean minimum summer flow was 0.6 $L \cdot s^{-1}$, or 0.02 cfs. Individual pools are sometimes isolated by stretches of exposed gravel during the lowest flows, which may reach 0.3 $L \cdot s^{-1}$, or 0.01 cfs (about 5 gallons per minute). The peak winter flow was 1.4 $m^3 \cdot s^{-1}$ (50 cfs). The temperature regime was similar to the other two streams, with an annual mean of 9.7°C. Daily temperatures ranged from a minimum of 1.6°C to a maximum of 16.1°C. Diel ranges were from 0.5° to 1.5°C (Hall and Lantz 1969).

Logging Treatments

Access roads for logging were constructed into Deer Creek and Needle Branch between March and August 1965, providing a 1-year period to evaluate the effects of road-building alone. Logging in Deer Creek began in May 1966 and was completed in November. In Needle Branch, logging was accomplished between March and August 1966.

Deer Creek was harvested in three patches of about 25 ha each. A streamside buffer, mainly red alder in this watershed, was left along all streams accessible to anadromous fish (Fig. 1.2). Trees were felled and then yarded to uphill landings by high-lead cable yarding systems. The cable system could lift one end of the log off the ground, but soil disturbance occurred as the logs were dragged to the tower. High-lead cable yarding was common practice during the study period. Although felling and yarding were completed in Deer Creek by November, the logs remained on the watershed in large decks at the landings, the result of another potential disruption to the study that occurred in the year of logging. A sharp decrease in the price being paid for logs in 1966 significantly threatened the integrity of the study. The timber had been bid on in 1964, a time of record prices, by a logging operator who planned to sell the logs to an independent mill. When it came time to harvest in 1966, the operator faced a substantial loss if forced to sell the timber that year. However, had the trees not been felled and yarded in 1966, the entire design of the study would have been compromised by having the two watersheds cut in different years. All subsequent comparisons between the response of Needle Branch and Deer Creek would have been open to question. Because the area was subject to the standard USFS timber sale regulations, there was no formal way to insure that the harvest would take place. Fortunately for all concerned, the USFS allowed the timber to be decked in the watershed during 1966 and sold the following year, when prices improved. This accommodation by both the agency and the operator was a significant contribution to the overall success of the study.

The Needle Branch watershed was completely clearcut, except for a small area (13 ha, or about 18%) in the upper northeast corner that had been cut in

Fig. 1.2 Timber harvest units in Deer Creek, with streamside vegetation buffers

1956. The earlier timber harvest was not considered to have had an effect on water quality or salmonid resources. The logging was carried out in accordance with normal practices on private lands at that time. No streamside vegetation buffers were left, and no effort was made to protect the stream during logging activity (Fig. 1.3). The majority of the watershed was logged by high-lead cable yarding to uphill landings, but a portion of the lower watershed was yarded with a tractor. In many places, logs were yarded across and through the stream, breaking down the banks and leaving extensive amounts of needles and larger debris in the stream channel. In September, the section of the channel accessible to anadromous fish was cleared of debris, primarily by a crew that used chain saws and manually removed material. A tractor was used in a small segment of stream at the lower end of the watershed.

The logging slash was burned on Needle Branch in October 1966. The fire was particularly hot, especially in the upper canyon region, where conditions allowed an intense firestorm to develop. Because logs remained on the watershed in Deer Creek, slash burning was delayed there. One unit was burned in May 1967, one in 1968, and the third in August 1969. Owing to the delay and consequent regrowth of vegetation, all three of the fires on Deer Creek were substantially cooler than the one at Needle Branch. Fuel oil was sprayed to

Fig. 1.3 View of the main channel of Needle Branch in fall 1966, after the debris was cleared from the channel and logging slash was burned

brown the vegetation before burning the last unit, but even that failed to produce a hot fire and much debris remained on the ground.

Reforestation in Deer Creek was accomplished by hand-planting following slash burning. The herbicides 2,4-D and 2,4,5-T were sprayed from a helicopter on two units to reduce competition from brush and promote survival of planted Douglas-fir. In Needle Branch, attempts were made to broadcast Douglas-fir seed from a helicopter. This technique was tried twice, in January 1967 and December 1967, using seed coated with endrin to deter seed-eating birds and mammals. However, the attempts met with limited success and hand-planting was necessary. Sampling was carried out to evaluate the fate of endrin in stream water and in the food chain (Marston et al. 1969; Moore et al. 1974).

Methods

Water Resources

A variety of data were collected on physical and biotic factors during both the pre and posttreatment periods. Compound broad-crested concrete weirs were installed on Deer Creek and Flynn Creek to measure streamflow and sediment production (Fig. 1.4). Needle Branch was gauged with a compound broad-crested concrete weir and H-type flume. The three stations were operated by the USGS, and measurements began in September 1958. Stage

Fig. 1.4 Broad-crested compound concrete weir with stilling well on Flynn Creek

measurements were recorded by a pen-and-chart recorder. Records were reduced manually. Punched-tape recorders were added later and used primarily for backup measurements. Streamflow and sediment data were published by the USGS in basic data reports.

Continuous recordings of water temperatures are available from the three USGS gauging stations from 1958 through 1973. In addition, the Oregon State University School of Forestry established a network of 18 Partlow® recording thermographs in March 1964. Flynn Creek had a thermograph at the gauging station, Needle Branch had five thermographs along the stream and one at the gauging station, and Deer Creek had 10 units along the stream and one at the gauging station. After 1969 the network of Partlow® thermographs was gradually reduced as some of the thermographs were used in special studies. The temperature data at each gauging station were published by the USGS in basic data reports. The network data were analyzed by Brown and Krygier (1970).

Precipitation measurements were made near the mouth of each watershed. Both precipitation intensity and amounts were recorded by weighing rain gauges. Annual precipitation values have been published (Harris 1977) data on precipitation intensity were used only for some stormflow analyses.

Water quality samples were collected for suspended sediment and chemical analyses for pre and posttreatment periods. (See Chapters 4 and 12 for respective discussions.) Water quality samples were collected for dissolved oxygen analyses at several sites and included intragravel measurements taken from standpipes driven into the gravel (Moring 1975a).

Biological Resources

A number of biological parameters were measured as part of the AWS. Based on information current at the time, the major effect on salmonids was expected to be an increase in fine sediment in spawning gravel that would decrease their survival in the gravel and as they emerged from the spawning beds into the stream. Thus, much of the effort, particularly by the Oregon State Game Commission, was focused on this aspect of the salmonids' life history (Phillips 1971). These researchers developed a unique trap to estimate the survival of juvenile coho salmon from egg deposition to emergence from the gravel (Phillips and Koski 1969).

Data on fish populations came primarily from fish traps located at the lower section of each study stream (Fig. 1.5), though additional sampling was done in the streams with small seines and electrofishing gear. The original traps were constructed of wood. Rotating screens directed downstream migrants into a holding box, but tended to accumulate woody debris and required nearly continuous cleaning during storm events. There were two major floods during the winter of 1964–1965, both estimated to be in the range of a 100-year return period. The wooden fish traps were substantially

Fig. 1.5 Two-way fish trap on Deer Creek

damaged during these high flows and were subsequently replaced by concrete structures. The Deer Creek trap was replaced in 1965, Needle Branch in 1966, and Flynn Creek in 1967.

Another objective of the AWS was to determine whether energy utilized by coho salmon for growth was originally derived from terrestrial or aquatic plants. The general approach was to determine food habits of the juvenile salmon and then to investigate the food of their principal insect prey groups (Chapman et al. 1961; Chapman and Demory 1963). Most of the macroinvertebrate analysis was done during the pretreatment period. Related studies were carried out on production by algae attached on the gravel, material forming one of the bases of the food chain for salmonids (Hansmann et al. 1971; Hansmann and Phinney 1973).

Results from the AWS showed that timber harvest practices could significantly affect water quality and salmonid populations. As a consequence, the U.S. Fish and Wildlife Service funded additional work to determine the short-term effects of timber harvesting practices on these resources. Twelve additional streams in the Oregon Coast Range were included in this expanded study. Observations were made during the summer one year before and one year after timber harvest. The primary measures used were estimates of population size of juvenile salmonids, stream temperature, surface and intragravel dissolved oxygen, and the proportion of fine sediment in spawning gravel. The purpose was to ascertain the effectiveness of streamside vegetation buffers in maintaining water quality and salmonid habitat during and following logging (Moring and Lantz 1974).

Training and Educational Opportunities

The AWS provided a unique opportunity for training and education. The study was one focus of "Forest Land Uses and Stream Environment," a three-day conference held at Oregon State University in 1970 (Krygier and Hall 1971). The meeting attracted more than 500 participants from numerous agencies and disciplines. A large number of graduate students in several departments at Oregon State University received an introduction to multidisciplinary research in their work on the three watersheds, and they also made substantial contributions to the project. The USFS used the Alsea site to conduct training sessions for foresters, engineers, and technicians (Moore 1971). The training was based on the recognition of the significance of very small headwater streams as habitat for resident and anadromous salmonids (heretofore unrecognized at this scale) and the interrelationships of land management practices and the fish resource (Lantz 1971; Moring 1975b; Hall et al. 1987). Such recognition resulted in interdisciplinary approaches for land management activities that consider all resources in management scenarios. These management practices may have been a forerunner to the current emphasis on ecosystem management.

Paired Watershed Studies in the United States

This brief summary is provided to put the AWS in context with other watershed studies in the U.S. The first paired watershed study in the United States was the Wagon Wheel Gap study in Colorado started in 1909. The USDA Forest Service and the U.S Weather Bureau monitored hydrological and meteorological conditions on paired watersheds; a clearcut watershed compared to an undisturbed watershed. The study lasted 16 years, and additional funding to continue the monitoring was continually sought, but to no avail. The study showed that timber harvesting increased water yields in the central Rocky Mountains (Ice and Stednick, 2004)

In 1934 the Coweeta Hydrologic Labortory was established in North Carolina by the USDA Forest Service as the first long-term forest hydrologic research facility in the United States. Early studies at Coweeta focused on how land management affects the hydrologic cycle and included studies on the effects of mountain farming, woodland grazing, and unrestricted logging. Additional work addressed comparisons of water resource changes from partial and clearcut harvesting, alternative road designs, and the use of cable logging.

The Fraser Experimental Forest in Colorado was established in 1937 by the USDA Forest Service and was oriented toward timber and water production resulting from forest management. The first paired watershed study was the Fool Creek experiment, where logging created variously sized openings to look at the effect of opening size on snow pack accumulation and streamflow. Pretreatment monitoring started in 1945, with the adjacent East St. Louis Creek serving as the control watershed. Fool Creek watershed manipulation began in 1955. The Fool Creek study was one of the first to look at the effects of timber harvesting activities on sediment yields.

The H. J. Andrews Experimental Forest in western Oregon was established by the USDA Forest Service in 1948. Paired watersheds were used to study timber harvesting and road practices. Watershed 2 serves as a control. Watershed 1 was clearcut between 1962 and 1966. Watershed 3 was partially harvested with road construction in the watershed. This study showed the importance of protecting the stream channel and adjacent vegatation to minimize changes in water quality.

The Fernow Experimental Forest was established in the Appalachians of central West Virginia in 1951. In response to a severe drought in the region, the original purpose of the Fernow Experimental Forest was to investigate the opportunity to increase water yields from forest watersheds. Studies at the Fernow showed that water yield did increase after harvesting but recovered rapidly. Research also looked at water quality and how timber harvesting and forest roads can affect sediment in streams.

In 1955, the USDA Forest Service established the Hubbard Brook Experimental Forest to evaluate the effects of forest management on water yield and quality and flood flow. One of the first experiments at Hubbard Brook was harvesting trees. Unlike typical commercial forest harvests, the trees were left on site, and the area was repeatedly sprayed with herbicide to prevent vegetation regrowth. Nitrate concentrations significantly increased in surface waters, and there was a concern that forest productivity could not be sustained due to nutrient losses. While this study is not representative of the effects of commercial forest harvesting on water quality, it did contribute to our understanding of nutrient cycling processes in forest watersheds.

Again, the Alsea Watershed Study was the first long-term watershed study in the nation to simultaneously consider the effect of timber harvesting on water quantity and quality, fish habitat, and fish populations. The paired watershed study approach was used to assess the effects of clearcutting with and without streamside management zones.

Summary

Looking back, it seems remarkable that the AWS was able to run its course and achieve a modicum of success, particularly because it required the use of national forest lands outside an experimental forest, limiting the control over timing and methods of logging. There has been only one comparable long-term study of logging impacts on salmonids on the West Coast, in Carnation Creek, British Columbia (Hartman and Scrivener 1990). That study had considerably more administrative support and a much larger budget. The Alsea study, even though it cost more than $1 million, was modestly budgeted for most of its existence relative to the scope of the work. Many of the researchers were involved in other projects at the same time and were able to devote only part time to the watershed study. Questions about the general design of the study have been debated. A long-term case study such as this one has certain shortcomings, and other approaches have been suggested (Hall et al. 1978). However, in spite of setbacks, the AWS made some substantial contributions to basic science and to resource management and should be counted a success. Findings from the AWS were used to help develop state forest practice regulations in the Pacific Northwest (Ice 1991). For example, most state regulations now require the use of buffer strips along fish-bearing streams to protect stream banks during yarding, to keep logging slash out of streams, to provide shade, and to provide a source of large wood for stream habitat into the future.

The succeeding chapters in the first part of this volume describe results of the original AWS by subject area: streamflow (Chapter 2), stream temperature (Chapter 3), sediment (Chapter 4), and salmonid populations and habitat (Chapter 5). Reports from the original AWS on effects of timber harvesting on water and salmonid resources often identified the potential for future studies (Brown and Krygier 1970, 1971; Harris 1977; Hall et al. 1987; Ice 1991; Stednick 1991).

Capitalizing on this potential, scientists began independent research efforts in fisheries and water resources on the AWS watersheds in 1988. The latter chapters of this volume describe results of post-AWS studies: the New Alsea Watershed Study (Chapter 7), streamflow (Chapter 9), water quality (Chapters 10 and 11), sediment (Chapter 12), and salmonid populations and habitat (Chapters 13, 14, and 15). Flynn Creek was designated a Research Natural Area by the USFS in 1977 (Chapter 8). It remains undisturbed, and will continue to serve as a control watershed. The New Alsea Watershed Study provides a unique opportunity to evaluate the effects of land management activities on water and salmonid resources in the context of long-term study (Chapters 16 and 17).

Literature Cited

Anonymous. 1954. Alsea meeting (minutes), Oregon State Committee on Natural Resources. February 17. Alsea, OR.

Brown, G.W., and Krygier, J.T. 1970. Effects of clear-cutting on stream temperature. Water Resour. Res. 6:1133–1139.

Brown, G.W., and Krygier, J.T. 1971. Clear-cut logging and sediment production in the Oregon Coast Range. Water Resour. Res. 7:1189–1198.

Chapman, D.W., Corliss, J.F., Phillips, R.W., and Demory, R.L. 1961. The Alsea Watershed Study. Misc. Paper 110. Oregon State Univ. Agricultural Experimental Station, Corvallis, OR. 52pp.

Chapman, D.W., and Demory, R.L. 1963. Seasonal changes in the food ingested by aquatic insect larvae and nymphs in two Oregon streams. Ecology 44:140–146.

Corliss, J.F. 1973. Soil Survey of Alsea Area, Oregon. USDA Soil Conservation Service and Forest Service, USDI Bureau of Land Management, in cooperation with Oregon Board of

Natural Resources and Oregon Agricultural Experiment Station, Superintendent of Documents, Washington, DC. 82pp.

Corliss, J.F., and Dyrness, C.T. 1964. Soil and Vegetation Survey, Alsea Area, Oregon. USDA Soil Conservation Service in cooperation with Oregon State Univ., USDI Bureau of Land Management, and USDA Forest Service, Portland, OR.

Corliss, J.F., and Dyrness, C.T. 1965. A detailed soil-vegetation survey of the Alsea area in the Oregon Coast Range, pp. 457–483. In: C.T. Youngberg, editor. Forest-Soil Relationships in North America. Oregon State Univ. Press, Corvallis, OR.

Hall, J.D., Brown, G.W., and Lantz, R.L. 1987. The Alsea Watershed Study: a retrospective, pp. 399–416. In: E.O. Salo and T.W. Cundy, editors. Streamside Management: Forestry and Fishery Interactions. Univ. of Washington Inst. of Forest Resour., Seattle, WA.

Hall, J.D., and Lantz, R.L. 1969. Effects of logging on the habitat of coho salmon and cutthroat trout in coastal streams, pp. 355–375. In: T.G. Northcote, editor. Symposium on Salmon and Trout in Streams. Univ. of British Columbia, H.R. MacMillan Lectures in Fisheries, Vancouver, BC.

Hall, J.D., Murphy, M.L., and Aho, R.S. 1978. An improved design for assessing impacts of watershed practices on small streams. Internationale Vereinigung für theoretische und angewandte Limnologie Verhandlungen 20:1359–1365.

Hansmann, E.W., Lane, C.B., and Hall, J.D. 1971. A direct method of measuring benthic primary production in streams. Limnol. Oceanogr. 16:822–826.

Hansmann, E.W., and Phinney, H.K. 1973. Effects of logging on periphyton in coastal streams of Oregon. Ecology 54:194–199.

Harris, D.D. 1977. Hydrologic changes after logging in two small Oregon coastal watersheds. Water-Supply Paper 2037. U.S. Geological Survey Washington, DC. 31pp.

Hartman, G.F., and Scrivener, J.C. 1990. Impacts of forest practices on a coastal stream ecosystem, Carnation Creek, British Columbia. Can. Bull. Fish. Aquat. Sci. 223. 148pp.

Ice, G.G. 1991. Significance of the Alsea watershed studies to development of forest practice rules, pp. 22–28. In: The New Alsea Watershed Study. Technical Bulletin 602. National Council of the Paper Industry for Air and Stream Improvement, Inc., New York, NY.

Ice, G.G. and Stednick, J.D. 2004. Forest watershed research in the United States. Forest History Today. Fall/Winter 2004. pp.16–26.

Krygier, J.T., and Hall, J.D., editors. 1971. Forest Land Uses and Stream Environment. Proceedings of a Symposium, October 19–21, 1970. Oregon State Univ., Corvallis, OR. 252pp.

Lantz, R.L. 1971. Guidelines for Stream Protection in Logging Operations. Oregon State Game Comm., Portland, OR. 29pp.

Marston, R.B., Tyo, R.M., and Middendorff, S. 1969. Endrin in water from treated Douglas-fir seed. Pesticides Monit. J. 2:167–171.

McKernan, D.L., Johnson, D.R., and Hodges, J.I. 1950. Some factors influencing the trends of salmon populations in Oregon. Trans. N. Amer. Wildl. Conf. 15:427–449.

Moore, D.G., Hall, J.D., and Hug, W.L. 1974. Endrin in forest streams after aerial seeding with endrin-coated Douglas-fir seed. Research Note PNW-219. USDA Forest Service, Pacific Northwest Forest and Range Experiment Station, Portland, OR. 14pp.

Moore, S.T. 1971. Forest Service views on the Alsea Watershed Study, pp. 250. In: J.T. Krygier and J.D. Hall, editors. Forest Land Uses and Stream Environment. Continuing Education Publications. Oregon State Univ., Corvallis, OR.

Moring, J.R. 1975a. The Alsea Watershed Study: effects of logging on the aquatic resources of three headwater streams of the Alsea River, Oregon. Part II. Changes in environmental conditions. Fish. Res. Rep. 9. Oregon Dept. of Fish and Wildlife, Corvallis, OR. 39pp.

Moring, J.R. 1975b. The Alsea Watershed Study: effects of logging on the aquatic resources of three headwater streams of the Alsea River, Oregon. Part III. Discussion and recommendations. Fish. Res. Rep. 9. Oregon Dept. of Fish and Wildlife, Corvallis, OR.

Moring, J.R., and Lantz, R.L. 1974. Immediate effects of logging on the freshwater environment of salmonids. Final Rep. AFS-58. Oregon Wildlife Commission, Portland, OR. 101pp.

Moring, J.R., and Lantz, R.L. 1975. The Alsea Watershed Study: effects of logging on the aquatic resources of three headwater streams of the Alsea River, Oregon. Part I. Biological studies. Fish. Res. Rep. 9. Oregon Dept. of Fish and Wildlife, Corvallis, OR. 66pp.

Phillips, R.W. 1971. Effects of sediment on the gravel environment and fish production, pp. 64–74. In: J.T. Krygier and J.D. Hall, editors. Proceedings of a symposium: Forest Land Uses and Stream Environment. Oregon State Univ., Corvallis, OR.

Phillips, R.W., and Koski, K.V. 1969. A fry trap method for estimating salmonid survival from egg deposition to fry emergence. J. Fish. Res. Board Can. 26:133–141.

Stednick, J.D. 1991. Purpose and need for reactivating the Alsea Watershed Study, pp. 84–93. In: The New Alsea Watershed Study. Technical Bulletin 602. National Council of the Paper Industry for Air and Stream Improvement, Inc., New York, NY.

Chapter 2
Effects of Timber Harvesting on Streamflow in the Alsea Watershed Study

John D. Stednick

The Alsea Watershed Study was the nation's first long-term watershed study to simultaneously consider the effects of timber harvesting on water and water related resources (fish habitat and fish populations) (Brown 1972). The study began in 1957 as a cooperative effort between Oregon State University (then Oregon State College) and other federal and state agencies to address the effects of integrated land management on the stream environment (Harr and Krygier 1972; Moring 1975; Harris 1977). The Alsea River Basin, in the Oregon Coast Range, was selected because of the diversity of land ownership, active timber harvesting, and its close proximity to the university. The initial goal to assess these potential effects at the large watershed level proved to be too ambitious and was reduced to three small watersheds in the Alsea River Basin. The final selection of the watersheds reflected similar geographic location, exposure, elevation, and land ownership of the participants, namely the USDA Forest Service and Georgia Pacific Company (now Plum Creek Timber Company), a private timber company.

The temperate coniferous forest in the western United States typically consists of well-developed overstories and understories. The overstory plant community of the temperate coniferous forest is dominated by Douglas-fir (*Pseudotsuga menziesii*) and western hemlock (*Tsuga heterophylla*). The understory consists of vine maple (*Acer circinatum*), red alder (*Alnus rubra*), salmonberry (*Rubus spectabilis*), rhododendron (*Rhododendron macrophyllum*), and others (Meehan 1991). The temperate coniferous forest of the Pacific Northwest of the United States extends from central Alaska to central California, including the Coastal Range of Alaska, British Columbia, Washington, Oregon, and California (Chamberlin et al. 1991).

Rain is the dominant form of precipitation in the temperate coniferous forest of the Pacific Northwest of the United States and drives the hydrology of small forested streams (Chamberlin et al. 1991). The climate of the Alsea watersheds is a maritime climate with mild temperatures, winter precipitation, and

John D. Stednick
Department of Forest, Rangeland, and Watershed Stewardship, Colorado State University, Fort Collins, CO 80523

J. D. Stednick (ed.), *Hydrological and Biological Responses to Forest Practices.*   19
© Springer 2008

a summer drought. Approximately 90% of the annual precipitation of 2500 mm occurs from October through April (Fig. 2.1). Precipitation events during the winter months generally occur as slow-moving, low-intensity frontal systems. These frontal storms usually occur frequently over the season, so precipitation volumes are up to an order of magnitude different between the wet and dry seasons. Convective storms are the primary cause of precipitation events during the summer and early fall. These storms are generally short in duration, but can often be of moderate to high intensity.

The largest precipitation events of the year occur in the winter on a soil mantle that is close to saturation, leaving most of the precipitated moisture available for runoff. This results in streamflow events that are 1000 to 5000 times larger than those observed in the summer for similar sized storms (Harr 1976) (Fig. 2.2). Increased streamflow resulting from the processes described previously causes the greatest contribution to annual water yield to occur during the wet winter months (Chamberlin et al. 1991).

Thick vegetation will also result in high rates of evapotranspiration during the growing season. These high rates of evapotranspiration contribute to the drying of the soil mantle, resulting in increased soil storage capacity during the summer months, which in turn contributes to the lack of streamflow response to summer precipitation events (Harr 1976; Hewlett and Helvey 1976).

Three watersheds were selected for the study: Flynn Creek, Deer Creek, and Needle Branch. Data on streamflow, sediment yield, temperature, and nutrients were collected during the study and compared to relations developed during the pretreatment period (1959–1966). The effects of treatment on the parameters of interest were evaluated in posttreatment (1967–1973) (Harr and Krygier 1972; Moring 1975; Harris 1977). Watershed elevations range from 135 to 490 m with mean slopes of 35 to 40%. Soils are derived from the Tyee sandstone formation: 80% of the soils are Bohannon and Slickrock series. Bohannon soils are stony,

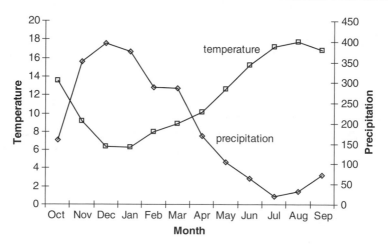

Fig. 2.1 Mean monthly temperature (°C) and precipitation (mm) as measured at Tidewater, OR

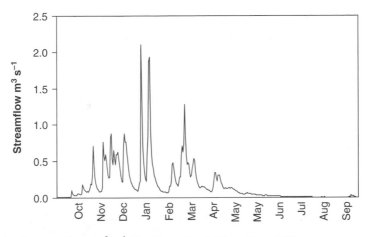

Fig. 2.2 Daily streamflow ($m^3 s^{-1}$) for Flynn Creek Water Year 1972

Fig. 2.3 Harvest unit in Deer Creek shortly after logging

generally less than 60 cm deep, and derived from sandstone residuum. Slickrock soils are derived from sandstone colluvium and range in depth to 140 cm. Rates of infiltration and percolation are high, and overland flow on undisturbed forest soil has never been observed.

Table 2.1 Summary of treatments and areas for each watershed

	Needle Branch	Flynn Creek	Deer Creek
Total area, ha	71	202	303
Area in roads 1965, ha[1]	3.6	0	11
Percent in roads[1]	5	0	4
Logged area 1966, ha	58	0	77
Percent logged	82[2]	0	25
Burned area 1966, ha	58	0	23
Percent burned	82	0	8

[1] Includes landings, road cutbanks and fill slopes, and tractor skid trails.
[2] In 1956, 13 ha in the headwaters of Needle Branch were logged.

Fig. 2.4 Needle Branch harvest unit with no streamside buffer left

Before treatment, vegetation consisted of various amounts of red alder and 120-year old Douglas-fir. Pure stands of Douglas-fir covered about 76% of Needle Branch and 17% of Deer Creek. Alder covered 30% of Flynn Creek. The remainder supported mixed stands of Douglas-fir and red alder (Harr and Krygier 1972).

Flynn Creek (202 ha) served as the control watershed, Deer Creek (303 ha) was harvested in three small patchcuts, with uncut forest left along the stream channels of 15 to 30 m wide (Brown 1972) (Fig. 2.3). The total area harvested in Deer Creek was 77 ha or 25% of the watershed area (Table 2.1). Needle Branch (71ha) was nearly completely clearcut with no streamside vegetation left (Fig. 2.4). Approximately 18% of upper Needle Branch watershed was harvested in 1956 (see Fig. 1.1).

Logging roads were constructed into Deer Creek and Needle Branch between March and August 1965 and were mostly located on ridgelines.

Roads were separated from logging for only one season. Logging began in March 1966 and was completed by November 1966. Most logging was done by high-lead yarding, but tractor skidding was done on the lower part of Needle Branch. As typical for the period, logging slash was burned after logging. The slash on Needle Branch was dry and resulted in a very hot fire in October 1966. Due to a depressed log market, logs were temporarily stored in Deer Creek landings and logging was not completed until summer 1969. One unit was burned in May 1967, one in 1968, and the lower unit in August 1969, but the vegetation regeneration resulted in cool fires.

Methods

Hydrometeorologic data were collected on all three systems for 15 years beginning in water year 1959 (October 1958). Data were collected for 7 years before logging (1959–1965 water years), 1 year during logging (1966), and 7 years posttreatment (1967–1973).

Measurements of precipitation, streamflow, sediment transport, and water temperature near the mouths of the watersheds before, during, and after logging provided the data needed to evaluate the potential effects of logging on streamflow. Streamflow data, sediment concentrations, and water temperature data collected by the U.S. Geological Survey (USGS) at the gauging stations are published in their basic data reports. Precipitation data were collected in forest openings near the mouths of the watershed by Oregon State University personnel. The original data were collected in English units, and converted to metric for this volume.

Precipitation

Weighing-type rain gauges (Belfort®) located near the gauging stations were serviced weekly. Precipitation data were reduced to daily values and compiled. Precipitation data allowed comparison between watersheds and comparison to the long-term record at Tidewater, Oregon. Double-mass analysis of cumulative precipitation suggested no change in areal distribution of precipitation after logging, thus streamflow changes are the result of logging and not precipitation differences (Harris 1977). Precipitation data were published through February 1968 (Harris 1977). In the process of compiling the AWS historical records, the remaining precipitation data (through September 1973) were located at Oregon State University and reduced (Table 2.2). The precipitation data records were marked "corrected" until February 1968. No documentation of this "correction" was located. Precipitation data records from February 1968 to the end of the study were not "corrected". The original reporting of the precipitation data (Harris 1977) suggested that the wettest and driest years

Table 2.2 Annual precipitation (mm) for Tidewater, OR and the three study watersheds for all years

Water year	Tidewater	Flynn Creek	Needle Branch	Deer Creek
1959	2599	2634	2940	2769
1960	2074	2090	2332	2244
1961	2560	2689	2790	2827
1962	2149	2176	2300	2248
1963	2224	2123	2223	2236
1964	2333	2422	2454	2525
1965	2309	2390	2344	2495
1966	2263	2249	2127	2347
1967	2396	2249	2127	2347
1968[1]	2383	2996	2990	2964
1969	2577	2262	2350	2260
1970	2301	2401	2551	2702
1971	2834	3317	3637	3429
1972	2901	3042	2952	2780
1973[2]	1808	1139	1077	1128

[1] Data from Harris 1977 through February 1968; unpublished data were compiled for the remaining years
[2] Data from January to September 1973

were 1972 and 1973, respectively, based upon precipitation data from Tidewater. The additional Alsea precipitation data indicated that the wettest year was 1971. Because the data were collected at the low point of the basin, they probably do not represent average precipitation over the basin. The frontal storm systems that generate most of precipitation would have orographic effects (i.e., increased precipitation with increased elevation) (Harris 1977).

Streamflow

The USGS built stream gauging stations at each watershed outlet in 1958. Broadcrested compound V-notch concrete weirs were built on Deer Creek and Flynn Creek (Fig. 2.5). Because of the smaller watershed area and stream channel size, Needle Branch had a smaller compound V-notch crest with vertical concrete walls. Each concrete weir had concrete cutoff walls built into the stream bank to prevent water short-circuiting of the control structure. The weirs are connected to the stilling well with two inlet pipes, one each for low flow and high flow conditions. The gauging house on the stilling well had a Leopold-Stevens® A-35 recorder that recorded stage at a 1:0.1 scale. Stream-flow measurements were made by the USGS and the stage-discharge relation frequently updated. Discharge measurements for high and medium flows were typically done with Price® or pygmy current meters, while low flows were

Fig. 2.5 Broad crested compound V-notch weir on Deer Creek

measured volumetrically (Harris 1977). Discharge records were considered to be good to excellent for all three stations.

Stream gauges were operated daily by Oregon State Game Commission personnel, with funding from Oregon State University, and serviced at intervals by the USGS (Moring 1975). Streamflows were converted from gauge heights to streamflow by hand and are part on the USGS streamflow records. During 1963–1965, six additional streamflow gauges were established in Deer Creek by the Oregon State University, School of Forestry, to monitor streamflow upstream at two locations on Deer Creek proper and on four tributaries (two with timber harvesting) (Table 2.3). Only some of these streamflow records were located, and are currently stored at the Oregon State University Forest Research Laboratory. This later study assessed the effects of logging and logging with roads on peak flows (Harr et al. 1975).

Table 2.3 Watershed characteristics for Deer Creek subbasins (adapted from Hall and Krygier 1967; Harr et al. 1975)

	I	II	III	IV	V	VI
Watershed area (ha)	3.4	56	40	16	12	231
Area logged (%)	0	30	65	90	0	25
Area in roads (%)*	0	3	12	0	5	5

* Includes landings, road cutbanks and fill slopes, and tractor skid trails (Harr et al., 1975).

Streamflows typically are low during the early fall months. As winter precipitation increases, the soil mantle becomes wet and responds to individual winter precipitation events. Most precipitation events occur as rain, and snowfalls on the Oregon Coast are relatively rare, short-lived, and add little water to the annual budget. The dry mantle storms are easily separated from the wet mantle storms. As winter storms decrease, the soil mantle drains and streamflow decreases to low flow conditions. There were no records of zero streamflow during the Alsea Watershed Study period.

Results

The principal method used to assess the effects of logging on water resources was to develop pretreatment relations between the treatment watersheds (Needle Branch and Deer Creek) and the control watershed (Flynn Creek). Regression equations were developed to estimate values of dependent variables (treatment watersheds) from values on the independent variable (Flynn Creek) (Table 2.4). Prediction limits at the 95% confidence interval were used to assess treatment departures (Harris 1977).

Selected streamflow characteristics were used to assess the effects of logging on the stream regimen. Annual runoff was used as the total amount of water leaving the watershed. Peak flows and three-day high flows represented the instantaneous peak flows and three-day high flow volumes. Low flows were daily flows in August and September (Harris 1977). Results presented in this chapter follow the format and methods of the earlier work (notably Harris 1977).

Annual Runoff

Generally, annual water yield increases following timber harvest, due to decreased evapotranspiration and interception on the harvested site, coupled with any physical disturbances caused by timber harvesting. This increase is

Table 2.4 Prediction equations derived from pretreatment streamflows (after Harris 1977).

Site	Prediction Equation	r^2 value
	Annual runoff (mm)	
Needle Branch	= (0.91) (Flynn) + 91.7	$r^2 = 0.80$
Deer Creek	= (1.03) (Flynn) − 124.9	$r^2 = 0.97$
	Peak flow ($m^3 s^{-1} km^{-2}$)	
Needle Branch	= (0.93) (Flynn) +0.227	$r^2 = 0.90$
Deer Creek	= (0.92) (Flynn) +0.100	$r^2 = 0.99$
	Low flow ($m^3 s^{-1}$)	
Needle Branch	= (0.245) (Flynn) − 0.585	$r^2 = 0.77$
Deer Creek	= (1.91) (Flynn) + 0.460	$r^2 = 0.88$

generally observed immediately following timber harvest, and decreases as vegetation recovers (Harr 1976; Hewlett and Helvey 1976; Chamberlin et al. 1991; Stednick 1996). The soil mantle is closer to saturation during the summer with no vegetation to transpire moisture from the soil. This leads to higher runoff during the beginning of the wet season because there is less soil moisture deficit to make up in the soil. As vegetation recovers following timber harvest, more soil moisture is transpired during the summer months. This leads to an increasingly dry soil mantle for the fall storms, which in turn leads to lower levels of runoff following precipitation events. This period of lower runoff continues until the soil moisture deficit has been satisfied. Ultimately the recovery of the soil moisture deficit in the summer leads to decreasing annual water yield, and a return to preharvest conditions. An analysis of annual water yield studies from paired watershed studies suggests that at least 20% of the watershed needs to be harvested to be detected using streamflow monitoring methods and a key factor governing changes in annual water yield is the proximity of harvest to streamflow source areas (Stednick 1996).

The mean annual runoff was approximately 1920 mm per year for all three watersheds (Harris 1977). Flynn Creek annual runoff ranged from 1195 to 2785 mm. After logging on Needle Branch, the mean runoff of 2353 mm was 483 mm or 26% greater than the predicted runoff (Harris 1977) (Fig. 2.6). Annual water yield increases were 20% to 31% greater than predicted in the posttreatment period. Water yield increases tended to increase with increased annual precipitation. Annual water yields were variable over time and did not suggest a return to pretreatment water yield levels.

On Deer Creek, the actual mean runoff of 1952 mm after logging was 64 mm or 3% greater than the predicted runoff of 1888 mm (Harris 1977) (Table 2.5)

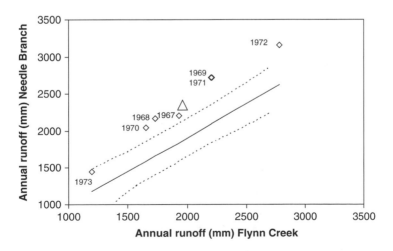

Fig. 2.6 Annual water yield regression between Flynn Creek and Needle Branch, and observed annual water yields after harvesting for the seven posttreatment years

Table 2.5 Annual runoff (mm) for all watersheds for all years

Water year	Flynn Creek	Needle Branch	Deer Creek
1959	1996	2135	1997
1960	1833	1767	1722
1961	2370	2173	2277
1962	1688	1565	1632
1963	1757	1596	1643
1964	1961	1912	1902
1965	2212	2052	2171
Prelogging mean	1973	1886	1907
1966 (logging)	1721	1734	1710
1967	1924	2209	1849
1968	1727	2173	1764
1969	2202	2716	2106
1970	1650	2045	1706
1971	2208	2717	2300
1972	2784	3162	2694
1973	1195	1446	1244
Post-logging mean	1956	2353	1952

and was not significantly different (Fig. 2.7). A covariance analysis using precipitation data at Tidewater and annual runoff for Needle Branch indicated that there was a significant difference in prelogging and post-logging runoff; the relations between hydrologic characteristics of watersheds before and after logging are significantly different; and the slope of the regression lines before and after logging are parallel. For Deer Creek, the analysis showed no

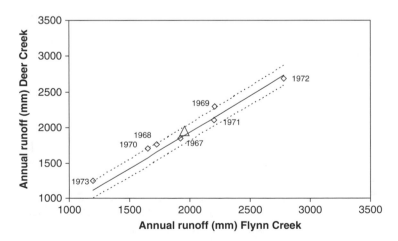

Fig. 2.7 Annual water yield regression between Flynn Creek and Deer Creek, with observed annual water yield after harvesting for the seven posttreatment years

significant difference between the prelogging and post-logging regression lines (Harris 1977).

Departures from the prediction equation (actual minus observed) were plotted over time. Needle Branch had significant and consistently positive increases in water yield after harvesting (Fig. 2.8). There was no discernible pattern in the increased annual water yields over time. The lowest increase was in the driest year (1973) and the highest increases in the wetter years. A similar plot for Deer Creek shows that water yield increases were not observed for every posttreatment year. Three of the seven posttreatment years had annual water yields less than predicted by Flynn Creek (Fig. 2.9).

Peak Flows

Peak flows result from the combination of incoming precipitation, interception, and the movement of water through the subsurface soil. The temperate coniferous forest environment generally exhibits seasonality in the runoff hydrograph, with peak flows occurring predominately in the wet winter season. Peak flows are important hydrologic characteristics because they are often responsible for moving large quantities of sediment in a river system, and are responsible for channel form. In the human environment, peak flows of various sizes are the driving design variables for culverts at road crossings, and in-stream structures.

The controlling hydrologic factor in temperate coniferous forest environments is rainfall, leading to large streamflows in the winter months when the

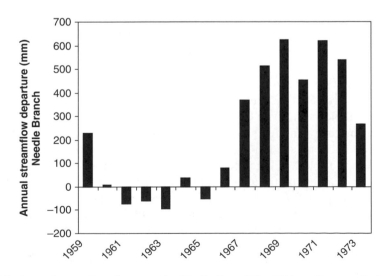

Fig. 2.8 Annual streamflow departure for Needle Branch from Flynn Creek prediction

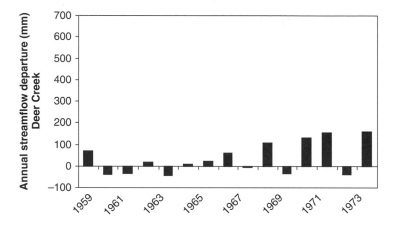

Fig. 2.9 Annual streamflow departure for Deer Creek from Flynn Creek prediction

majority of rainfall occurs, and low flows in the summer when little rain occurs. The highest annual peak flows generally occur during the winter months when precipitation is highest, the soil is generally near saturation, and the vegetation is not transpiring at peak levels (Harr 1976). Decreased evapotranspiration due to less vegetative cover causes the soil to be wetter than during pre-harvest conditions, resulting in earlier saturation of the soil mantle, and potentially higher peak flows (Harr 1976). This effect is not generally observed in the winter when the soil moisture is fully recharged in both harvested and unharvested watersheds. The timing of peak flows is dependent on the site-specific impacts of the particular timber harvest. The increase in fall soil moisture associated with decreased evapotranspiration has the greatest increase on peak flows with a one- to five-year recurrence interval (Harr 1976). Larger peak flows are not as susceptible to change by timber harvest, since the amount of precipitation during these storms will exceed increased soil moisture due to timber harvest (Harr 1976).

The evaluation criterion for peak flows was selected as flows greater than 0.55 $m^3 s^{-1} km^{-2}$ (or 50 $ft^3 s^{-1} mi^{-2}$) (Harris 1977). In the prelogging period there were 15 peak flows on Flynn Creek above the threshold and 16 peak flows in the post-logging time period (Table 2.6). On Needle Branch, the mean peak flow increased to 1.19 $m^3 s^{-1} km^{-2}$ or 20% greater than the predicted mean of 1.0 $m^3 s^{-1} km^{-2}$ (Table 2.4).Three peak flow events in the post-logging period were outside (or greater than) the 95% confidence interval (Fig. 2.10). The mean of all posttreatment peak flows was within the regression confidence intervals.

After logging on Deer Creek, the actual mean of the peak flows increased 0.02 $m^3 s^{-1} km^{-2}$ or 2% greater than the predicted mean of 0.86 $m^3 s^{-1} km^{-2}$ (Table 2.5) (Harris 1977). Two peak flows were outside the 95% regression confidence intervals and the posttreatment mean was within the confidence intervals (Fig. 2.11). The mean of all posttreatment peak flows was within the

Table 2.6 Peak flows ($m^3 s^{-1} km^{-2}$) on all three watersheds for the pre and posttreatment periods

Water year	Date	Flynn Creek	Needle Branch	Deer Creek
1959	Jan 9	0.74	0.89	0.77
1959	Jan 27	0.59	0.77	0.66
1960	Feb 9	0.60	0.81	0.63
1961	Nov 24	1.09	1.34	1.06
1961	Feb 10	0.90	1.13	0.99
1961	Feb 13	0.66	0.73	0.71
1962	Nov 22	0.64	1.17	0.76
1962	Dec 19	0.57	0.73	0.63
1962	Dec 20	0.60	0.69	0.61
1963	Nov 26	0.91	1.13	0.98
1964	Jan 19	0.88	1.13	0.90
1964	Jan 25	0.56	0.69	0.59
1965	Dec 1	0.60	0.69	0.65
1965	Dec 22	1.26	1.30	1.21
1965	Jan 28	1.92	2.02	1.88
Prelogging mean		0.84	1.02	0.87)
1967	Jan 27	0.98	1.34	0.98
1968	Feb 19	0.63	1.01	0.75
1969	Dec 4	0.63	0.97	0.65
1969	Dec 10	0.60	0.97	0.65
1969	Jan 7	0.57	0.81	0.67
1970	Jan 18	0.70	1.01	0.71
1970	Jan 23	0.60	0.89	0.73
1970	Jan 27	0.63	0.77	0.68
1971	Dec 30	0.81	1.25	1.06
1971	Jan 16	0.71	1.25	0.70
1971	Jan 25	0.57	0.89	0.69
1972	Jan 11	1.95	2.59	1.83
1972	Jan 20	1.67	1.93	1.61
1972	Mar 2	0.67	1.01	0.69
1973	Dec 21	0.78	1.46	1.07
1973	Dec 27	0.59	0.89	0.56
Post-logging mean		0.82	1.19	0.88

regression confidence interval, indicating no significant increase in peak flows after harvesting.

Many studies have shown that few changes in peak flows occur as a result of timber harvest, even clearcutting (Harris 1977; Harr 1980; Harr et al. 1982). This evidence suggests that changes in peak flows are not as important as were once thought, especially since the small to average peak flows, not the larger channel forming flows, are those most affected by timber harvest.

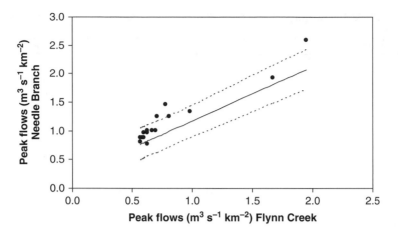

Fig. 2.10 Regression for peak flows between Flynn Creek and Needle Branch. Observed peak flows in posttreatment years are identified

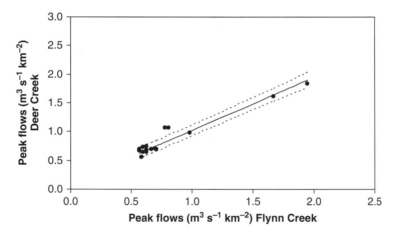

Fig. 2.11 Regression for peak flows between Flynn Creek and Deer Creek. Observed peak flows in posttreatment years are identified

Three-Day High Flow Runoff

The mean three-day high flow runoff for Flynn Creek in the pretreatment period was 114 mm compared to 116 mm, or 2 mm greater in the posttreatment period (Harris 1977). Deer Creek runoff in the post-logging period was 2.5 mm greater and Needle Branch was 31 mm greater. On Needle Branch, the predicted mean of the three-day high flow was significantly greater, 121 mm after logging compared to an actual mean of 150 mm. On Deer Creek, the predicted mean was 116 mm compared to the actual mean of 117 mm and not statistically different after harvesting.

Storm Hydrograph Changes

There has been continuing speculation about the influence of road building and clearcutting on the magnitude and frequency of peak flows (Harr et al. 1975). In the Western Cascade Mountains of Oregon, average peak flows in the fall from a 100% clearcut watershed increased $0.1 \, m^3 \, s^{-1} \, km^{-2}$, but winter peak flows were largely unchanged (Rothacher 1970). A similar pattern of smaller increases was noted when 25% of a watershed was harvested.

Deer Creek was divided into subwatersheds in an attempt to examine the effects of roading, clearcutting, and roading and clearcutting on streamflow (Harr et al. 1975). During the rainy season in western Oregon, storm runoff occurs under conditions of both recharging (fall season) and recharged (winter season) soil moisture conditions. Since the largest effects of road building and clearcutting on streamflow are expected to occur in the fall, data for this season were separated from the remainder of the rainy season. For Deer Creek, storm events were arbitrarily separated by date (September through November) and by antecedent moisture conditions as expressed as a baseflow of $0.038 \, m^3 \, s^{-1} \, km^{-2}$ (Harr et al. 1975).

There were few storm events suitable for the analysis of the effects of roads on peak flows because roads were separated from clearcutting for only one year. Study results were variable and a significant change in peak flow was only detected in Deer Creek subwatershed III, where roads occupied 12% of the total watershed area and 64% was logged (Harr et al. 1975). This became a management "rule-of-thumb" where no watershed should have more than 12% of its area in roads. This is a misrepresentation of the study results (Harr, personal communication 1996).

Needle Branch had 82% of the watershed harvested and 5% of the area was in roads. Roads had no detectable effect on storm hydrograph volume. Again, roads were only separated from logging by one year, and few storm events of sufficient magnitude were available for the analysis of the effects of roads on peak flows.

After logging, changes in total streamflow volume generally increased with increased watershed area harvested. Only Needle Branch had statistically significant increases in hydrograph volumes (Harr et al. 1975). Most increases were largest in the fall, when maximum differences in soil moisture content existed between cut and uncut watersheds. No consistent change in time to peak was noted among the watersheds (Harr et al. 1975).

Low Flows

Daily mean low flows measured in August and September in Needle Branch immediately after logging were higher than expected from the prelogging

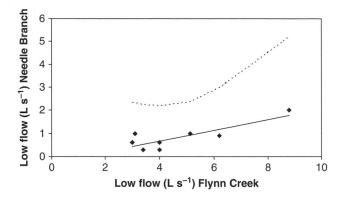

Fig. 2.12 Regression between low flows on Flynn Creek and Needle Branch and posttreatment low flow observations.

relation (Fig. 2.12), but generally decreased each subsequent year toward the prelogging relation (Table 2.7) (Harris 1977). Recalculation of the low flow regression between Flynn Creek and Deer Creek showed a different regression (Fig. 2.13) than presented earlier (Harris 1977). The mean low flows measured on Deer Creek were significantly lower than predicted.

Table 2.7 Minimum daily flow (L s^{-1}) for all three watersheds for the period of record

Water year	Flynn Creek	Needle Branch	Deer Creek
Prelogging period			
1959	7.08	1.42	14.16
1960	3.96	0.57	8.78
1961	5.10	0.57	9.63
1962	5.66	0.57	9.91
1963	5.95	0.85	13.31
1964	6.23	0.85	12.18
1965	3.40	0.28	6.80
Mean	5.38	0.85	10.76
Logging period			
1966	3.12	1.13	6.51
Post-logging period			
1967	2.55	0.57	5.10
1968	8.78	1.98	13.59
1969	5.10	1.13	9.06
1970	3.96	0.57	7.36
1971	6.23	0.85	11.33
1972	3.40	0.28	5.10
1973	3.96	0.28	5.38
Mean	4.81	0.85	8.21

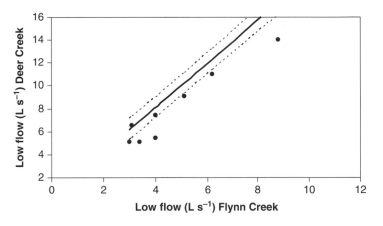

Fig. 2.13 Regression between low flows on Flynn Creek and Deer Creek and posttreatment low flow observations

Given the importance of low flows particularly as related to fish habitat and connectivity of pools, a subsequent analysis of low flows used the number of low flow days. The number of daily low flows less than $0.01 \ m^3 \ s^{-1} \ km^{-2}$ decreased for Needle Branch after logging when compared to Flynn Creek (Harr and Krygier 1972). The number of low flow days in Deer Creek decreased in two of the five posttreatment years (Harr and Krygier 1972).

Summary

Timber harvesting on Needle Branch increased annual water yield up to 31% over pretreatment conditions. The increases in annual water yields were greater in the wet years, and the posttreatment period of record did not suggest a hydrologic recovery or return to pretreatment water yields. Patch cutting with streamside vegetation in Deer Creek increased water yield by 3%. Timber harvesting did not increase mean peak flows on either treated watershed when compared to Flynn Creek. On Deer Creek, two of 16 peak flows were outside the confidence interval, and on Needle Branch three of 16 peak flows were outside the confidence interval. Daily low flows were increased on Needle Branch, and suggested a return to pretreatment conditions over time. The low flow response on Deer Creek showed streamflows lower than the pretreatment period. In general, additional research could be done on the effects of timber harvesting on streamflow responses.

This study was instrumental in illuminating the physical processes governing the hydrology of the temperate coniferous forest of the Pacific Northwest, and the changes in hydrologic process following timber harvest. The Alsea Watershed Study results, especially the effect of timber harvesting on water resources in Needle Branch is often cited as typical of forest management practices. It must be remembered that it was part of a study designed to have measurable responses. The understanding of hydrological processes as affected by timber harvesting with this study better afforded the development of best management practices (BMPs) designed to prevent or minimize adverse water resource damage.

Literature Cited

Brown, G.W. 1972. The Alsea Watershed Study. Pacific Logging Congress. Loggers Handbook 32:13–15, 127–130.

Chamberlin, T.W., Harr, R.D., and Everest, F.H. 1991. Timber harvesting, silviculture, and watershed processes, pp. 181–205. In: W.R. Meehan, editor. Influences of Forest and Rangeland Management on Salmonid Fishes and their Habitats. Amer. Fish. Soc. Spec. Publ. 19.

Hall, J.D., and Krygier, J.T. 1967. Progress Report to Federal Water Pollution Control Administration: Studies on Effects of Watershed Practices on Streams, May 1, 1963 through April 30, 1967. Oregon State Univ. Agric. Exp. Sta. and Forest Res. Lab, Corvallis, OR.

Harr, R.D. 1976. Hydrology of small forest streams in western Oregon. General Technical Report GTR-PNW-55. USDA Forest Service Portland, OR. 19pp.

Harr, R.D. 1980. Streamflow after patch logging in small drainages within Bull Run municipal watershed, Oregon. Research Paper RP-PNW-268. USDA Forest Service, Pacific Northwest Forest and Range Experiment Station, Portland, OR. 16pp.

Harr, R.D., Harper, W.C., Krygier, J.T., and Hsieh, F.S. 1975. Changes in storm hydrographs after road building and clear-cutting in the Oregon Coast Range. Water Resour. Res. 11:436–444.

Harr, R.D., and Krygier, J.T. 1972. Clearcut logging and low flows in Oregon coastal watersheds. Forest Research Lab. Research Note 54. Oregon State Univ., Corvallis, OR. 3pp.

Harr, R.D., Levno, A., and Mersereau, R. 1982. Streamflow changes after logging 130-year-old Douglas-fir in two small watersheds. Water Resour. Res. 18:637–644.

Harris, D.D. 1977. Hydrologic changes after logging in two small Oregon coastal watersheds. Water-Supply Paper 2037. U.S. Geological Survey Washington, DC. 31pp.

Hewlett, J.D., and Helvey, J.D. 1976. Effects of forest clear-felling on the storm hydrograph. Water Resour. Res. 6:768–782.

Meehan, W.R., editor. 1991. Influences of Forest and Rangeland Management on Salmonid Fishes and Their Habitats. American Fisheries Society Special Publ. 19, Bethesda, MD. 622pp.

Moring, J.R. 1975. The Alsea Watershed Study: effects of logging on the aquatic resources of three headwater streams of the Alsea River, Oregon. Part II. Changes in environmental conditions. Fish. Res. Rep. 9. Oregon Dept. of Fish and Wildlife, Corvallis, OR. 39pp.

Rothacher, J. 1970. Increases in water yield following clearcut logging in the Pacific Northwest. Water Resour. Res. 6:653–658.

Stednick, J.D. 1996. Monitoring the effects of timber harvest on annual water yields. J. Hydrol. 176:79–95.

Chapter 3
Stream Temperature and Dissolved Oxygen

George G. Ice

Stream temperature and the concentration of dissolved oxygen (DO) in water are important and traditional measures of water quality, especially for coldwater fish species such as salmonids. Sanitary engineers have used these measures for decades to assess the health of rivers where industrial wastes are released. One of the major advances in water-quality management was the development of oxygen–sag curve equations and their use in modeling protection needs for streams. The Alsea Watershed Study (AWS) showed that timber harvesting and yarding could depress DO when fresh slash was deposited in a stream. Removal of streamside shade was also found to increase stream temperatures, contributing to concerns about how forest practices could potentially influence water quality and the health of fish. Equally important, the AWS also identified management techniques that could minimize negative temperature and DO impacts, thus protecting aquatic organisms.

Stream Temperature and Fish Habitat

Water temperature is one of the most important factors affecting habitat quality for fish (Lantz 1971). Temperature influences a fish in three important ways: by directly influencing physiological rates, by affecting interspecific competition and fish pathogens, and by controlling biochemical rates and gas solubilities.

Because fish are poikilothermic, or "cold-blooded," their internal body temperature must adjust to their environment and their metabolism is closely tied to temperature. Within a range of temperatures, fish can be active and capture food to maintain activity and provide for growth. Too low a temperature reduces activity so that sufficient food cannot be obtained. Too high

George G. Ice
National Council for Air and Stream Improvement, Inc., Corvallis, OR 97339
gice@wcrc-ncasi.org

J. D. Stednick (ed.), *Hydrological and Biological Responses to Forest Practices.*
© Springer 2008

a temperature can result in a metabolism too rapid to supply sufficient energy for the fish, especially when food is limited (Warren 1971). For very high stream temperatures, direct mortality can result. Different species of fish and different life stages of these fish can withstand higher or lower temperatures. Salmon are among the fish most sensitive to increases in stream temperature. Bjornn and Reiser (1991) provided temperature ranges considered favorable for migration, spawning, incubation, and rearing for different salmon species. Table 3.1 summarizes temperatures recommended for salmonid fish that occur in the Alsea Watershed, including Chinook salmon (*Oncorhynchus tshawytscha*), coho salmon (*O. kisutch*), cutthroat trout (*O. clarkii*), and steelhead (*O. mykiss*).

If species are acclimated to elevated temperatures, their ability to survive high water temperatures is improved, but there are limits. Brett (1952) determined lethal temperatures (both high and low) for chum salmon (*O. keta*) that were acclimated at various temperatures. No species of salmon can survive extended exposure to water temperatures in excess of 25°C (77°F). Fluctuations in stream temperatures may be just as stressful as prolonged elevated temperatures. In fact, salmonids may acclimate more slowly to cooling water temperatures than to warming temperatures (Lantz 1971).

Salmon are also susceptible to diseases that are influenced by stream temperature. An EPA report summarized the temperature ranges in which common diseases result in mortality of infected fish (Fryer and Pilcher 1974). In general, higher stream temperatures result in higher mortality to salmonids.

Although lethal and optimum temperatures for fish production are often considered, the influence of stream temperature on development, migration patterns, and interspecific competition is less often recognized. Both the AWS and, later, the Carnation Creek Study in British Columbia, detected more rapid emergence and development of salmon fry due to increased stream temperatures (Hartman and Scrivener 1990). This led to the early migration of some smolts out of these headwater stream systems. Some researchers believe that smolts that migrate out of headwater streams too early are at a competitive disadvantage. Interspecific competition and temperature have been studied for a number of species, including steelhead and redside shiner (*Richardsonius balteatus*) (Reeves et al. 1987). Temperature influence on

Table 3.1 Recommended temperatures for salmonids reported by Bjornn and Reiser (1991) in °C

Species	Migration	Spawning	Incubation	Rearing (preferred)	Upper[1] Lethal
Chinook salmon	10.6–19.4	5.6–13.9	5.0–14.4	12–14	26.2
Coho salmon	7.2–15.6	4.4–9.4	4.4–13.3	12–14	26.0
Cutthroat trout	–	6.1–17.2	–	–	22.8
Steelhead	–	3.9–9.4	–	10–13	23.9

[1] Species acclimated to 20°C

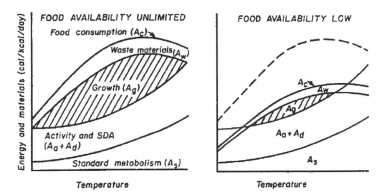

Fig. 3.1 Fish growth in response to varying temperatures and food supplies (Warren 1971)

interspecific competition is an important consideration in bull trout (*Salvelinus confluentus*) habitat, particularly where there is competition with brook trout (*S. fontinalis*) (Hillman and Essig 1998). Research is underway to better understand how changes in stream temperature will influence interspecific competition.

One anomaly associated with stream temperature is the existence of fish in streams where water temperatures occur that are considered lethal for those species. In some cases, fish are believed to avoid these adverse conditions through use of thermal refuges such as cool tributaries, thermally stratified pools, and cool effluent groundwater in streambed interstices. Refuge to warmer streambed interstices during cold winter periods may be important to fish survival in some regions, and there is concern about filling-in of gravels with fines. Changes in food supply may also shift temperature requirements. Warren (1971) presented theoretical curves for fish response to varying temperatures and food supplies (Fig. 3.1) which showed that when food supply is abundant, fish can survive and grow at temperatures that are limiting for low food availability conditions. Where instream primary production is increased due to removal of riparian shade or to an increase in nutrients, food supplies to fish may ultimately increase. This type of response was observed for the high salmonid productivity in streams exposed following the eruption of Mt. St. Helens (Bisson et al. 1988).

Response of Stream Temperature to Forest Management in the Alsea Watershed Study

Recording thermographs were used during the AWS to measure stream temperature. Locations and number of recording thermographs varied over the years. Moring (1975) reported that "in most years prior to 1969, six units were positioned along Needle Branch, 11 units were placed along Deer Creek, and one unit was located at the stream gauging station in Flynn Creek." Temperature-monitoring stations changed after 1969 to two units in Deer Creek, six units in Needle Branch, and one unit in Flynn Creek. Other temperature monitoring was conducted as part of biological monitoring, including intragravel water temperature measurements to assess the environment in salmon redds.

Prior to harvesting, the three Alsea Watershed Study streams had stream temperature patterns that were very similar (Table 3.2). Annual mean water temperatures recorded during the prelogging period were 9.7°C for Flynn Creek, 9.6°C for Deer Creek, and 9.6°C for Needle Branch. Diurnal fluctuations were similar for these three streams. Minimums and maximums recorded were 2.2°C to 16.7°C for Flynn Creek, 1.1°C to 16.1°C for Deer Creek, and 1.7°C to 16.1°C for Needle Branch.

Following logging, Deer Creek and Needle Branch both showed increases in temperature. However, those changes were dramatically larger for Needle Branch, which lacked the riparian buffer provided in the Deer Creek watershed. Maximum diurnal fluctuations prior to harvesting and site preparation were 3.9°C for Flynn Creek, 4.5°C for Deer Creek, and 3.3°C for Needle Branch. In 1967, after harvesting, Deer Creek had an increase in maximum stream temperature of only 1.7°C above the pretreatment maximum. A maximum diurnal temperature fluctuation of 7.3°C was observed after treatment. In contrast to these modest changes, Needle Branch experienced large increases in maximum stream temperature and in diurnal temperature fluctuation.

Harvesting and site preparation had both immediate and more prolonged effects on stream temperature for Needle Branch. Prior to harvesting (1959–1965), monitoring at the gauging station on Needle Branch had never shown a stream temperature greater than 16.1°C (August 1961). The immediate impact of the prescribed burn was a rapid increase in temperature, particularly in the upper stretches of Needle Branch. During slash burning, stream temperatures rose from 13°C to above 28°C in the upper canyon of Needle

Table 3.2 Comparison of water temperatures (°C) of the Alsea Watershed for pretreatment (1959–1965), posttreatment (1966–1973)[1], and New Alsea Watershed Study (NAWS)[2]

	Flynn Creek			Deer Creek			Needle Branch		
	Pre	Post	NAWS	Pre	Post	NAWS	Pre	Post	NAWS
Annual mean	9.7	–	–	9.6	–	–	9.6	–	–
Max. Daily Range	3.9	3.4	2.7[2]	4.5	7.3	4.1[2]	3.3	12.0	2.7[2]
Minimum	2.2	1.7	5.2	1.1	2.2	2.2	1.7	1.3	2.2
Maximum	16.7	15.0	16.1	16.1	17.8	16.7	16.1	26.1	16.1

[1] U.S. Geological Survey. Water years 1959–1973. Water resources data – Oregon.
[2] Values estimated from daily minimum/maximum chart records.

Branch. Mortality was observed for juvenile coho salmon, cutthroat trout, and reticulate sculpin (*Cottus perplexus*) (Hall and Lantz 1969).

In 1966, after harvesting but before site preparation, Needle Branch at the gauging station had a maximum stream temperature of 22.8°C. In 1967, following burning and removal of debris in the stream channel, the stream temperature maximum at the gauging station was 26.1°C (July). A maximum diurnal temperature fluctuation at one of the upstream gauges of 15.6°C and a maximum temperature of 29.5°C were measured in Needle Branch (15.6°C increase over the 1965 maximum for the same station) (Brown and Krygier 1970). In comparison, Flynn Creek had a preharvest-period maximum stream temperature of 16.7°C (August and September 1961) but experienced a maximum temperature of only 14.4°C in 1966 and 15.0°C in 1967. Figure 3.2 shows the temperature pattern on the days of annual maximum temperature for Needle Branch before and after treatment, compared to the temperatures observed for Flynn Creek. It is clear that harvesting of the forest canopy in Needle Branch opened up the stream to increased solar radiation and warming. The prescribed burn and stream cleanup further exposed Needle Branch.

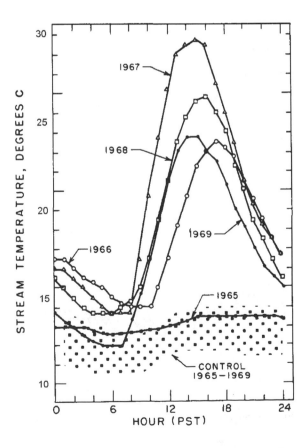

Fig. 3.2 Comparison of temperature pattern for the days of annual maximum temperature of Needle Branch vs. Flynn Creek, before and after treatment (Brown and Krygier 1970)

Deer Creek Water Year 1993
Maximum and Minimum Daily Temperatures

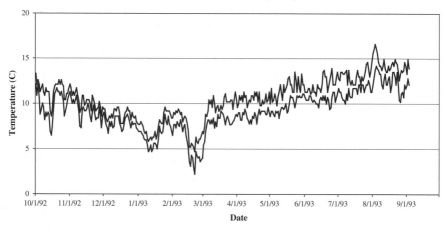

Flynn Creek Water Year 1993
Maximum and Minimum Daily Temperatures

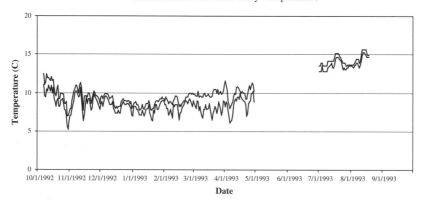

Needle Branch Water Year 1993
Maximum and Minimum Daily Temperatures

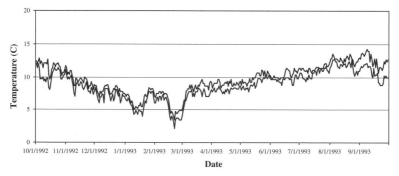

Fig. 3.3 Stream temperatures of Needle Branch, Flynn Creek, and Deer Creek in water year 1993

Intragravel temperatures were also elevated in Needle Branch, although the effects were muted compared to those observed for the surface waters (Ringler and Hall 1975). Intragravel water temperatures tended to lag behind those observed in the surface waters and there was considerable spatial variation in absolute temperatures and diurnal fluctuations. Monitoring was focused on winter and spring, when temperature had a direct influence on embryos and alevins. Temperature effects in the summer, which exceeded those measured in the winter, could influence macroinvertebrate populations in these types of streams.

Temperatures returned to near pretreatment conditions at the end of the AWS (Moring 1975). In 1997, during the New Alsea Watershed Study (NAWS), Needle Branch had a maximum stream temperature of 16.1°C, which is exactly the same as the pretreatment maximum (Table 3.2). Figure 3.3 shows temperatures for Flynn Creek, Deer Creek, and Needle Branch in 1993. Maximum temperatures are near the pretreatment values and diurnal fluctuations appear to have also returned to near pretreatment levels. For example, during 1993, Needle Branch experienced a maximum temperature of 14.1°C, which was less than the maximum seen for Flynn Creek (15.6°C).

Discussion of Temperature Response

The immediate heating of water during the prescribed burn is consistent with some studies, but contrasts with others where riparian areas did not burn. For example, stream temperature monitoring during a brown-and-burn site-conversion project on the Oregon Coast showed no increase in stream temperature associated with the burn (Ice 1980). However, fires in forested watersheds have resulted in rapid elevation of stream temperature from both heat of combustion and (more importantly for long-term response) from the removal of shade. After the fire on the Entiat Experimental Forest in Washington State, Berndt (1971) concluded that a drop in streamflow resulted from "vaporization of water from live stream surface ventilated by strong convection currents ..." although the downstream temperature recorder showed "no drastic immediate change ..." In assessing the effects of fire on stream temperature, Tiedemann et al. (1969) concluded that "water temperature has been shown to change markedly regardless of how shade is removed." Helvey et al. (1976) reported summer stream temperature increases following wildfires, and Levno and Rothacher (1969) reported significant stream temperature increases following prescribed burning and debris clean-out of the stream.

The AWS showed that exposure of small headwater streams to solar radiation is certainly a dominant process for stream temperature increases. This is particularly useful information to managers because maintenance of shade can be used as a management tool to avoid temperature increases.

However, other mechanisms that influence stream heating include air temperature increases associated with removal of shade within the watershed; soil warming and its effects on groundwater temperatures; channel widening as a result of debris flows or aggradation, channel meandering, or channel-type change; and changes in streamflow. All of these influence stream temperature response to greater or lesser degrees.

The role of shade in minimizing increases in stream temperatures has been shown in numerous paired watershed studies comparing streams with and without buffers. Yet, some continue to argue that shade from vegetation does not influence stream temperature (Larson and Larson 1996). Beschta (1997) showed that shade from vegetation is important in regulating stream temperatures in forested watersheds.

The temperature of a stream is constantly moving toward equilibrium with the temperature of the atmosphere to which it is exposed. Sullivan et al. (1990) showed that basin air temperature can be used to predict stream temperatures. Removal of shade in a watershed can increase air temperatures near streams, potentially shifting the rate and magnitude of stream temperature response. The width of riparian buffer needed to maintain the microclimate near a stream is an area of active research. Preliminary results show that streams are not only influenced by air temperature but that they in turn influence humidity and air temperature in their vicinity.

Some have argued that warming of soil exposed to direct solar radiation following harvesting can increase soil-water temperature, thus influencing stream temperatures as groundwater flows into the stream. This mechanism has been suggested by Hewlett (1979) and Hartman and Scrivener (1990). In the Carnation Creek Watershed Study in Canada, Hartman and Scrivener speculated that "upslope cutting and post-logging slash burning may increase stream temperatures during the winter by causing groundwater, which is replaced during the first autumn rains, to be warmer." Todd (1959) found that one of the most conservative properties of groundwater is temperature; annual variations under ordinary conditions are almost negligible. The insulating qualities of the earth's crust damp out the extreme temperature variations found at ground surface. Studies have shown that the annual range of the earth's temperature at a depth of 9 m may be expected to be less than $0.5°C$. Analysis of thousands of records of groundwater temperature in the United States revealed that the temperature of groundwater occurring at a depth of 9 to 18 m will generally exceed the mean annual air temperature by $1.1°C$ to $1.7°C$.

As with most heating mechanisms, special conditions can lead to a response. For example, Ice (1980) monitored a shallow spring in an open field in the Oregon Coast Range and found that this source ranged from $10.0°C$ to $12.3°C$ over a season, about four times the range reported by Todd. Temperatures for this spring followed a pattern similar to but delayed from the air temperature recorded at the site. Nevertheless, $2°C–3°C$ is still a relatively small temperature range and at least a portion of this temperature pattern would occur with or without a forest cover. Changes in groundwater temperature are an unlikely source of increased maximum stream temperatures. Summer groundwater sources, when stream temperatures are at their maximum, are generally deep. Also, paired watershed studies have shown only minimal responses to clearcutting where buffers were used to shade streams.

Streamflow is also an important variable. Small, shallow, slow-moving streams with large surface areas are more susceptible to rapid heating (Brown 1969; Brown 1972b). In the AWS, it was the upper sections of Needle Branch, with shallow, low flows, that experienced the greatest increases in temperature (Brazier and Brown 1973).

Although small streams may be more susceptible to heating, they also recover from elevated temperatures more rapidly. Andrus and Froehlich (1991) studied the recovery of coastal Oregon riparian forest streams after disturbance (including the AWS streams). They concluded that "within ten years, stream shading was similar to that provided by developed forests . . . "

Channel widening is a potentially serious and long-lasting process that leads to increased stream temperatures. A debris torrent in Watershed 3 of the H. J. Andrews Experimental Forest in Oregon resulted in elevated temperatures along a scoured section of the stream (Levno and Rothacher 1969). McSwain (1987) found that, in the Elk River Basin of coastal Oregon, diurnal stream temperature fluctuations increased as the proportion of the basin covered with landslides increased (aggradation causing increased width-to-depth ratios). In areas of even greater numbers of landslides, this trend reversed as flow came to be subsurface. Holaday (1992) reported that the 1964–1965 floods in Oregon caused significant damage to riparian vegetation in the Steamboat Creek Basin, resulting in elevated stream temperatures. The 1996 floods in Oregon caused extensive channel scouring and deposition, which can be expected to increase insolation of streams (Robison 1997). A recent stream temperature Total Maximum Daily Load (TMDL) for the Sucker/Grayback Watershed in southwest Oregon (Park and Boyd 1998) addressed not only shade requirements but also riparian conservation reserves to reestablish favorable (narrow) channel morphology.

An additional stream temperature consideration, which was at least partially addressed in the AWS, was the transport of energy downstream and the potential for cumulative stream temperature effects. Even for Needle Branch with no buffer, elevated temperatures in the upper headwater reaches did not continue to increase downstream and instead were lower at the main gauging station. Holaday (1992) showed that temperature changes in headwater tributaries may have little influence on mainstem stream temperatures. Zwieniecki and Newton (1999) found that small increases in stream temperature through buffered clearcuts returned to the normal temperature trend line within 150 m downstream.

The AWS taught us that we could harvest trees in a watershed without causing a major increase in stream temperature if riparian shade was retained. More subtle heating processes may occur, but studies of areas with histories of basin-wide harvesting patterns show that well-buffered streams maintain temperatures near those found for unmanaged and undisturbed forests (Moring 1975; McSwain 1987; Holaday 1992).

Dissolved Oxygen and Fish Habitat

The concentration of DO in forest streams is important for aquatic organisms, including fish. A number of early forest watershed studies revealed the potential for depressed DO concentrations when harvesting occurred near streams (Krammes and Burns 1973; Feller 1974; Moring 1975). The most influential of these studies was the AWS. During this study, surface and intragravel DO concentrations for Needle Branch were reduced to 2.5 and $1.3\,\text{mg}\,\text{L}^{-1}$, respectively, following harvesting on the watershed (Hall and Krygier 1967; Moring 1975). These results led to forest practice rules that maintain shade and keep organic debris out of streams (Ice 1989; Ice et al. 1989; Ice et al. 1994). Rules were developed that required that a minimum amount of shade be retained near streams and that equipment and logging slash be kept out of streams. Forest practice rules require that any slash that does reach the channel must be removed and placed above the high-water line of the channel.

The "atmosphere" of water is important in determining its suitability for aquatic organisms. For aerobic organisms, the most important gas in water is oxygen, which is necessary for respiration to convert sugars and other materials into energy (see equation below).

$$C_2H_{12}O_6 + 6O_2 \rightarrow 6CO_2 + 6H_2O$$

respiration (oxidation of sugar)

Dissolved oxygen is considered a "sparingly" soluble gas because the amount in solution is so small. The volumetric proportion of oxygen in the atmosphere is about 21%, while the concentration of oxygen in water is well below 1% (by weight). The problem that fish have in obtaining sufficient oxygen is also increased because diffusion of oxygen is much slower in water than in the atmosphere.

Fish obtain oxygen by passing water across their gills. Gills are a series of finely dissected surfaces with a thin membrane (only two cells) covering blood vessels. Oxygen diffuses across the cells and is picked up by the blood (Keeton 1967).

Just as temperature affects all fish activity, so does DO. Bjornn and Reiser (1991) reported studies showing reduced swimming speeds for salmonids when DO dropped below 6.5 to 7.0 mg L^{-1}. They also reported relations between DO concentration and salmonid embryo survival and salmonid weight gain. Brett and Blackburn (1981) found a critical oxygen concentration of around 4.0 mg L^{-1} at 15°C for fingerling coho and sockeye salmon growth. Thurston et al. (1981) showed a synergistic effect between DO and NH_3. For rainbow trout the toxicity of NH_3 increased with decreasing DO.

Unlike most water quality standards, for which baseline values are difficult to determine, the baseline criteria for DO can be readily established. Surface water

DO concentrations for steep, turbulent forest streams in nutrient-poor systems (typical of the Pacific Northwest) are at or near saturation most of the year. The DO saturation concentration is a function of the water temperature and barometric pressure. The solubility of oxygen is inversely related to water temperature (Maron and Prutton 1958). As the barometric pressure increases, the partial pressure of oxygen also increases, which raises the saturation concentration. Forest operations can modify dissolved oxygen by increasing temperature and consequently decreasing DO solubility, by introducing organic matter that is decomposed by microorganisms (with the consumption of DO), and by inhibiting the relative rate of oxygen input to the water (reaeration) by modifying streamflow.

The introduction of organic matter (such as logging slash) can create a demand for oxygen. This demand for oxygen is called the Biochemical Oxygen Demand (BOD) to reflect both an inorganic and biological demand for oxygen. Work by Ponce (1973, 1974) and by Berry (1975) provided information on how much BOD various forest slash components exert. The rate of oxidation increases as temperature increases. When we consider the oxygen load imposed by various slash-material sizes, it becomes apparent that needles, leaves, and other fine materials provide the greatest oxygen loads for short periods. This is because fine material is oxidized at a much faster rate than is large material. When slash is utilized for food by aquatic organisms, oxygen is also utilized and a deficit between the DO saturation concentration and actual concentration occurs. Mathematically, the BOD concentration L, in mg L^{-1}, is related to the oxygen deficit D, in mg L^{-1}, by the equation:

$$-\frac{\partial L}{\partial t} = \frac{\partial D}{\partial t}$$

When a deficit in DO exists, there is a disequilibrium and a tendency for more molecules of oxygen to move from the atmosphere into solution. This process is known as reaeration which is a function of the oxygen deficit in the solution and a reaeration rate constant K_2. Reaeration in small turbulent Oregon forest streams has been found to be very rapid and is a function of energy dissipation and stream depth (Ice 1978, 1990; Ice and Brown 1978).

In addition to fines, which exert a rapid BOD on streamwater, large organic material can dam streams and cause deep, quiescent pools where more organic material is in contact with water and reaeration is less rapid. Stream water mixing is inhibited by seeping or leaking debris jams with coarse and fine slash and sediment.

Dissolved Oxygen Response in the Alsea Watershed

During the AWS, a worst-case scenario for DO impacts occurred at Needle Branch. In spring 1966 nearly the entire watershed was clearcut down to the stream edge, and then yarded using a high-lead cable system. The stream was

choked with sediment and organic debris and exposed to sunlight. In September
the stream channel was cleaned of organic debris, and the watershed was site-
prepared by broadcast burning in October. This further exposed the stream to
sunlight (see discussion on temperature response) but also lowered the BOD
load on the stream. None of the best management practices (BMPs) currently
considered standard for fish-bearing streams, such as riparian protection zones,
immediate stream cleaning, or precluding the use of fire in and around streams,
were implemented. Surface and intragravel DO concentrations were monitored
in Needle Branch before and after the treatment.

The effect of timber harvesting on the DO content of the surface water in Needle
Branch the first summer after harvesting (1966) is presented in Fig. 3.4. The data,

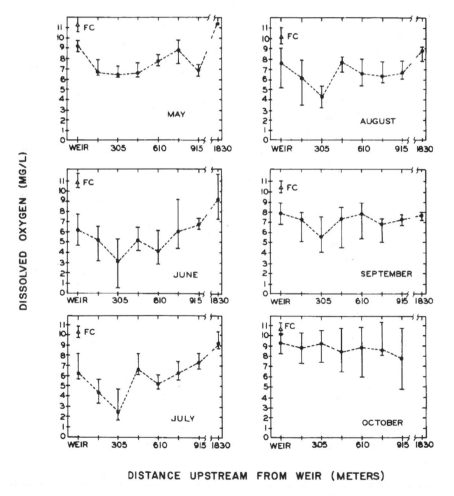

Fig. 3.4 Surface dissolved oxygen levels at eight stations in Needle Branch and at the Flynn
Creek weir, May through October 1966

showing the monthly mean and range of surface DO, were taken twice weekly at eight locations upstream of the weir in Needle Branch and at the weir in Flynn Creek. These results show clearly that the surface DO concentrations were depressed in Needle Branch during the summer months of 1966. Not only did DO concentrations in Needle Branch fall markedly below the levels in the control stream, Flynn Creek, but they also approached or went below values considered limiting to growth of juvenile coho salmon (about 4 to 6 mg L^{-1}). In July, the surface DO levels actually went below lethal levels for juvenile coho salmon (Moring 1975). In subsequent summers after 1966, mean monthly surface DO levels in Needle Branch remained lower than the values for the control stream, Flynn Creek, but never again approached the limiting values that were recorded during summer 1966. The differences in surface DO levels in the summers after 1966 are primarily attributed to the increased stream temperatures in Needle Branch. The limiting and lethal values of DO observed in Needle Branch during summer 1966 are attributed to the combination of the effects of increased stream temperature, reduced reaeration rates, and an increased BOD from logging slash (Hall and Lantz 1969).

Intragravel DO levels were also monitored in Needle Branch during the winter and spring using both permanent and temporary standpipes (Moring 1975; Ringler and Hall 1975). The results from this monitoring are shown in Fig. 3.5. These results show that posttreatment average intragravel DO levels fell below pretreatment levels and below limiting values for optimum fry survival. Intragravel dissolved oxygen dynamics are more complex than surface dissolved oxygen dynamics. Incorporation of organic material into the stream gravel (Garvin 1974), effluent groundwater and surface water exchange, and water temperature all play a role. It is difficult to identify the contribution of these factors to the reduction in DO observed in the streambed. Ringler and Hall (1975) attributed these reductions to the entrainment of both fine organics and inorganics in stream gravels. This would increase intragravel BOD, reduce intragravel water velocities, and reduce mixing of depleted intragravel waters with aerated surface waters. They also noted great spatial variations and that DO reductions seen with standpipes in redds appeared to be less than those observed for permanent standpipes. All these results are clouded, however, by reductions in intragravel DO observed in the control stream, Flynn Creek, and throughout the period of monitoring. The permanent standpipes used to collect intragravel DO samples may have somehow promoted clogging of intragravel sites, leading to a long-term reduction in DO immediately around these locations.

As a consequence of the DO concentrations observed during the AWS, a series of studies was conducted at Oregon State University to determine how DO concentrations in forest streams could be protected in surface and in intragravel water (Holtje 1971; Ponce 1973; Berry 1975; McGreer 1975; Ice 1978; Skaugset 1980). Slow-moving, sun-exposed streams with low reaeration rates and prolonged contact with organic material are susceptible to oxygen deficits. Methods have been developed to predict potential stream temperature changes related to harvesting (Brown 1972a; McGurk 1989; Sullivan et al. 1990), and there are

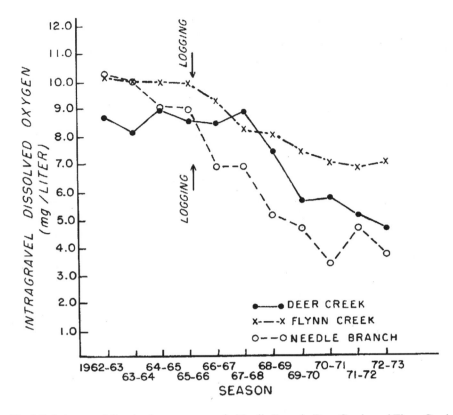

Fig. 3.5 Intragravel dissolved oxygen means in Needle Branch, Deer Creek, and Flynn Creek for winter and spring, 1962 to 1973 (Moring 1975)

some data on potential oxygen demand from slash (Atkinson 1971; Servizi et al. 1971; Ponce 1973, 1974; Schaumburg 1973; Berry 1975). Streams with low reaeration rates are particularly susceptible to depressed DO concentrations and need to be identified (Ice 1990). Even though many small forest streams are important for fish spawning and rearing, they tend to be quite different than streams traditionally studied for reaeration rates. Small forest streams can have very low flows in the summer as well as irregular channel geometries.

Because fines (e.g., needles and leaves) are difficult to remove from a stream and because their BOD is largest when first introduced, the best approach to controlling DO is to keep slash out of the stream. If slash is introduced and causes a DO deficit, then debris that impounds flow, traps fine debris, and inhibits mixing can be removed. An integrated evaluation of possible DO problems needs to be made because forest organic matter also has benefits for streams. Forest organic matter is a major food source for stream organisms; modifies stream habitat by increasing pool volume, habitat diversity, and cover; and helps to stabilize streambanks (Bisson et al. 1987).

Although this discussion has focused on surface DO, it is the intragravel DO that is critical during embryo development. Biochemical Oxygen Demand may be an even more important variable for intragravel DO because the reaeration process is not active in the streambed, and mixing of reaerated water from the surface may be inhibited. Of equal concern are unrealistic water-quality standards for intragravel DO. Skaugset (1980) found no significant difference in intragravel DO depressions for undisturbed, partially harvested, and completely harvested watersheds. He did find that increases in inorganic fines were associated with decreased intragravel DO.

Changes are now being proposed in forest practices that emphasize long-term sustainability of environmental values. These sustainable practices incorporate an increased awareness of the role of large woody debris (LWD) in providing stream structure and habitat for fish (Swanson and Lienkaemper 1978; Wilson 1986; Bisson et al. 1987; Gregory 1987). Changes have been made in forest practices rules for Washington, Oregon, and Idaho requiring that trees be left to provide not only shade but also future sources of LWD for streams (Ice 1989; Ice et al. 1994). This reverses a trend in stream "enhancement" activities that had resulted in removal or modification of both fresh slash and LWD in streams (Froehlich 1975). In some cases, windthrow has resulted in a rapid delivery to streams of trees left for shade and LWD recruitment, with consequent decline in water quality and criticism of the stream management zone policies (Mussallem and Lynch 1980; DeWalle 1983; Steinblums et al. 1984). These rule changes could renew concerns about DO in managed forest streams.

The Alsea Watershed Study taught forest hydrologists that DO concentrations could be depressed severely in small forest streams with massive accumulations of fresh slash and exposure of the stream to sunlight. Practices that avoid these conditions will maintain DO concentrations near saturation in surface waters of turbulent forest streams. Intragravel DO can also be depressed with incorporation of fresh organic and inorganic material into the gravels and with heating of the stream. Management solutions are readily available to avoid these circumstances and have been incorporated into state forest practice rules.

Literature Cited

Andrus, C.W., and Froehlich, H.A. 1991. Riparian forest development after logging or fire in the Oregon Coast Range: wildlife habitat and timber value. National council of the Paper Industry for Air and Stream Improvement, Inc., New York.

Atkinson, S.W. 1971. BOD and Toxicity of Log Leachate. M.S. Thesis. Oregon State Univ., Corvallis, OR. 58pp.

Berndt, H.W. 1971. Early effects of forest fire on streamflow characteristics. Res. Note PNW 148. USDA Forest Service, Portland, OR. 9pp.

Berry, J.D. 1975. Modeling the Impact of Logging Debris on the Dissolved Oxygen Balance of Small Mountain Streams. M.S. Thesis. Oregon State Univ., Corvallis, OR. 163pp.

Beschta, R.L. 1997. Riparian shade and stream temperature: an alternative perspective. Rangelands 19(2):25–28.

Bisson, P.A., Bilby, R.E., Bryant, M.D., Dolloff, C.A., and coauthors. 1987. Large woody debris in forest streams in the Pacific Northwest: past, present, and future, pp. 143–190. In: E.O. Salo and T.W. Cundy, editors. Streamside Management: Forestry and Fishery Interactions. Univ. of Washington Inst. of Forest Resour. Contrib. 57, Seattle, WA.

Bisson, P.A., Sullivan, K., and Nielsen, J.L. 1988. Channel hydraulics, habitat use, and body form of juvenile coho salmon, steelhead, and cutthroat trout in streams. Trans. Amer. Fish. Soc. 117:262–273.

Bjornn, T.C., and Reiser, D.W. 1991. Habitat requirements of salmonids in streams, pp. 83–138. In: W.R. Meehan, editor. Influences of Forest and Rangeland Management on Salmonid Fishes and their Habitats. Amer. Fish. Soc. Spec. Publ. 19.

Brazier, J.R., and Brown, G.W. 1973. Buffer strips for stream temperature control. Res. Pap. 15. Forest Research Laboratory, Oregon State Univ., Corvallis, OR. 9pp.

Brett, J.R. 1952. Temperature tolerance in young Pacific salmon, genus *Oncorhynchus*. J. Fish. Res. Board Canada 9:265–323.

Brett, J.R., and Blackburn, J.M. 1981. Oxygen requirements for growth of young coho (*Oncorhynchus kisutch*) and sockeye (*O. nerka*) salmon at 15 degrees C. Can. J. Fish. Aquat. Sci. 38:399–404.

Brown, G.W. 1969. Predicting temperatures of small streams. Water Resour. Res. 5:68–75.

Brown, G.W. 1972a. The Alsea Watershed Study. Pacific Logging Congress. Loggers Handbook 32:13–15, 127–130.

Brown, G.W. 1972b. An improved temperature prediction model for small streams. WRRI 16. Water Resources Research Institute, Oregon State Univ., Corvallis, OR. 24pp.

Brown, G.W., and Krygier, J.T. 1970. Effects of clear-cutting on stream temperature. Water Resour. Res. 6:1133–1139.

DeWalle, D.R. 1983. Wind damage around clearcuts in the Ridge and Valley Province of Pennsylvania. J. Forestry 81:158–159.

Feller, M.C. 1974. Initial Effects of Clearcutting on the Flow of Chemicals through a Forest Watershed Ecosystem in Southwestern British Columbia. Ph.D. Thesis. University of British Columbia, Vancouver, BC.

Froehlich, H.A. 1975. Accumulation of large debris in forest streams before and after logging. In: Logging Debris in Streams, Notes for a Workshop. Oregon State Univ., Corvallis, OR. 10pp.

Fryer, J.L., and Pilcher, K.S. 1974. Effects of temperature on diseases of salmonid fishes. U.S. Environmental Protection Agency Research Series EPA 600/3-73-020. Washington, DC. 114pp.

Garvin, W.F. 1974. The intrusion of logging debris into artificial gravel streambeds. WRRI 27. Water Resources Res. Inst. Oregon State Univ., Corvallis, OR. 88pp.

Gregory, S.V. 1987. Aquatic production – some good news and some bad news, pp. 28–34. In: Managing Oregon's Riparian Zone for Timber, Fish and Wildlife. National Council of the Paper Industry for Air and Stream Improvement, Inc., New York, NY.

Hall, J.D., and Krygier, J.T. 1967. Progress Report to Federal Water Pollution Control Administration: Studies on Effects of Watershed Practices on Streams, May 1, 1963 through April 30, 1967. Oregon State Univ. Agric. Exp. Sta. and Forest Res. Lab, Corvallis, OR.

Hall, J.D., and Lantz, R.L. 1969. Effects of logging on the habitat of coho salmon and cutthroat trout in coastal streams, pp. 355–375. In: T.G. Northcote, editor. Symposium on Salmon and Trout in Streams. Univ. of British Columbia, H.R. MacMillan Lectures in Fisheries, Vancouver, BC.

Hartman, G.F., and Scrivener, J.C. 1990. Impacts of forest practices on a coastal stream ecosystem, Carnation Creek, British Columbia. Can. Bull. Fish. Aquat. Sci. 223. 148pp.

Helvey, J.D., Tiedemann, A.R., and Fowler, W.B. 1976. Some climatic and hydrologic effects of wildfire in Washington State, pp. 201–222. In: Proceedings 15th Annual Tall Timbers Fire Ecology Conference. Tall Timbers Research Station, Tallahassee, FL.

Hewlett, J.D. 1979. Forest Water Quality. An Experiment in Harvesting and Regenerating Piedmont Forest. School of Forest Resources, Univ. of Georgia, Athens, GA. 22pp.

Hillman, T.W., and Essig, D. 1998. Review of Bull Trout Temperature Requirements: A Response to the EPA Bull Trout Temperature Rule. Idaho Division of Environmental Quality, BioAnalysts, Inc., Boise, ID. 55pp.

Holaday, S.A. 1992. Summertime Water Temperature Trends in Steamboat Creek Basin, Umpqua National Forest. M.S. Thesis. Oregon State Univ., Corvallis, OR. 128pp.

Holtje, R.K. 1971. Reaeration in Small Mountain Streams. Ph.D. Thesis. Oregon State Univ., Corvallis, OR. 154pp.

Ice, G.G. 1978. Reaeration in a Turbulent Stream System. Ph.D. Thesis. Oregon State Univ., Corvallis, OR. 182pp.

Ice, G.G. 1980. Immediate Stream Temperature Response to a Prescribed Burn. Spec. Rept. West Coast Regional Center, Natl. Council of the Paper Industry for Air and Stream Improvement, Inc, Corvallis, OR.

Ice, G.G. 1989. Guidelines and approaches for forest riparian management: state forest practice rules, pp. 94–98. In: Proc. 1989 Natl. Convention, Soc. American Foresters, Bethesda, MD.

Ice, G.G. 1990. Dissolved oxygen and woody debris: detecting sensitive forest streams, pp. 333–345. In: S.C. Wilhelms and J.S. Gulliver, editors. Air and Water Mass Transfer. Amer. Soc. Civil Eng., New York, NY.

Ice, G.G., Beschta, R.L., Craig, R.S., and Sedell, J.R. 1989. Riparian protection rules for Oregon forests. Proceedings of the California Riparian Systems Conference. USDA Forest Service Gen. Tech. Report PSW-110:533–536, Berkeley, CA.

Ice, G.G., and Brown, G.W. 1978. Reaeration in a turbulent stream system. WRRI 58. Water Resour. Res. Inst. Oregon State Univ., Corvallis, OR. 96pp.

Ice, G.G., Megahan, W.F., McGreer, D.J., and Belt, G.H. 1994. Streamside management in northwest forests, pp. 293–298. In: Managing Forests to Meet Peoples' Needs. Proc. 1994. Nat. Convention Soc. Amer. Foresters, Bethesda, MD.

Keeton, W.T. 1967. Biological Science. W.W. Norton & Co., New York, NY.

Krammes, J.S., and Burns, D.M. 1973. Road construction on Caspar Creek Watershed. 10-year report on impact. Research Paper PSW-93. USDA Forest Service, Pacific Southwest Forest and Range Experiment Station, Berkeley, CA. 10pp.

Lantz, R.L. 1971. Influence of water temperature on fish survival, growth and behavior, pp. 182–193. In: J.T. Krygier and J.D. Hall, editors. Forest Land Uses and Stream Environment. Oregon State Univ., Corvallis, OR.

Larson, L.L., and Larson, S.L. 1996. Riparian shade and stream temperature: a perspective. Rangelands 18(4):149–152.

Levno, A., and Rothacher, J. 1969. Increases in maximum stream temperatures after slash burning in a small experimental watershed. Research Note PNW-110. USDA Forest Service, Pacific Northwest Forest Range and Experiment Station, Portland, OR. 7pp.

Maron, S.H., and Prutton, C.F. 1958. Principles of Physical Chemistry. The MacMillan Co, New York, NY. 789pp.

McGreer, D.J. 1975. Stream Protection and Three Timber Falling Techniques-a Comparison of Costs and Benefits. M.S. Thesis. Oregon State Univ., Corvallis, OR. 92pp.

McGurk, B.J. 1989. Predicting stream temperature after riparian vegetation removal, pp. 157–164. In: Proc. Calif. Riparian Systems Conf. General Technical Report PSW-110. USDA Forest Service Berkeley, CA.

McSwain, M.D. 1987. Summer Stream Temperatures and Channel Characteristics of a Southwestern Oregon Coastal Stream. M.S. Thesis. Oregon State Univ., Corvallis, OR. 106pp.

Moring, J.R. 1975. The Alsea Watershed Study: effects of logging on the aquatic resources of three headwater streams of the Alsea River, Oregon. Part II. Changes in environmental conditions. Fish. Res. Rep. 9. Oregon Dept. of Fish and Wildlife, Corvallis, OR. 39pp.

Mussallem, K.E., and Lynch, J.A. 1980. Controlling nonpoint source pollution from commercial clearcuts, pp. 669–681. In: Symposium on Watershed Management 1980. Amer. Soc. Civil Eng., New York, NY.

Park, C., and Boyd, M. 1998. Sucker/Grayback total maximum daily load (TMDL). Oregon Dept. Environmental Quality, USDA Forest Service, Portland, OR. 33pp.

Ponce, S.L. 1973. The Biochemical Oxygen Demand of Douglas-fir Needles and Twigs, Western Hemlock Needles and Red Alder Leaves in Stream Water. M.S. Thesis. Oregon State Univ., Corvallis, OR. 141pp.

Ponce, S.L. 1974. The biochemical oxygen demand of finely divided logging debris in stream water. Water Resour. Res. 10:983–988.

Reeves, G.H., Everest, F.H., and Hall, J.D. 1987. Interactions between the redside shiner (*Richardsonius balteatus*) and the steelhead trout (*Salmo gairdneri*) in western Oregon: the influence of water temperature. Can. J. Fish. Aquat. Sci. 44:1603–1613.

Ringler, N.H., and Hall, J.D. 1975. Effects of logging on water temperature and dissolved oxygen in spawning beds. Trans. Amer. Fish. Soc. 104:111–121.

Robison, E.G. 1997. Storm of '96: small channel impacts in forested areas of western Oregon, pp. 23–28. In: A. Laenen and J.D. Ruff, editors. The Pacific Northwest Floods of February 6–11, 1996. Amer. Inst. Hydrology, Portland, OR.

Schaumburg, F.D. 1973. The influence of log handling on water quality. EPA R2 73 085. Office of Res. and Management, U.S. Environmental Protection Agency. 114pp.

Servizi, J.A., Martens, D.W., and Gordon, R.W. 1971. Toxicity and oxygen demand of decaying bark. J. Water Pollution Control Federation 43:278–292.

Skaugset, A.E. 1980. Fine Organic Debris and Dissolved Oxygen in Streambed Gravels in the Oregon Coast Range. M.S. Thesis. Oregon State Univ., Corvallis, OR. 96pp.

Steinblums, I.J., Froehlich, H.A., and Lyons, J.K. 1984. Designing stable buffer strips for stream protection. J. Forestry 82:49–52.

Sullivan, K., Tooley, J., Doughty, K., Caldwell, J., and coauthors. 1990. Evaluation of Prediction Models and Characterization of Stream Temperature Regimes in Washington, WA. Dept of Natural Resources. Olympia, WA. 245pp.

Swanson, F.J., and Lienkaemper, G.W. 1978. Physical consequences of large organic debris in Pacific Northwest streams. General Technical Report PNW-69. USDA Forest Service, Pacific Northwest Forest and Range Experiment Station Portland, OR. 12pp.

Thurston, R.V., Phillips, G.R., Russo, R.C., and Hinkins, S.M. 1981. Increased toxicity of ammonia to rainbow trout (*Salmo gairdneri*) resulting from reduced concentrations of dissolved oxygen. Can. J. Fish. Aquat. Sci. 38:982–988.

Tiedemann, A.R., Conrad, C.E., Dieterich, J.H., Hornbeck, J.W., and coauthors. 1969. Effects of fire on water. USDA Forest Service Gen. Tech. Rep., WO-10, Washington, DC.

Todd, D.K. 1959. Groundwater Hydrology. John Wiley & Sons, New York, NY. 336pp.

Warren, C.E. 1971. Biology and Water Pollution Control. W.B. Saunders Co., Philadelphia, PA.

Wilson, A. 1986. Fish and Wildlife Riparian Areas on Private Forest Land in Oregon: A Report to the Regional Forest Practice Committees. Oregon Dept. Forestry, Salem, OR.

Zwieniecki, M.A., and Newton, M. 1999. Influence of streamside cover and stream features on temperature trends in forested streams of western Oregon. Western J. Applied Forestry 14:106–113.

Chapter 4
Forest Practices and Sediment Production in the Alsea Watershed Study

Robert L. Beschta and William L. Jackson

The Alsea Watershed Study (AWS) was initiated in 1958 to determine the effects of forest practices upon the water quality and fishery resources of small watersheds in the Oregon Coast Range. At that time there was an increasing body of evidence indicating sediment increases to streams could significantly impact fisheries (Cordone and Kelley 1961). Since then, various results from the AWS have been reported by Brown and Krygier (1971), Harris and Williams (1971), Harris (1973, 1977), and Beschta (1978) based on stream-flow and suspended sediment data collected by the U.S. Geological Survey. However, each study differed somewhat in its analytical approach and the period of record that was analyzed. In this chapter we summarize published results and conclusions concerning the effects of road building, logging, and slash burning on sediment production during the AWS. We also discuss these results in terms of the processes and conditions that contributed to the observed changes in sediment production following these forest practices.

Factors Affecting Sedimentation in the Oregon Coast Range

Sedimentation involves the erosion, transport, and deposition of inorganic soil material. The erosion phase may initially involve soil, rock, and woody debris from watershed hillslopes that are eventually deposited in stream channels. For forested mountain watershed in the Oregon Coast Range, sedimentation processes are primarily initiated during rainfall associated with frontal storms moving inland from the Pacific Ocean, particularly during the months of November through March. Rain-on-snow conditions, although relatively infrequent, can be especially important for increasing hillslope erosion since

Robert L. Beschta
Department of Forest Engineering, Oregon State University, Corvallis, OR 97331
robert.beschta@oregonstate.edu

J. D. Stednick (ed.), *Hydrological and Biological Responses to Forest Practices.*
© Springer 2008

condensation and snowmelt provide additional moisture to forest soils during a rainstorm.

For undisturbed Coast Range watersheds, precipitation readily infiltrates into forest soils and is routed to topographic depressions and channels as subsurface flow. Thus, coastal streams respond rapidly to rainfall inputs (Harr 1976). The capability of large storms to affect instream sediment concentrations and yields during rainfall is further influenced by watershed conditions, processes that deliver upland sediments to streams, and characteristics of the stream network. Hence, instream sediment concentrations can be highly varied in time and space.

Since the kinetic energy of rainfall is largely absorbed by forest vegetation and its underlying litter, surface erosion from undisturbed forested watersheds is typically uncommon and the primary mechanism for delivering watershed sediments to coastal streams involves mass erosion of hillsides (soil creep, landslides) and sometimes the erosion of streambanks. Mass erosion process such as debris flows and rotational failures can vary greatly as to size, frequency of occurrence, and their potential to deliver sediment to streams since, in turn, they are influenced by a variety of factors such geology, topography, vegetation, and distance to channels.

Once hillslope sediments are delivered to the stream network, they are transported or deposited in relation to particle size and stream discharge. Coarse-grained sediments from hillslope erosion are generally prone to becoming deposited within the channel. During subsequent storms, these sediments may be mobilized and transported downstream as bedload. Thus, a temporal delay often occurs between the time when coarse sediments are delivered to a channel and when they are actually measured as export from the mouth of a watershed. This situation contrasts from that of smaller sized sediments (clay, silts, and fine sands) that are efficiently transported as suspended load and relatively quickly routed downstream.

Streamflow interacts with the gradient and morphology of a channel to locally define flow hydraulics (e.g., velocity, depth, bed shear stress, stream power). Of particular importance for forested streams is the role of streamside vegetation and large woody debris in affecting local channel morphology. Large wood from fallen trees along a channel or that delivered to a stream system by mass soil movements, as well as the root systems of riparian trees and shrubs, can contribute significant hydraulic roughness to a channel. Where the instream loading of large wood is relatively high and debris jams have formed, these structural features may locally decrease stream gradients and cause considerable volumes of sediment to become deposited upstream of the debris accumulation. In other instances, large wood may divert flows laterally and increase channel sinuosity. The net result of large wood in stream channels is typically an increase in the variability of channel dimensions and local gradients, thus influencing the characteristics of pools (scour features) and riffles (depositional features).

Land use activities such as road building, logging, and slash burning practices are capable of affecting sedimentation processes in several critical ways. First, if such practices expose mineral soils to raindrop impact and reduce the infiltration capacities of forest soils, surface runoff and erosion may occur where formerly these mechanisms were absent. Second, if forest practices reduce the inherent stability of a slope, this may increase the frequency and magnitude of landslides or other types of mass soil movements. Finally, if increases in the magnitude and frequency of peak flows were to occur, this would increase the erosive potential of a stream and its capacity to transport bed and bank materials.

Study Design

The AWS was designed as a classic paired watershed study. Three geographically close watersheds with similar elevations, aspects, soils, and forest vegetation were selected for study. The Flynn Creek watershed, with a drainage area of $2.02 \, km^2$, became the "control" or reference watershed and remained untreated throughout the study. After a pretreatment calibration period, clearcut logging and slash burning occurred on the Needle Branch watershed (area $= 0.71 \, km^2$) and roading, patchcut logging, and burning occurred on the Deer Creek watershed (area $= 3.03 \, km^2$).

The collection of sediment and hydrologic data began in October 1958, the start of water year (WY) 1959, and continued for 15 years. After a seven-year period of pretreatment data collection (WYs 1959–65) during which all three watersheds were unaffected by forest operations, watershed treatments were begun. Logging roads were constructed in the Deer Creek and Needle Branch watersheds between March and August 1965. Many of the roads in the Deer Creek watershed were mid-slope roads and thus had a potential to reroute intercepted subsurface flows and locally alter hillslope stability. In contrast, only relatively minor amounts of roading actually occurred on the Needle Branch watershed. Here, roads that provided access to landings were constructed outside the topographic boundaries of the watershed or were located along ridge tops comprising the watershed divide.

Logging on both Deer Creek and Needle Branch watersheds occurred between March and November 1966 with slash burning in October of 1966. Approximately 25% of Deer Creek was clearcut and involved three harvest units, each 25 ha in size. In the lowermost harvest unit, a buffer strip of forest vegetation was left for stream protection purposes. Approximately 82% of Needle Branch was clearcut and burned. Stream protection buffers were not left along any of Needle Branch. After felling, much of the downed timber in Needle Branch was yarded across the channel to ridge-top landings. This yarding contributed to high loadings of slash along lower hillslopes and within the stream channel. As a result, the watershed experienced a particularly hot slash burn. Posttreatment sediment and hydrologic data were collected during

WYs 1966–73. More detailed descriptions of the watersheds, forest practices, and overall experimental design are provided in Brown and Krygier (1971), Moring and Lantz (1975), and Beschta (1978).

Concrete weirs, in conjunction with water level recorders, were used to monitor streamflow at the mouth of each research watershed throughout the period of study. Suspended sediment samples were collected using depth-integrating DH-48 samplers. During periods of measurable suspended sediment transport, samples were usually collected on a daily basis with more frequent sampling during storms. Daily estimates of suspended sediment discharge were calculated by summing the product of streamflow and suspended sediment concentration for discrete time intervals within a given day. The number of intervals depended upon variations in flow and concentration that occurred during a particular day. Annual sediment yields were, in turn, calculated by summing the daily estimates of suspended sediment discharge over a water year. Daily streamflow and suspended sediment discharge for each of the three experimental watersheds have been published in Water Resources Data for Oregon (U.S. Geological Survey).

During the study, it was noted that the weir pools trapped some of the bedload in transport by each stream (Harris 1977). The volume of trapped sediment was estimated by the method of average-end areas based on cross-sectional surveys conducted at the end of each water year. This volume was prorated by month on the basis of the suspended-sediment load amounts (Harris and Williams 1971). Harris estimated that between 1–4% of the total sediment yield occurred as unmeasured bedload. However, subsequent studies suggested that bedload transport may represent a substantially higher proportion of the total load (Beschta et al. 1981).

In their analysis of sediment yields, Brown and Krygier (1971) accounted for the effects of flow increases that occurred after harvesting. To do so, they developed average flow durations for both pre and posttreatment periods. They then used the relationships between daily streamflow and daily suspended sediment concentration to calculate "normalized" annual sediment yields.

Two general methods were used to analyze the effects of forest management practices on suspended sediment yields and concentrations during the AWS. For the first, regression analysis was used to develop pretreatment relationships between annual sediment yields and flow-weighted sediment concentrations for Flynn Creek (control watershed) and corresponding yearly values for Deer Creek and Needle Branch (treatment watersheds). During the posttreatment period, measured sediment concentrations and yields at Flynn Creek thus became the basis for estimating sediment concentrations and yields for the treated watersheds, using the established pretreatment regression equations. These estimates were then compared to the measured values for the treated watersheds. Statistically significant differences ($p \leq 0.05$) between predicted and measured suspended sediment yields (or concentrations) in the posttreatment period were assumed to be attributable to the effects of treatment.

A second method of data comparisons involved double mass analysis, whereby the cumulative yield of sediment over time for the treated watersheds is plotted against the cumulative yield of sediment for the control watershed. Any change in sediment yield in the treated watershed compared to that of the control watershed will show as a change in the slope of the cumulative yield relation.

Annual Sediment Yields

Annual suspended sediment yields for the three research watersheds (Table 4.1) indicated considerable year-to-year variability during both the pre and posttreatment periods. In reporting annual sediment yields, Beschta (1978) noted the overriding importance of several large storms as primary contributors to total sediment yields from the unharvested control watershed. Specifically, two large storms with peak flows occurring on January 28, 1965 and January 11, 1972 accounted for 36% of the total 15-year suspended sediment yield from the Flynn Creek watershed.

During the pretreatment period, WY 1965 sediment yields were dominated by the January 28, 1965 storm. Suspended sediment yields in WY 1965 for the three watersheds were approximately 3–4 times the average yield for the pretreatment period and approximately an order of magnitude larger than the lowest annual yields for the pretreatment period.

During the posttreatment period, sediment yields from Flynn Creek and Deer Creek were dominated by the January 11, 1972 storm. Sediment yields for WY 1972 were approximately four times the mean annual yield for the

Table 4.1 Annual sediment yields ($t\,mi^{-2}$) for the study watersheds (Harris 1977, Table 8). Flynn Creek is the control (untreated) watershed

Water Year	Flynn	Needle Branch	Deer
1959	88	59	91
1960	65	41	91
1961	338	186	340
1962	136	141	118
1963	114	117	162
1964	226	184	213
1965	1270	430	1070
Average (prelogging period)	320	165	298
1966 (logging period)	291	368	746
1967	131	905	218
1968	67	490	87
1969	142	515	161
1970	121	232	147
1971	189	415	211
1972	1103	519	1411
1973	58	132	131
Average (post-logging period)	263	458	338

posttreatment period. However, WY 1972 yields for Needle Branch were less dramatically influenced by the January 11 storm and were only slightly increased above the mean annual yield for the posttreatment period.

The pattern of increases in annual measured sediment yields following road building, logging, and slash burning on Deer Creek (Fig. 4.1) and Needle Branch (Fig. 4.2) are presented as deviations from expected yields based on the pretreatment regression analysis with Flynn Creek. Brown and Krygier (1971) found that suspended sediment yield was significantly increased ($p \leq 0.05$) for Deer Creek in WY 1966, the first year following road construction. They also indicated a significant increase ($p \leq 0.05$) in sediment yield for Deer Creek in WY 1967, immediately following the effects of both roading and logging, with sediment yields returning to prelogging levels in 1968 and 1969.

Harris (1977) reported that post-logging annual sediment yields increased an average of 25% ($67\,t\,mi^{-2}$) for Deer Creek and 181% ($295\,t\,mi^{-2}$) for Needle Branch. He further noted that the average increase was not significant ($p > 0.05$) for Deer Creek but was significant for Needle Branch. Similarly, Beschta (1978) found that, while the average increase in sediment yields for the entire post-logging period on Deer Creek was not significant ($p > 0.05$), significant increases did occur in three of the post-logging years (WYs 1966, 1967, and 1972). Although the average sediment yield increase for the post-logging period on Needle Branch was significant ($p \leq 0.05$) (Beschta 1978), the annual sediment yield for WY 1966, following road construction but prior to logging, was not.

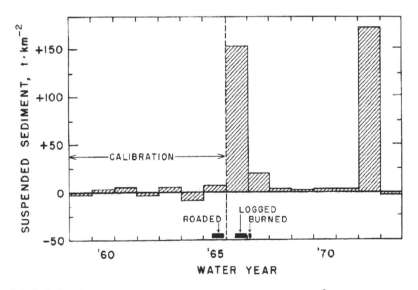

Fig. 4.1 Relative changes in annual suspended sediment yield ($t\,km^{-2}$) after road construction, 25% patchcut logging, and site preparation (burning) for the Deer Creek watershed (Beschta 1978). Reproduced from Beschta 1978, with permission from the American Geophysical Union

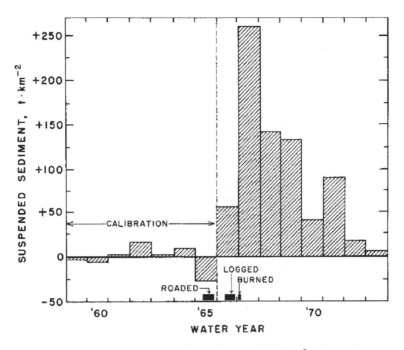

Fig. 4.2 Relative changes in annual suspended sediment yield (t km^{-2}) after road construction and 82% clearcut logging, and site preparation (burning) for the Needle Branch watershed (Beschta 1978). Reproduced from Beschta 1978, with permission from the American Geophysical Union

Overall, these analyses provide a relatively clear picture regarding the effects of road building and logging on annual sediment yields. Road building alone caused a significant increase in sediment yield for Deer Creek in WY 1966 as well as in WY 1972. Patchcut logging on Deer Creek did not result in a measurable increase in annual sediment yield. For Needle Branch, the relatively minor amount of road building on the watershed did not increase WY 1966 sediment yields over that predicted from the pretreatment relation with Flynn Creek. However, the combination of clearcut logging, yarding, and slash burning on Needle Branch resulted in five consecutive years of significant ($p \leq 0.05$) increases in measured sediment yield. The initial increase was the greatest and was followed by a general decline toward pretreatment levels during the five-year period (Fig. 4.2).

Sediment Concentrations

Because sediment concentrations represent a mass per unit volume of water, it is conceivable that sediment yields could increase due to streamflow increases alone, even though suspended sediment concentrations remained unchanged.

Thus, the appropriate metric for assessing the effects of forest operations on sediment is to test for changes in the relation between suspended sediment concentration and streamflow. However, this analytical approach was complicated by the considerable variability in sediment concentration vs. stream discharge relationships for each of the three watersheds during the pre and posttreatment periods. Some of this variability is attributable to differences in storm characteristics, such as magnitude, frequency, and timing, during the pre and posttreatment periods. Even within a given storm system that moved across the Coast Range, there was no guarantee that each watershed received identical amounts or temporal distributions of precipitation.

To reduce the variability between streamflow and sediment concentration, Brown and Krygier (1971) attempted to develop "tighter" relationships between stream discharge and instantaneous sediment concentrations by utilizing only rising-limb data. However, a shift in the relationship between stream discharge and sediment concentration on the control watershed during the early portion of the posttreatment period (WYs 1966–69) made it difficult to detect statistically significant changes in sediment concentrations. Nevertheless, Brown and Krygier (1971) concluded that an increase in rising-limb suspended sediment concentration for Deer Creek in WY 1966, following road building, was significant at the 90% confidence level (*CL*). No statistically significant increases (90% *CL*) were found for WYs 1967–69 at Deer Creek. Similarly, the mean annual flow-weighted sediment concentrations following logging on Needle Branch, though large, were not significant (90% *CL*) for WY 1967 due to the shift in the posttreatment sediment rating curve on the control watershed.

Brown and Krygier (1971) also evaluated the effects of road building and logging at sub-watershed sampling sites on Deer Creek. Although one of four sub-watershed sites recorded a sediment concentration increase following road building, none of the sub-watershed sites recorded sediment concentration increases following logging.

Harris (1977) analyzed posttreatment data over a longer period than Brown and Krygier (1971) and reported maximum mean daily sediment concentrations for the prelogging, logging, and entire post-logging periods (Table 4.2). Harris (1977) also developed regression relations for mean annual flow-weighted sediment concentrations between control watershed and the treated watersheds as a basis for comparing annual flow-weighted sediment concentrations in the post-logging period (Fig. 4.3). He concluded that average sediment concentrations for Needle Branch increased significantly ($p \leq 0.05$) to 40 mg L^{-1} over the

Table 4.2 Maximum mean daily sediment concentration (mg L^{-1}) and date of occurrence for the study watersheds (Harris 1977). Flynn Creek is the control (untreated) watershed

Measurement Period (yrs)	Flynn Creek	Needle Branch	Deer Creek
Prelogging (1959–1965)	1,580 (1/28/65)	300 (1/28/65)	1,220 (1/28/65)
Logging (1966)	390 (12/27/66)	477 (12/27/66)	1,010 (12/27/66)
Post-logging (1967–1973)	1,530 (1/11/72)	1,260 (2/18/68)	2,450 (1/11/72)

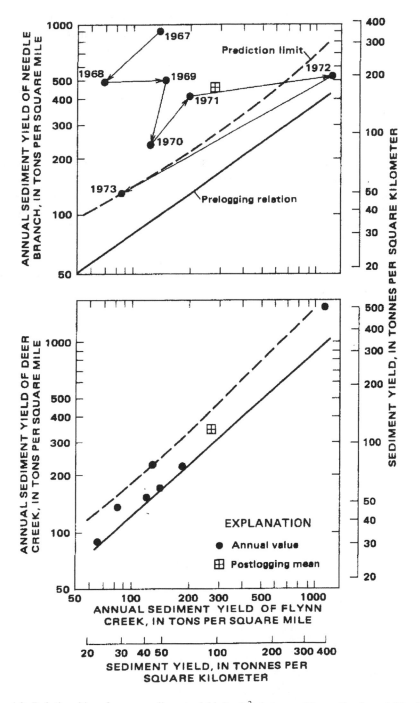

Fig. 4.3 Relationship of mean sediment yield $(t\,mi^{-2})$ between Flynn Creek and Needle Branch (top panel) and between Flynn Creek and Deer Creek (lower panel) (Harris 1977)

predicted value of $27 \, mg \, L^{-1}$ in the posttreatment period, with concentrations returning to pretreatment levels by the fifth year following road-building. Although average annual flow-weighted sediment concentrations on Deer Creek were greater than predicted values in the posttreatment period, none of the increases were significant ($p \leq 0.05$).

Beschta (1978) performed an analysis similar to that of Harris (1977), regressing flow-weighted mean annual suspended sediment concentrations of the treated watersheds against the control watershed and comparing posttreatment results to the pretreatment relation. Beschta (1978) concluded that the same three years that had significant increases in sediment yield on Deer Creek (WYs 1966, 1967, 1972) also had significant ($p \leq 0.05$) increases in suspended sediment concentration. He also reported significant ($p \leq 0.05$) increases in sediment concentrations for the posttreatment period on Needle Branch for WYs 1967, 1968, 1969, and 1971. The reason for the slight discrepancy between the sediment concentration results of Harris (1977) and Beschta (1978) is not readily apparent. It may be that Harris (1977) used concentrations that were modified slightly to account for sediment storage changes in weir pools immediately upstream from the sediment sampling locations, whereas Beschta (1978) calculated concentrations from published USGS daily values of suspended sediment yields and streamflows. Furthermore, Harris's (1977) statistical evaluation tested for increases and decreases in sedimentation following treatment (a two-tailed statistical test) whereas Beschta (1978) tested only for potential increases (a one-tailed test). If posttreatment increases occur, this latter approach is more likely to identify statistically significant differences at a given level of probability.

Discussion

In total, the various published analyses of sediment concentration data from the AWS illustrated a general pattern of suspended sediment concentration changes as a result of the imposed treatments. First, the data comprising stream discharge vs. sediment concentration relationships are highly variable. A major portion of this variability appears to be associated with differences in concentrations associated between rising and falling limbs (hysteresis) of storm hydrographs as well as differences in concentrations between fall/early winter runoff and late-winter/springtime runoff (See Chapter 12). Second, the effects of timber harvesting (including the combined effects of yarding and burning for slash disposal) on sediment concentrations are somewhat more difficult to detect than the effects of these same treatments on sediment yields. This is because, in part, sediment yield increases can stem from increases in both streamflow and sediment concentrations, whereas sediment concentrations tend to minimize the effects of any flow changes following harvest. Finally, suspended sediment concentrations were significantly ($p \leq 0.05$) increased for all posttreatment years that experienced significant increases in annual sediment yield, with the exception of WY1970 at Needle Branch. The degree of statistical significance associated with increases from Deer Creek is less clear and appears somewhat dependent upon the method of analysis. In general, sediment concentration increases for Deer Creek during the posttreatment period are substantially smaller than those for Needle Branch.

Overall, the Alsea Watershed Study found that road building, logging, and site preparation measurably increased sediment production even though background levels for these watersheds were highly variable. To explain the nature of the sedimentation effects measured during the AWS, it is necessary to postulate the effects of treatments on the controlling processes.

In the case of the AWS, several mechanisms of accelerated sedimentation are believed to have occurred. While road surfaces likely generated some surface runoff and erosion in both the Needle Branch and Deer Creek watersheds, the reported studies did not identify such processes as an important component of the increased sediment concentrations and yields. The increases in sediment yield and sediment concentration that occurred on Deer Creek appeared to correspond with road-associated mass failures, which occurred in WYs 1966 and 1972. Although there may have been some residual effect of the WY 1966 Deer Creek failures in causing increased sediment yields and concentrations in WY 1967, these effects were not significant ($p > 0.05$). For Deer Creek, sediment yields and concentrations were not significantly ($p > 0.05$) increased during five of the seven posttreatment years.

The high-severity slash burn on Needle Branch was thought to be responsible for reducing soil infiltration rates and causing surface runoff and erosion to occur (Brown and Krygier 1971). In addition, this fire likely contributed to localized toeslope sloughing and channel instabilities that could contribute to higher sediment yields. Whereas high-lead cable yarding may result in some soil compaction and contribute to localized surface runoff and erosion, felling and yarding were not identified as having an important effect on surface or mass erosion processes for either the Deer Creek or Needle Branch Watersheds. However, because logs were yarded across the Needle Branch channel and caused high levels of slash to accumulate in and near the stream, streambank disturbance from yarding in conjunction with the subsequent high-severity slash burn appears to have been a major contributor to the increased sediment yields from this watershed.

Sediment yield and concentration increases attributed to the combined effects of cable yarding and slash burning in Needle Branch occurred in six of the seven posttreatment years. A large increase the year immediately following treatment was followed by a general trend toward pretreatment conditions in subsequent years. There were no mass failures reported in Needle Branch during the posttreatment period that would have caused episodic pulses of sediment to channels such as occurred on Deer Creek. Based on observations during the posttreatment period, it is generally presumed that logging, and most particularly the intense slash burn, created soil and channel conditions in the Needle Branch watershed that initially favored surface erosion. Sediment concentration and yield increases following these treatments occurred primarily during large storms. Observations and results indicated that the recovery of vegetation on this watershed, initially deciduous and herbaceous species, was responsible for decreasing the posttreatment increases in erosion over time.

Sedimentation studies associated with the AWS ascertained the magnitude and duration of increased watershed sediment yields and instream suspended sediment concentrations associated with forest practices typically used in the 1960s on forested watersheds in the Oregon Coast Range. While results indicated that there was considerable variation in sediment production associated with individual forest practices, several conclusions emerged from the study. For example, minimizing road construction within a watershed, small clearcuts, forested stream buffers, and low-severity slash burns were practices that appeared to prevent increases in sediment production. Conversely, sidecast construction of mid-slope roads across headwalls and other over-steepened portions of a hillslope, clearcutting a large portion of a watershed, yarding logs across stream channels, and high-severity slash burning contributed to accelerated rates of erosion involving both surface erosion and mass failures. These watershed-scale studies were important in that they illustrated the extent to which various forest practices could affect erosion processes. In doing so, they provided an impetus for altering forest practices to help minimize potential adverse effects upon water quality, stream systems, and aquatic habitats.

Literature Cited

Beschta, R.L. 1978. Long-term patterns of sediment production following road construction and logging in the Oregon Coast Range. Water Resour. Res. 14:1011–1016.

Beschta, R.L., O'Leary, S.J., Edwards, R.E., and Knoop, K.D. 1981. Sediment and organic matter transport in Oregon Coast Range streams. WRRI-70. Water Resources Research Institute, Oregon State Univ., Corvallis, OR. 67pp.

Brown, G.W., and Krygier, J.T. 1971. Clear-cut logging and sediment production in the Oregon Coast Range. Water Resour. Res. 7:1189–1198.

Cordone, A.J., and Kelley, D.E. 1961. The influence of inorganic sediment on the aquatic life of streams. Calif. Fish and Game 47:189–228.

Harr, R.D. 1976. Hydrology of small forest streams in Western Oregon. General Technical Report GTR-PNW-055. USDA Forest Service Portland, OR. 19pp.

Harris, D.D. 1973. Hydrologic changes after clearcut logging in a small Oregon coastal watershed. U.S. Geological Survey J. Res. 1:487–491.

Harris, D.D. 1977. Hydrologic changes after logging in two small Oregon coastal watersheds. Water-Supply Paper 2037. U.S. Geological Survey Washington, DC. 31pp.

Harris, D.D., and Williams, R.C. 1971. Streamflow, sediment-transport, and water-temperature characteristics of three small watersheds in the Alsea River basin, Oregon. U.S. Geological Survey Circ. 642pp.

Moring, J.R., and Lantz, R.L. 1975. The Alsea Watershed Study: effects of logging on the aquatic resources of three headwater streams of the Alsea River, Oregon. Part I. Biological studies. Fish. Res. Rep. 9. Oregon Dept. of Fish and Wildlife, Corvallis, OR. 7pp.

U.S. Geological Survey. Water years 1959–1972. Water resources data - Oregon. Water Resources Division, Portland, OR.

Chapter 5
Salmonid Populations and Habitat

James D. Hall

The three streams included in the Alsea Watershed Study (AWS) are small headwater systems that were selected as typical of those supporting anadromous salmonids in the central Oregon Coast Range. Two principal salmonid species were present, the coho salmon *(Oncorhynchus kisutch)* and cutthroat trout *(O. clarkii)*. Two others, Chinook salmon *(O. tshawytscha)* and steelhead *(O. mykiss)* were occasionally seen in Deer Creek, the largest watershed, but received little emphasis in our work. The other common species in all three streams was the reticulate sculpin *(Cottus perplexus)*. The Pacific lamprey *(Lampetra tridentata)* and western brook lamprey *(L. richardsoni)* also occurred in low numbers in the streams.

The general study objective was to determine the effects of two patterns of clearcut logging on fish habitat and fish populations. The two logging patterns were harvest of a complete watershed compared with harvest of several smaller patches that included buffer strips along the stream. Ours was the first long-term watershed study to investigate fish populations. Another purpose of the study was to learn as much as possible about the basic biology and life history of the fish species occupying the watersheds and to better understand their habitat requirements. The long-term focus of the study and the census capability provided by two-way fish traps were substantial assets in pursuit of this objective.

The study design, description of the watersheds, and a brief description of the logging operations and watershed treatments are included in Chapter 1. The work on fish and fish habitat involved substantial contributions from biologists of the Oregon State Game Commission and from graduate students at Oregon State University. In reporting results of that work, I have used the plural to indicate these cooperative efforts.

James D. Hall
Department of Fisheries and Wildlife, Oregon State University, Corvallis, OR 97331

Methods

Habitat Characterization

The habitat features that we investigated were chosen based on changes expected after logging. At the time the study was planned, available information indicated that spawning gravel would receive the greatest impact. Hence, much of the effort was placed on characterizing the quality of gravel in the three streams before and after logging. We emphasized three measures of gravel quality: dissolved oxygen concentration, particle size composition, and permeability.

Dissolved Oxygen

Concentration of dissolved oxygen (DO) within the gravel matrix was determined by Winkler titration and measured from permanent standpipes (Terhune 1958) located in riffle areas in all three streams. There were 19 stations in Deer Creek, 11 in Flynn Creek, and 9 in Needle Branch, all located in areas that appeared to be suitable spawning sites for coho salmon. Measurements were made from a depth of about 25 cm throughout the year, with the most frequent sampling (biweekly) during spawning and incubation, November through June. The standpipes in Needle Branch were removed just prior to logging and replaced after the logging operation was completed. Supplemental measurements of the DO in gravel within actual salmon redds were made from temporary plastic standpipes. These were 1.25-cm diameter PVC pipe with the lower 8 cm of each pipe perforated (McNeil 1962). These temporary pipes were driven into redds approximately 4 weeks after spawning, and measurements were made until fry emergence was complete.

Gravel Composition

After all fry had emerged from those redds that had been marked for sampling, quality of the gravel was analyzed from three cores taken from each redd with a McNeil sampler (McNeil and Ahnell 1964). The core was 10 cm in diameter and 25 cm deep. The samples were washed through a series of Tyler sieves with mesh openings of 50.8, 25.4, 12.7, 6.35, 3.33, 1.65, and 0.833 mm. Volume of material in each size class was determined by displacement in water. During 1968, additional gravel samples were extracted from the same redds sampled with the McNeil sampler, and also from some other areas that had been used as redds in previous years. These samples were taken with a frozen-core sampler that preserved the vertical stratification in the sample (Ringler 1970). In addition to similar volumetric analysis, these samples were evaluated for content of organic material, as determined by loss on ignition (Ringler and Hall 1988).

The initial analyses of gravel composition (Hall and Lantz 1969; Moring 1975a) indexed quality as the percentage of total sample volume less than either 3.33 or 0.833 mm. Since that time, two other indices of gravel quality have been proposed as more accurately describing conditions related to survival of salmonid alevins in the gravel. These are geometric mean particle size (Platts et al. 1979) and the fredle index (Lotspeich and Everest 1981). For the reanalysis in this chapter, I calculated the geometric mean diameter by the method of moments (Lotspeich and Everest 1981). I used a modified fredle index recommended by Beschta (1982), calculated as the geometric mean divided by its standard deviation (Young et al. 1991b).

Permeability

The rate of water seepage through the gravel was estimated by measurement of permeability, using methods of Terhune (1958). These measurements were taken biweekly during winter and spring from the permanent standpipes used to sample for dissolved oxygen.

Fish Population Statistics

Two-way fish traps at the outlet of each watershed (See map and photo in Chapter 1) provided a nearly complete census of the two principal anadromous species, coho salmon and cutthroat trout. The traps were occasionally overtopped in extreme high flows, resulting in some loss of data. In addition to the floods, there were problems with the trap at Needle Branch from the beginning. For reasons that were never determined, female salmon often returned downstream through the trap immediately after they had been confined in the trap box. Consequently, beginning in the winter of 1961-1962, the upstream barrier was pulled out during the spawning season (usually November through January). We relied on frequent streamside surveys and redd counts to estimate the numbers of salmon spawning each year until the trap was rebuilt in the fall of 1966.

Adult Enumeration

Spawning adult salmon and trout that returned to the three watersheds were intercepted in the upstream trap, measured (fork length) and weighed, and passed above the trap. For most of the study period, mature female coho salmon were marked distinctively with Petersen disk tags in various color combinations so that the location of their spawning could be recorded. Beginning in 1967, we fitted some of the female salmon with a radio-frequency tag that could be detected with a simple FM receiver. The radio tag allowed us to locate a high proportion of spawning redds, because spawning fish could be found even when they were hidden by undercut banks or dense cover. When a redd could be associated with a tagged female, an estimate of egg deposition was made. This estimate was based on a relationship between female length, weight, and egg number for a sample of 92 mature coho salmon from the Fall Creek Hatchery on the Alsea River (Koski 1966).

Juvenile Populations

We estimated the rate of survival from egg deposition to emergence from the gravel (survival-to-emergence, or STE) in two ways: directly, by sampling emergence from fry traps; and indirectly, by combining population estimates with enumeration of emigrant fry moving through the downstream trap. These estimates were made only for coho salmon; it was not feasible to estimate STE for trout.

From 1964 through 1971, redds that could be unequivocally attributed to a marked female coho salmon were marked with flagging for later sampling. Approximately 30 days before predicted emergence, these marked redds were covered with a nylon-netting fry trap (Phillips and Koski 1969) that capped the redd and allowed enumeration of emergent fry. Traps were emptied several times each week and emergence success measured as total emerged fry divided by estimated egg deposition. Emergence success could then be related to physical characteristics of the gravel in individual redds.

It was also possible to construct an approximate estimate of overall STE for each stream for the years 1960 through 1968. Emergence was calculated as the sum of (1) a population estimate of resident fry made soon after all emergence was complete and (2) the number of newly emerged fry that had left the study area through the downtrap. Dividing this value by the estimated total fecundity of females spawning that year, I derived an estimate of STE. This estimate was available for only 2 years after logging, so the method did not provide a good comparison of pre- and post-logging emergence success.

I formulated another index of emergence that provided coverage for the entire study period. Research on the early emigrating fry by Au (1972) and Haak (1984) had shown that the numbers of these migrants were closely proportional to the numbers of spawning females in each watershed. The proportionality differed in each watershed, probably owing to different average distances between spawning locations and the downstream fish traps, and perhaps to other differences among the three watersheds, such as size. However, the inference was that these early migrants represented an approximately constant fraction of the emergent fry in each stream. Thus, the numbers of early migrating fry per spawning female could be used as a comparative estimate of STE within any one stream. Because it made maximum use of the data available, this ratio was the primary basis used for comparison of pre- and post-logging emergence success.

Periodic population estimates were made of juvenile coho salmon and cutthroat trout in the streams. From 1962 through 1974, trout estimates were made in August or September with mark-recapture techniques that incorporated electroshocking. The recapture sample was taken approximately 1 week after marking. For this analysis I reviewed the original field notebooks and recalculated all estimates. Consequently, some corrections were made to values previously published by Moring and Lantz (1975) and Hall et al. (1987).

During the period 1959 through 1968, estimates of the abundance of juvenile coho salmon were made by marking fish several times during the year with distinctive fin clips. Recaptures were tabulated as the fish moved through the downstream trap. From 1969 through 1974, estimates of juvenile salmon were made only in August or September, at the same time as those of trout. These estimates were not made every year for all three streams. Data on juvenile salmon are presented in Chapter 14.

Downstream movement of juveniles was monitored at inclined-plane traps (Wolf 1951) incorporated into the upstream trapping facilities. There are two principal periods of emigration of juvenile coho salmon: (1) during their first few months of life (mainly March through May) as early migrating fry (Chapman 1962), and (2) as smolts after about 1 year of residence in the stream (primarily February through May). A small fraction of the smolts ($<5\%$) remain in the stream an additional year and migrate at 2 years of age.

Abundance of outmigrating smolts was one of the primary measures of biological productivity of the watersheds. Juvenile coho salmon moved through the downstream trap during most months of the year, so it was necessary to develop a criterion for classifying a smolt. A small number of fish moved through the downstream traps in the fall, mostly during November and December. During the prelogging period, approximately 8–10% of all the fish leaving the watersheds from November through May moved in the period from November through January. More recent monitoring of the physiological characteristics and general appearance of wild juvenile coho salmon migrating from a nearby coastal Oregon stream (Rodgers et al. 1987) suggested that those fish moving early in the season may not be true smolts ready to enter salt water. The exact period when true smolt migration begins in AWS streams is still not known, but based on the work of Rodgers et al. (1987) and on analysis of the seasonal distribution of movement from the Alsea streams, I selected February 1 as the beginning point and ended the tally on May 31. Ideally, smolts should be defined by cohort and separated into ages 1 and 2. However, the relatively small proportion of age-2 fish and the small number of scale samples available made this distinction impractical and probably unnecessary.

Prelogging Conditions

Fish Populations

A major advantage of the AWS was the ability to monitor fish populations from the time of spawning until downstream migration of juveniles, and then to count adult returns to the watershed from these downstream migrants. This nearly total census capability made possible by the two-way fish traps provided significant new information about the biology of the principal species in the watersheds.

Productivity of Headwater Streams

One of the principal contributions of the prelogging study was to document the importance of small headwater streams to both anadromous and resident salmonids. For example, Needle Branch, the smallest of the watersheds (70 ha), with a stream that nearly dried in late summer, supported annual spawning populations of female coho salmon that numbered from 1 to 28 (Table 5.1). With accompanying adult males and jacks (precociously mature males), the total spawning migration in a winter season ranged from 5 to 117 for this small stream. Comparable totals for Flynn Creek ranged from 11 to 226 and for Deer Creek from 58 to 280.

Ecology of Salmonid Incubation

A major focus of the AWS was to increase understanding of the earliest period in the life history of anadromous salmonids, their incubation and emergence from the gravel. Most of our information on this topic came from the fry traps placed over coho salmon redds and from evaluation of gravel quality. There were two major conclusions from this work.

We found that STE for coho salmon was substantially lower than had been thought. The prelogging average STE for the three streams was approximately 35% (Table 5.2), compared to estimates of 60% or more from earlier work on other streams. Those studies (e.g., Briggs 1953) usually estimated survival only to hatching, whereas our fry traps measured the additional mortality that occurred as the fry attempted to emerge from the gravel.

We also found that survival was inversely related to the percentage of fine sediment in the gravel. More fine sediment resulted in decreased STE. This inverse relationship had been established in laboratory work, and McNeil and Ahnell (1964) had shown the influence of gravel quality on survival to hatching in pink salmon redds. However, work on coho salmon in the Alsea streams by Koski (1966) was the first to demonstrate in a naturally spawning population the quantitative influence of fine sediment on survival from egg deposition to

Table 5.1 Population statistics for coho salmon in the three streams of the Alsea Watershed Study (numbers of spawning females, emigrant fry, and smolts). Rows are arranged by year of spawning[a]. Numbers are rounded to reflect approximate level of accuracy. Dashes indicate no data available. An asterisk indicates the first of each life stage that could have been influenced by the logging operations in 1966

Year	Deer Creek (patchcut)			Flynn Creek (control)			Needle Branch (clearcut)		
	Females	Fry	Smolts	Females	Fry	Smolts	Females	Fry	Smolts
1958–59	–	–	2,970	–	–	1,230	–	–	190
1959–60	21	1,730	1,830	8	6,540	840	2[b]	190	430
1960–61	19	3,560	1,800	26	10,400	640	4[c]	1,920	200
1961–62	28	5,850	2,700	51	29,900	1,280	15[d]	12,400	430
1962–63	18	4,410	1,770	2	70	460	4[d]	5,770	280
1963–64	27	7,350	1,840	20	6,470	660	15[d]	14,200	150
1964–65	44	5,420	1,070	10	3,160	590	25[d]	19,400	240
1965–66	24	2,610	*2,080	11	2,750	880	28[d]	16,600	*240
1966–67	*56	*12,500	2,320	55	28,800	600	*19	*5,540	180
1967–68	23	7,240	1,810	10	4,460	350	15	3,680	140
1968–69	39	8,720	1,370	19	8,180	170	17	7,150	140
1969–70	8	1,690	690	5	2,110	140	1	10	70
1970–71	10	1,190	990	5	1,450	310	2	1,660	110
1971–72	36	3,940	1,820	18	4,640	391	18	8,070	340
1972–73	6	50	–	3	20	–	1	0	–

[a] For example, for year 1959–60 values represent spawning females in fall-winter 1959–60, emigrant fry in spring 1960, and smolts from February through May 1961.

[b] Though female coho salmon entered Needle Branch, they left the stream before they had spawned. Consequently 1,627 emigrant fry from the Flynn Creek down trap were stocked into Needle Branch between April 16 and May 6, 1960 (Chapman 1962). Approximate equivalent number of females indicated.

[c] Two females spawned successfully in Needle Branch, but for experimental purposes 1,577 emigrant fry from Flynn Creek were stocked between April 19 and May 4, 1961 (Chapman 1962).

[d] Estimated from spawning surveys because upstream trap was not in service.

Table 5.2 Survival to emergence estimated from trapping of redds during spring of the year indicated. Trapped redds with no emergence are omitted, as are a few with emergence timing suggesting that more than one female had spawned in that redd. Means are indicated for all years and the prelogging (1964–1966) and post-logging (1967–1971) periods, with standard error in parenthesis

Year	Deer Creek (patchcut)			Flynn Creek (control)			Needle Branch (clearcut)		
	Mean	Range	n	Mean	Range	n	Mean	Range	n
1964	68.0	58.1–77.5	4	20.4	1.1–35.6	6	29.2	3.8–54.9	6
1965	39.1	28.2–47.5	4	23.9	14.3–31.0	3	22.2	6.4–41.9	3
1966	18.4	6.2–30.6	2	22.9	0.73–48.0	4	53.4	30.5–82.0	4
Pre-	46.5 (7.04)		10	22.0 (3.92)		13	35.0 (6.26)		13
1967	41.9	13.7–61.8	6	44.0	27.6–60.3	2	20.1	4.2–41.5	5
1968	45.9	17.8–70.0	6	9.7	2.3–14.3	3	34.7	10.8–66.6	4
1969	24.2	1.3–38.1	4	34.2	3.9–77.8	6	48.9	13.0–77.0	5
1970	34.8	32.0–37.7	2	23.4	15.2–31.7	2	52.7	–	1
1971	39.6	–	1	33.9	0.70–53.0	3	no data	–	–
Post-	38.6 (4.25)		19	29.4 (6.19)		16	35.8 (6.45)		15
All years	41.3 (3.69)		29	26.1 (3.84)		29	35.4 (4.43)		28

emergence. This relationship existed both in natural redds (Fig. 5.1a), and in an experiment carried out in a laboratory stream channel at our field headquarters (Fig. 5.1b). Particularly in the natural redds, there was substantial variation in STE that was not explained by the percentage of fine sediment in the gravel.

Both the laboratory and field work confirmed that there was significant mortality in the period after hatching, during emergence from the gravel. Large amounts of fine sediment in spawning gravel impair survival in several

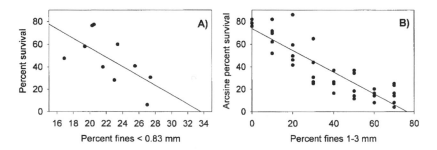

Fig. 5.1 (a) The relationship between percentage of fine sediments < 0.83 mm in gravel and survival of coho salmon to emergence for natural redds in Deer Creek, 1964–1966 ($r = -0.65$, $p < 0.05$),and (b) the relationship between the percentage of fine sediments between 1 and 3 mm and the arcsine transformation of percent survival of coho salmon alevins in replicated gravel troughs at the field laboratory ($r = -0.89$, $p < 0.001$). Details of the trough experiment are in Phillips et al. (1975)

ways. Decreased interchange of surface and intragravel water reduces dissolved oxygen levels in the intragravel water and slows the removal of metabolic wastes. In addition, at high sediment concentrations, fry can sometimes hatch successfully, but then be entrapped and die before reaching the gravel surface.

Early Juvenile Residence and Emigration

One feature of the life history of coho salmon is a substantial downstream migration of fry soon after they emerge from the gravel. Much of the early analysis of this phenomenon was carried out by D.W. Chapman as part of the AWS. His initial research focused on the role of social behavior in initiating this downstream movement, and he termed these early migrants "nomads." Chapman (1961, 1962) was the first to extensively document the repertoire of aggressive behavior exhibited by juvenile coho salmon. Dominance of large individuals over smaller ones was prominent, and the early migrating fry were on average smaller than the fry that remained upstream. Several experiments demonstrated that fry that had emerged early in the season and spent some time in the stream were more likely to remain there than were those just emerged from the gravel.

Two experiments involving transfer of downstream-migrating salmon fry from Flynn Creek to Needle Branch reinforced the conclusion that there is a behavioral component to this early migration. Of 1,627 migrants transferred in 1960, a year in which no salmon spawned naturally in Needle Branch, only 4% moved downstream through the fish trap in the first month after transfer. In 1961, when naturally produced fry were already present, 27% of the 1,577 fry placed in Needle Branch left the stream during the same period (Chapman 1962). Chapman concluded that "aggressive behavior is one factor causing the downstream migration of coho fry," and "it is likely that a part of the spring downstream emigration of coho fry ... is due to current displacement or to an innate migration urge." Chapman recognized the possibility of multiple influences in this migration, but the term "nomad" apparently caused many biologists and managers to discount the importance of these fry. Subsequent work by Au (1972) and Lindsay (1975) on the AWS found that these early migrating fry are a significant segment of the population.

Au (1972) conducted experiments in both laboratory troughs and in Needle Branch. He concluded that there is a developmental sequence in newly emerged fry. Their innate tendency to stay near the stream surface at night leads to significant downstream movement soon after emergence from the gravel. This initial phase is followed by settling behavior at night that leads increasing numbers of fry to take up residence. Au concluded that this behavior is a mechanism that serves to disperse fry from headwater spawning regions to rearing areas lower in the basin.

Lindsay (1975) conducted experiments that showed the ecological significance of the early emigrating fry. In the spring of 1972, he distinctively marked
all the emigrant fry from each of the three AWS streams. He found that many
of them survived in the larger systems downstream, but at substantially lower
rates than those fry remaining in the streams above the fish traps. By analyzing
earlier marking experiments he also confirmed that some of these fry survived
to adult size and returned to spawn in the stream in which they emerged. The
findings of these two investigators focus attention on the early migrating fry.
They suggest that these fry represent an additional source of production from
small headwater spawning tributaries, rather than simply a loss of less-fit
individuals.

Determinants of Smolt Abundance

A crucial problem in the management of natural populations of salmon is how
to determine the capacity of tributary streams to produce juveniles that survive
to go to the ocean as smolts. The long-term studies in the AWS produced a
substantial body of data relevant to this question. In each stream, numbers of
female spawners varied by at least an order of magnitude over the 15 years
of the study (Table 5.1). This variation provided evidence for the influence of
spawner abundance on the numbers of smolts produced.

Data from the AWS have been used by fishery managers to establish goals
for spawner abundance in coastal streams, particularly in Oregon. One analysis
is based on a Ricker curve of smolts produced versus number of spawning
females, illustrated here for Deer Creek (Fig. 5.2). There is an indication of an
upper limit in the number of smolts produced as spawner numbers reach about

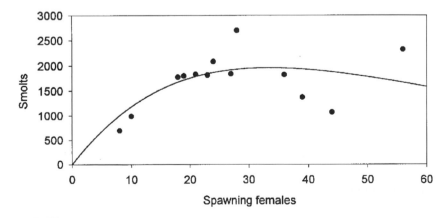

Fig. 5.2 The relationship between abundance of spawning females and the resulting number
of smolts produced in Deer Creek, 1961–1973. The curve is fitted to a Ricker equation of the
form: $y = a \cdot x \cdot e^{-(b \cdot x)}$, where $y =$ smolts, $x =$ spawning females, a = 160, and b = 0.030

30 (equivalent to about 15 spawners per kilometer). However, there are several reasons why spawner abundance above this level could be beneficial, and other reasons why these data should be used with caution when setting escapement goals:

- Owing to the direct proportionality between spawning females and early migrant fry (Au 1972), the abundance of downstream migrating fry would likely continue to increase with increasing number of spawning females, thereby increasing abundance of returning adults in the basin.

- Even though the AWS watersheds had not been noticeably disturbed since the great coastal fires of the middle 1800s, their current productivity for anadromous salmonids may not represent the historical potential. Two possible influences are suggested. Ongoing harvest of salmon over many years has substantially reduced the numbers of spawning fish returning to the watersheds, thereby depriving the stream systems of nutrient enrichment that would have resulted from return of these adults. Bilby et al. (1996) found that as much as 30% of the carbon and nitrogen in juvenile coho salmon in a coastal Washington stream may have been derived from the ocean and released back to the stream by decay of spawned-out adults. In addition, the historic decline of beaver because of widespread trapping has probably reduced the potential for coastal streams to produce coho salmon. Beaver ponds have been shown to provide superior habitat for juvenile coho salmon (Nickelson et al. 1992). Reversal of both these harvest trends could result in significant improvement in the productive potential of these small stream systems. As noted below, beaver populations have now returned to Deer and Flynn Creeks.

- The inherent productivity for juvenile coho salmon of other watershed basins along the coast will differ from that of the AWS streams based on differing geologic parent material, nutrient regime, food supply, and other factors.

Our attempts to relate smolt production in the AWS streams to environmental factors have met with little success. Other work had found relationships between low summer streamflow and abundance of returning adult coho salmon over large regions in Washington (Smoker 1955) and Oregon (Scarnecchia 1981). In the early years of the AWS there was a significant positive correlation between smolt production and summer streamflow for Deer and Flynn Creeks (but not Needle Branch). However, by the end of the study these relationships were nonexistent or statistically insignificant (Knight 1980). There were some marginally significant correlations between smolt numbers and winter flow, but they were not consistent. It seems likely that there are environmental limitations on production of smolts from these headwater drainages, but our work has not yet revealed specific relationships.

Post-logging Changes

There were substantial changes in the stream channel and in salmonid habitat after logging in Needle Branch, the watershed totally clearcut. Felling of timber along and into the stream, cable yarding of logs across the stream channel to uphill landings, and the subsequent hot slash fire all contributed to these changes. From the beginning of felling in mid-March 1966, considerable debris accumulated in the stream channel, forming small dams that slowed the flow of the stream (Fig. 5.3). In September the larger debris was removed from the channel. Crews used chain saws to cut a path through the debris that allowed the stream to flow freely again. In the process, they also removed any large wood that had been in the channel prior to the logging operation. The slash fire in October was especially intense in the upper canyon of Needle Branch, beyond about 550 m above the weir (Fig. 5.4). At a station 915 m above the weir, stream temperature was 13°C when the fire was set and rose rapidly to 28°C during the fire. Many cutthroat trout and reticulate sculpins were found dead in the stream beyond the 550-m point. This was above the major zone of rearing for coho salmon, so fewer of this species were affected by the rapid increase in temperature. Below 550 m, the slopes are not so steep and the fire was less intense near the stream. There was no significant short-term heating of the stream in this zone. During the subsequent winter there was substantial bank erosion that continued until the streambanks and hillslopes revegetated. These events

Fig. 5.3 Logs and debris that remained in the stream channel of Needle Branch, the clearcut stream, after merchantable timber had been removed during the logging operation in 1966

Fig. 5.4 Looking upstream into the canyon of Needle Branch at approximately 500 m above the fish trap during winter 1966–1967, after stream clearance and the slash fire

combined to leave the streambanks broken down, all undercut banks removed, and the stream channel devoid of cover and large wood.

In the Deer Creek watershed, the buffers left along the stream protected the stream channel from any significant influence of the adjacent clearcutting. No changes were noted in the stream during harvest and slash burning, except for the sedimentation that resulted from two road-related landslides that occurred in the winter of 1965–1966 (See Chapter 4).

There was one unexpected occurrence in the patchcut watershed. There had been no evidence of beaver in any of the basins in the prelogging period, but during the year of logging the upper meadow of Deer Creek was invaded by beavers. They built a number of dams that formed extensive ponds. We felt that their influence would confound analysis of any changes caused by logging, so we removed the dams and trapped the animals to keep this influence to a minimum until the study was completed. In recent years, beaver have returned to Deer Creek and are now also found in Flynn Creek.

Major results from the post-logging analysis are summarized in this chapter. Additional details are in Hall and Lantz (1969), Moring (1975a), Moring and Lantz (1975), and Hall et al. (1987).

Habitat Quality

Gravel Composition

Suspended sediment loads changed substantially in both the clearcut and patch-cut watersheds after logging. The road slides in the patchcut watershed during the winter of 1965–1966 caused significant increases in sediment load during the winters of 1965–1966 and 1966–1967. The logging itself did not appear to influence sediment loads in this stream. In the clearcut stream, increases in suspended sediment load were much larger, though they began a year later (in the winter of 1966–1967). These increases were caused primarily by removal of vegetation from the entire watershed, particularly the streambanks. Road building had less influence on sediment production in this watershed than in the patchcut basin (See Chapter 4).

The amount of fine sediment in spawning gravel increased in both the clearcut and patchcut watersheds (Table 5.3). For this chapter I reanalyzed the original data and have included the measures of geometric mean diameter (DG) and fredle index (FI). For statistical analysis I used a 3×5 factorial Analysis of Variance (ANOVA) with specified contrasts (Systat 1996). To provide approximately equal sample sizes in each cell, I used data from 1964 and 1965 for prelogging, and data from 1967 through 1969 for the post-logging period.

In Needle Branch, the clearcut stream, there were statistically significant post-logging increases in the percentages of fine sediments in both size categories, <0.83 mm ($p=0.005$) and <3.3 mm ($p=0.04$), when contrasted with the control stream. The changes in DG and FI were not quite significant ($p=0.13$ and 0.085 respectively). This result is in agreement with the views of

Table 5.3 Summary statistics (mean and standard error) for indices of gravel quality in salmon redds in the three streams. In this summary the prelogging period is 1964–1965 and the post logging period is 1967–1969. Number of redds in sample (n) is indicated. Some redds are included here for which emergence was not recorded

Stream	Period	Percent <0.83 mm	Percent <3.3 mm	Geometric mean (mm)	Fredle index
Deer Creek (patchcut)	Prelogging ($n=11$)	22.9 (1.10)	34.0 (2.21)	6.86 (0.70)	1.21 (0.13)
	Post-logging ($n=17$)	26.6 (0.97)	39.4 (1.30)	5.64 (0.39)	0.898 (0.055)
Flynn Creek (control)	Prelogging ($n=15$)	28.3 (1.30)	42.6 (1.89)	4.39 (0.38)	0.807 (0.071)
	Post-logging ($n=16$)	27.2 (1.36)	41.2 (1.54)	4.74 (0.32)	0.836 (0.049)
Needle Branch (clearcut)	Prelogging ($n=11$)	27.3 (0.87)	40.3 (1.09)	4.80 (0.24)	0.834 (0.040)
	Post-logging ($n=22$)	31.5 (0.85)	44.4 (0.88)	4.10 (0.21)	0.617 (0.032)

Beschta (1982) and Young et al. (1991b) that the percentage of fine sediments is expected to be the most sensitive measure of management-caused change in gravel composition. Although fine sediment also increased in the gravels of the patchcut stream, the variability was greater there and those changes were not statistically significant for any category, when contrasted with the control stream.

Stream Temperature

There was a substantial increase in temperature of the surface water in Needle Branch, the clearcut stream. In the year after logging, the maximum temperature observed in the watershed was 30°C at a point about 300 m above the fish trap. This value is above the recognized lethal limit for juvenile coho salmon and cutthroat trout (Bjornn and Reiser 1991). Temperatures were lower both above and below this point, and there was substantial diurnal variation, so that the lethal temperature levels were experienced for only a short period each day. This variation, and the likely influx of some cooler groundwater into the channel, allowed juvenile salmon and trout to survive, though they undoubtedly experienced significant stress. Rapid revegetation of the riparian zone moderated high temperatures in subsequent years. By 1973 the stream temperature regime in Needle Branch had returned to prelogging levels. There were only minor increases in stream temperature in the patchcut watershed (See Chapter 3).

Temperature of the intragravel water also increased in the clearcut stream, and the mean intragravel temperatures were similar to those of the surface water (Ringler and Hall 1975). Peak temperatures at 25 cm in the gravel lagged peaks in surface water by 1–6 hours, reflecting differences in the rate of interchange of surface and intragravel water.

Dissolved Oxygen

The most striking post-logging change in dissolved oxygen occurred in the surface water of Needle Branch, the clearcut stream. The influx of fine debris during logging, the ponding of stream water by the larger debris, and the increased stream temperature caused by opening the canopy all combined to substantially decrease the level of DO during the summer of logging. The lowest concentration observed was $0.6 \, mg \, L^{-1}$ on June 27, 1966, at a station about 300 m above the gauging weir. In a 300 m reach centered on that station, oxygen levels were too low to support salmon and trout during late June and most of July (Hall and Lantz 1969). Juvenile coho salmon placed in live boxes in that reach survived only 8–40 minutes. Levels of surface DO increased substantially when the stream was cleared of debris in September. However, they remained somewhat below the levels of the control stream for the next few summers owing mainly to increased stream temperatures that reduced the solubility of oxygen.

Dissolved oxygen in intragravel water was also reduced in the clearcut stream. The most significant reduction occurred during the late spring and early summer of logging. The mean value on June 30 was 1.3 mg L^{-1}, compared to 7.2 mg L^{-1} in the control stream (Hall and Lantz 1969). The presence of a blanket of debris over the stream gravel was thought to be the main cause of the decline, both by reducing DO in surface water and restricting interchange of surface and intragravel water. Significance of this decline to fish incubation was minor, however, because almost all the fry had emerged from the gravel before oxygen reached dangerous levels. During subsequent winters, DO in the gravel was also reduced in Needle Branch, but to a lesser extent. Little change was seen in the patchcut watershed immediately after logging. However, there was a long-term decline in intragravel DO in all three streams from 1967 through 1973 (Moring 1975a), suggesting a shortcoming in the standpipe monitoring methods. More details are provided in Chapter 3.

Permeability

Permeability is another measure of the quality of the intragravel environment as substrate for rearing salmonid alevins. It is an index to the rate of movement of water through pore spaces in the gravel. This process affects survival of alevins by delivery of oxygen and removal of waste products. Permeability decreased substantially after logging in the clearcut basin, whereas there was no apparent change in the patchcut or control basins (Moring 1975a, 1982). This measurement provides further evidence of degradation in the quality of the gravel environment in the clearcut watershed.

Fish Populations

Fry Emergence

As indexed by emergence in the fry traps, there was no demonstrable change in percentage of survival from egg deposition to emergence after logging in either Needle Branch or Deer Creek (Table 5.2). However, the power of this analysis was low, constrained by the substantial variability in survival rate as measured by the fry traps. Given the evidence of a reduction in gravel quality, I explored other estimates of STE.

Two other estimates of emergence success were available, both of which suggest that STE decreased after logging in the clearcut stream, Needle Branch. The estimate that uses resident fry plus emigrant fry (Table 5.4a) indicates a large reduction in emergence success for the first two post-logging years, the only years for which these data are available. The estimate based on emigrant fry per spawning female (Table 5.4b) includes the entire study period. The reduction in fry per female also suggests a substantial post-logging reduction in survival in

Table 5.4 Two estimates of survival to emergence for coho salmon: (a) Survival rate calculated as (number of emigrant fry + estimated number of June resident fry)/estimated number of eggs deposited. This estimate is available for only two years post-logging, 1967 and 1968; (b) Survival indexed as emigrant fry per spawning female. Sample size (n) refers to number of years for which observations are available. For both measures the ratio of post-logging to prelogging means is shown

(a) Fry survival rate				
		Deer Creek	Flynn Creek	Needle Branch
Prelogging	mean (n)	0.357 (7)	0.452 (7)	0.588 (6)
	range	0.257–0.559	0.231–0.837	0.396–1.0
Post-logging	mean (n)	0.281 (2)	0.315 (2)	0.188 (2)
	range	0.256–0.306	0.307–0.323	0.153–0.222
Ratio Post/Pre		0.787	0.697	0.320

(b) Fry per female				
		Deer Creek	Flynn Creek	Needle Branch
Prelogging	mean (n)	175 (7)	389 (7)	737 (7)
	range	83–272	33–818	95–1,442
Post-logging	mean (n)	172 (7)	340 (7)	321 (7)
	range	7.7–315	8.0–524	0–829
Ratio Post/Pre		0.989	0.874	0.436

the clearcut watershed. The appropriate statistical analysis of these data uses a plot of emigrant fry and spawning females (Fig. 5.5). The slopes of the lines in prelogging and post-logging periods provide an index of STE. For the control watershed, the slopes for the periods before and after logging (363 and 380 respectively) did not differ ($p = 0.57$). In contrast, there was a significant decrease in the post-logging slope in the clearcut watershed (672 and 357 respectively, $p = 0.038$). This index suggests an approximately 50% reduction in survival to emergence after logging. The same analysis was not available for the patchcut stream because the prelogging regression was not significant.

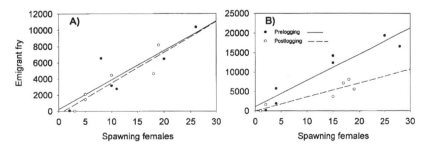

Fig. 5.5 Pre and post-logging regressions of number of spawning females and resultant emigrant fry for (a) Flynn Creek, and (b) Needle Branch. The slopes of the regression lines represent an index of survival to emergence

Emergent Survival and Gravel Quality

With the newer indices to gravel quality (DG and FI), I expected to find that additional data beyond that published earlier (Hall and Lantz 1969) would improve the correlation between gravel quality and STE. Instead, at least on Deer Creek, the inverse relationship between STE and percentage fines, significant in the early years, progressively worsened as more years of data accumulated (Table 5.5). The same regression for the post-logging period became positive, a most unexpected result. In contrast, the same relationship on Flynn Creek, not significant early, improved substantially over time. In Needle Branch, no significant correlation was found between STE and any measure of gravel quality during the entire study period.

Contrary to expectation, DG and FI never performed better than percentage of fines in accounting for variation in STE in the natural redds. In the laboratory experiment (Fig. 5.1b), DG did produce a slightly better correlation with emergence than did percentage of fines ($r^2 = 0.83$ compared with 0.79). In the natural redds, for the periods that produced statistically significant linear regressions, FI

Table 5.5 Correlation coefficients between various measures of gravel quality and survival to emergence (STE) for Deer Creek and Flynn Creek during a number of time periods (none of the coefficients was statistically significant for Needle Branch). As a common comparison, the value for STE and percent fines <0.833 mm (P83) is shown for all periods. When other correlations were higher, the highest is shown in addition, along with some others of interest. P33 = percent fines <3.33 mm; DG = geometric mean diameter; FI = fredle index; Arc (STE) = arcsine transformation; Log (STE) = natural logarithm. Correlations with percent fines are expected to be negative, those with DG and FI to be positive. Asterisks after the correlation coefficient indicate level of significance: $*p < 0.05$, $**p < 0.01$

Years	Deer Creek				Flynn Creek		
	Variables	Correlation	n		Variables	Correlation	n
1964–66	STE – P83	−0.65*	10		STE – P83	−0.47	13
	STE – P33	−0.66*					
	STE – DG	0.52					
	STE – FI	0.51					
1964–67	STE – P83	−0.41	16		STE–P83	−0.43	15
1964–68	STE – P83	−0.20	22		STE– P83	−0.51*	18
	STE – P33	−0.35			Arc (STE – P83)	−0.53*	
1964–71	STE – P83	−0.20	29		STE – P83	−0.56**	29
(all years)	STE – DG	0.29			Log (STE) – P83	−0.66**	
	STE – FI	0.31			Arc (STE) – DG	0.44*	
					Arc (STE) – F1	0.48**	
1967–71	STE – P83	0.16	19		STE-P83	−0.64**	16
(post-logging)	Log (STE) – P83	0.26			Log(STE) – P83	−0.86**	

was generally better than DG in accounting for variation. Clearly, we were unable to account for significant sources of variation in STE in the natural redds.

Fry Emigration

As noted earlier (Table 5.4b), the numbers of emigrant fry per female decreased substantially in Needle Branch after logging. The numbers of females spawning in Needle Branch (and in the other two streams) were approximately equal before and after logging. Thus there was a large decrease in the total number of fry leaving Needle Branch after logging. The post-logging annual average was about 37% of the prelogging value, compared to 84% and 114% for Flynn Creek and Deer Creek, respectively.

Juvenile Density

The most significant change in the fish populations after logging was a substantial reduction in the late-summer abundance of juvenile cutthroat trout in Needle Branch, the clearcut stream (Fig. 5.6). Prelogging biology and abundance of the trout had been documented by Lowry (1964, 1965). During the post-logging period (1966–1974) the average abundance of juveniles in both Deer and Flynn Creeks was higher than during the prelogging period, but in Needle Branch, total numbers averaged only 37% of the prelogging value. The numbers of age-1 and older fish were also reduced by almost the same percentage. The reduction was somewhat less if indexed as grams per square meter (50% of the prelogging mean), reflecting an increase in average size of trout in the post-logging period.

There is a small falls about midway up the Needle Branch watershed that forms a barrier to upstream migration of trout. This falls provided an unusual opportunity to document the contribution made by cutthroat trout residing above a migration barrier, both to populations in downstream areas and also to the downstream migration of juveniles and smolts. Prior to logging, estimates of juvenile population size for both salmon and trout were made between the fish trap and this small barrier falls approximately 870 m above the fish trap. After logging was completed, we found large numbers of trout above the falls, so we extended our estimates to include trout in this reach. For 4 years (1968–1971) we marked and recaptured trout in about 500 m immediately above the first falls. Distinctive marks on these fish showed that substantial numbers of juvenile trout moved from above the falls to the section below the falls.

Thus, the large reductions in trout numbers that we saw in the lower reach were buffered by an influx of juvenile trout from above the barrier falls. In several of the post-logging years, unique marks indicated that the majority of trout in the lower study reach had originated above the falls. Without this influx from above, the numbers of cutthroat trout in the study reach below

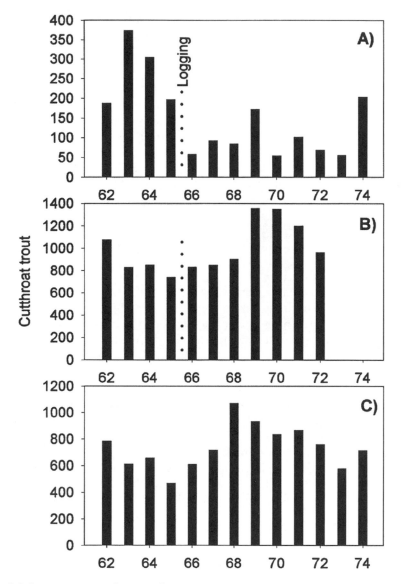

Fig. 5.6 Late-summer estimates of population size of cutthroat trout in the three study streams, 1962–1974; (a) Needle Branch (clearcut), (b) Deer Creek (patchcut), and (c) Flynn Creek (control). Beginning of logging influence is marked by dotted line

the falls during late summer would have been exceptionally low. An increase in numbers in 1974 suggested that recovery might have begun, but later sampling indicated that not to be true (See Chapter 14). Changes in the stream accompanying clearcut logging resulted in a long-term decrease in the productivity of the watershed for cutthroat trout.

There were also indications of an adverse effect on the reticulate sculpin population in the clearcut stream, at least in the short term. Age-0 fish did not survive in the stream during the year of logging, and age-1 fish were almost nonexistent (Krohn 1968). Numbers of sculpins moving through the downstream trap in the first 6 years following logging were only 16% of the prelogging average (Moring 1981).

Smolt Migrations

The great majority (approximately 85–95%) of salmonids of smolt size moving through the downstream traps were coho salmon. A number of changes occurred in migration of this species after logging, but substantial natural variation made evaluation of these changes difficult.

One major change in the clearcut watershed was a shift toward earlier migration of juvenile salmon for the first 4 years after logging. In the year when this was most extreme, 1968–1969, nearly 70% of the total migration occurred prior to February 1, the range during the 4 years being 29–68% and the mean 41%. Comparable figures for Flynn Creek were a range of 3–18% and a mean of 12%. The increased temperatures in Needle Branch were probably the impetus for this early migration. A shift to earlier migration of coho salmon smolts was also noted after logging in the Carnation Creek, British Columbia, watershed, and temperature was implicated as the cause (Holtby 1988). By the 1970–1971 migration year in Needle Branch, when stream temperatures had returned toward their prelogging levels, migration timing returned to its earlier pattern.

There did not seem to be a similar shift in migration timing of juvenile and smolt cutthroat trout. For the same 4 years in Needle Branch, the percentage of total migrants <275 mm that moved prior to February 1 ranged from 3.3% to 14.8% and averaged 11.1%. Comparable figures for Flynn Creek were a range of 1.7% to 16% and a mean of 7.4%.

It was difficult to determine the influence of logging on the numbers of coho salmon smolts leaving the watersheds for the sea. In both the clearcut and patchcut watersheds, the numbers of smolts leaving the watersheds after logging were lower than those during the prelogging period (Fig. 5.7). The numbers of smolts leaving the clearcut watershed were 63% of the prelogging average, and those from the patchcut were 79% of the prelogging value (Table 5.1). However, the number for the control watershed was only 50% of the prelogging average. Of additional concern, there appeared to be a long-term downward trend in the production of smolts from all three watersheds over the 15-year period of the study. This trend was particularly strong in Flynn Creek. Thus it was not possible to use the "control" watershed as a reference for production of juvenile coho salmon, and no conclusion about change in smolt abundance after logging is possible for this species.

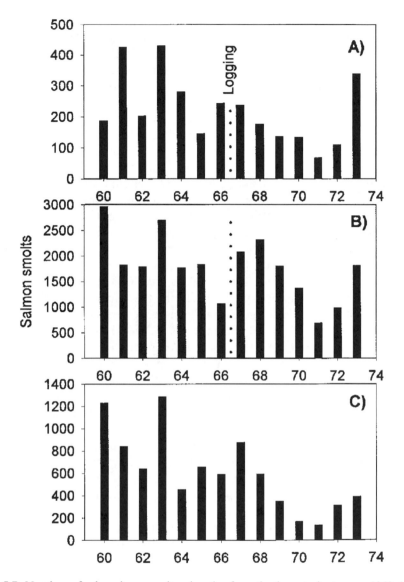

Fig. 5.7 Numbers of coho salmon smolts migrating from the three study streams, 1960–1973; (a) Needle Branch, (b) Deer Creek, and (c) Flynn Creek. Year indicated is spring of migration. Logging influence on smolt migration was 1 year later than on late-summer trout populations

Using the same period of smolt enumeration (February 1–May 31), we determined that there was no significant post-logging change in the downstream migration of smolt-sized cutthroat trout (150–275 mm) from the clearcut watershed. In the post-logging period, the mean was 24 compared with the prelogging mean of 37. Numbers of trout in this size range migrating from this small stream were low and variable, ranging from 17 to 85 in the prelogging period and from 11 to 36 after logging.

There was, however, a notable increase in numbers of trout smaller than 150 mm passing through the downstream trap after logging. Beginning in 1966–1967 and continuing through 1970–1971, more cutthroat trout were counted through the downstream trap than were estimated to have been present in the previous late summer in the section below the barrier falls. Based on recapture of fish uniquely marked above the falls from 1968 through 1971, we were able to show that the source of the additional fish was the population above the falls. Most of these fish moving from above the falls and through the down-trap were <125 mm, and many were <100 mm. Thus, they were not contributing directly to the sea-going population of cutthroat trout, but they still represented a significant contribution from an isolated stream segment above a barrier falls to populations in downstream rearing areas. Such extensive downstream movement as we found from above the migration barrier in Needle Branch is unusual for stream salmonids (Northcote 1992, 1997). In most populations above waterfalls, behavioral mechanisms restrict downstream movement. It may be that the severe environmental conditions in the stream after logging contributed to this unusual behavior. Analysis of this phenomenon is ongoing.

Evaluation

The changes in fish habitat that occurred after logging in the clearcut watershed were dramatic, especially in the first year or two. In spite of these major changes, there were few equally clear responses from the fish populations. Only the long-term depression of the cutthroat trout population and the reduction in numbers of early migrating salmon fry showed changes large enough to be considered statistically significant. In particular, the coho salmon showed considerable resilience in the face of dramatic shifts in habitat quality. The fact that they survived at all in the low-oxygen, high-temperature regime during the summer of logging and the one immediately after suggests that they possess considerable ability to withstand habitat perturbation, at least in the short term.

The general downward trend in numbers of coho salmon smolts, particularly evident in the control watershed, contributed significantly to our inability to evaluate the effects of logging on the salmon populations. There has been much speculation about possible causes, but little direct evidence is available to explain this trend. There is evidence that the relationship between streamflow and sediment load changed in the control watershed after the major storms of December 1964 and January 1965 in the Pacific Northwest. Brown and Krygier (1971) noted an increased production of sediment in Flynn Creek in the year after these events, raising the possibility of additional channel storage of sediment that could have influenced the productive capacity of that stream.

The conflicting results in reanalysis of relationships between gravel quality and survival to emergence measured with the fry traps were disappointing. In his thorough critique of research on STE, Chapman (1988) suggested some reasons that studies like ours might fail to find conclusive relationships where they would be expected. Young et al. (1990) also provided suggestions for improvement that should help future researchers avoid some of the pitfalls of our work. Among the sources of error that they note are (1) failure to sample gravel from the egg pocket, (2) error in estimating fecundity from female length, and (3) interannual variation in egg viability or percentage of fertilization. Young et al. (1991a) noted another potential source of error, which is bias associated with various types of substrate samplers. However, they showed that the McNeil sampler (the one we used in the AWS) most often yielded samples representative of the true substrate composition.

Another factor hampering our ability to draw conclusions in the biological phase of the study was the large natural variation in most of the fish population parameters. Many others have noted that this variation makes it difficult to detect changes in salmonid abundance that might be related to particular land-use treatments (Pella and Myren 1974; Hall and Knight 1981; Platts and Nelson 1988; House 1995) (See Chapter 15).

Part of the failure to achieve clear unambiguous results might be attributed to the pioneering nature of the project. It was the first attempt to include studies of effects on fish populations and fish habitat in a paired watershed analysis. For example, had the work begun later, we would have placed more emphasis on description of the stream channel features, particularly large wood. More recent work has been able to take advantage of experience gained in the AWS. Among the advances have been improvements in study design (Hall et al. 1978), better definition of causal mechanisms (Wilzbach et al. 1986), and the ability to demonstrate clear changes in fish populations and their habitat caused by logging (Murphy et al. 1986; Hartman and Scrivener 1990; Hicks et al. 1991).

One conclusion seems clear: given the significant stream protection afforded by streamside buffers in the patchcut watershed, fish populations in that basin were relatively little affected by the logging operation. The buffer strips, however, did not provide protection from the sediment that resulted from the two road-related landslides that occurred in the headwaters during the winter prior to logging.

The AWS contributed substantially to knowledge of the life history and biology of coho salmon and cutthroat trout. Our findings on the importance of small headwaters streams for both coho salmon and cutthroat trout resulted in an increased impetus for protection of the stream channels and riparian areas of streams as small as these. In addition, many recommendations were made for improved protection of fish habitat during logging operations (Lantz 1971; Moring 1975b). Results from the study were used in formulating early Oregon Forest Practice Rules, after passage of the Forest Practices Act by the Oregon legislature in 1971 (See Chapter 6). Also of significance is the baseline of data that provided the impetus for establishment of Flynn Creek as a Research

Natural Area by the U.S. Forest Service in 1977 (See Chapter 8), and formed the basis for the New Alsea Watershed Study.

Acknowledgment This chapter benefited from comments by a number of people, chief among them Drs. Donald W. Chapman and Thomas G. Northcote. Dr. Ken Rowe provided advice on statistical analysis. I am grateful for the assistance and guidance from the late Dr. Jim Krygier, who was of immense help over the years that I was involved in the AWS.

Literature Cited

Au, D.W.K. 1972. Population Dynamics of the Coho Salmon and its Response to Logging in Three Coastal Streams. Ph.D. Thesis. Oregon State Univ., Corvallis, OR. 245pp.

Beschta, R.L. 1982. Comment on "Stream system evaluation with emphasis on spawning habitat for salmonids" by M. A. Shirazi and W. K. Seim. Water Resour. Res. 18:1292–1295.

Bilby, R.E., Fransen, B.R., and Bisson, P.A. 1996. Incorporation of nitrogen and carbon from spawning coho salmon into the trophic system of small streams: evidence from stable isotopes. Can. J. Fish. Aquat. Sci. 53:164–173.

Bjornn, T.C., and Reiser, D.W. 1991. Habitat requirements of salmonids in streams, pp. 83–138. In: W.R. Meehan, editor. Influences of Forest and Rangeland Management on Salmonid Fishes and their Habitats. Amer. Fish. Soc. Spec. Publ. 19.

Briggs, J.C. 1953. The behavior and reproduction of salmonid fishes in a small coastal stream. Fish Bull. 94. Calif. Dept. Fish and Game. 66pp.

Brown, G.W., and Krygier, J.T. 1971. Clearcut logging and sediment production in the Oregon Coast Range. Water Resour. Res. 7:1189–1198.

Chapman, D.W. 1961. Factors Determining Production of Coho Salmon (*Oncorhynchus kisutch*) in Three Oregon Streams. Ph.D. Thesis. Oregon State Univ., Corvallis, OR. 227pp.

Chapman, D.W. 1962. Aggressive behavior in juvenile coho salmon as a cause of emigration. J. Fish. Res. Board Canada 19:1047–1080.

Chapman, D.W. 1988. Critical review of variables used to define effects of fines in redds of large salmonids. Trans. Amer. Fish. Soc. 117:1–21.

Haak, R.J. 1984. Causal Influences and Logging Effects on Emigration of Juvenile Coho Salmon from Natal Streams. B.S. Thesis. Oregon State Univ., Corvallis, OR. 40pp.

Hall, J.D., Brown, G.W., and Lantz, R.L. 1987. The Alsea Watershed Study: a retrospective, pp. 399–416. In: E.O. Salo and T.W. Cundy, editors. Streamside Management: Forestry and Fishery Interactions. Univ. of Washington Inst. of Forest Resour., Seattle, WA.

Hall, J.D., and Knight, N.J. 1981. Natural Variation in Abundance of Salmonid Populations in Streams and its Implications for Design of Impact Studies. EPA–600/S3-81-021. U.S. Environmental Protection Agency, Corvallis, OR. 85pp.

Hall, J.D., and Lantz, R.L. 1969. Effects of logging on the habitat of coho salmon and cutthroat trout in coastal streams, pp. 355–375. In: T.G. Northcote, editor. Symposium on Salmon and Trout in Streams. Univ. of British Columbia, H.R. MacMillan Lectures in Fisheries, Vancouver, BC.

Hall, J.D., Murphy, M.L., and Aho, R.S. 1978. An improved design for assessing impacts of watershed practices on small streams. Internationale Vereinigung für theoretische und angewandte Limnologie Verhandlungen 20:1359–1365.

Hartman, G.F., and Scrivener, J.C. 1990. Impacts of forest practices on a coastal stream ecosystem, Carnation Creek, British Columbia. Can. Bull. Fish. Aquat. Sci. 223. 148pp.

Hicks, B.J., Hall, J.D., Bisson, P.A., and Sedell, J.R. 1991. Responses of salmonids to habitat changes, pp. 483–518. In: W.R. Meehan, editor. Influences of Forest and

Rangeland Management on Salmonid Fishes and their Habitats. Amer. Fish. Soc. Spec. Publ. 19.

Holtby, L.B. 1988. Effects of logging on stream temperatures in Carnation Creek, British Columbia, and associated impacts on the coho salmon (*Oncorhynchus kisutch*). Can. J. Fish. Aquat. Sci. 45:502–515.

House, R. 1995. Temporal variation in abundance of an isolated population of cutthroat trout in western Oregon, 1981–1991. N. Amer. J. Fish. Manage. 15:33–41.

Knight, N.J. 1980. Factors Affecting the Smolt Yield of Coho Salmon (*Oncorhynchus kisutch*) in Three Oregon Streams. M.S. Thesis. Oregon State Univ., Corvallis, OR. 105pp.

Koski, K.V. 1966. The Survival of Coho Salmon (*Oncorhynchus kisutch*) from Egg Deposition to Emergence in Three Oregon Coastal Streams. M.S. Thesis. Oregon State Univ., Corvallis, OR. 84pp.

Krohn, D.C. 1968. Production of the Reticulate Sculpin (*Cottus perplexus*) and its Predation on Salmon Fry in Three Oregon Streams. M.S. Thesis. Oregon State Univ., Corvallis, OR. 78pp.

Lantz, R.L. 1971. Guidelines for Stream Protection in Logging Operations. Oregon State Game Comm., Portland, OR. 29pp.

Lindsay, R.B. 1975. Distribution and Survival of Coho Salmon Fry after Emigration from Natal Streams. M.S. Thesis. Oregon State Univ., Corvallis, OR. 41pp.

Lotspeich, F.B., and Everest, F.H. 1981. A new method for reporting and interpreting textural composition of spawning gravel. Research Note RN-PNW-369. USDA Forest Service, Pacific Northwest Forest and Range Experiment Station, Portland, OR. 11pp.

Lowry, G.R. 1964. Net Production, Movement, and Food of Cutthroat Trout (*Salmo clarki clarki* Richardson) in Three Oregon Coastal Streams. M.S. Thesis. Oregon State Univ., Corvallis, OR. 72pp.

Lowry, G.R. 1965. Movement of cutthroat trout, *Salmo clarki clarki* (Richardson) in three Oregon coastal streams. Trans. Amer. Fish. Soc. 94:334–338.

McNeil, W.J. 1962. Variations in the dissolved oxygen content of intragravel water in four spawning streams of southeastern Alaska. Special Scientific Report-Fisheries No. 402. USDI Fish and Wildlife Service. 15pp.

McNeil, W.J., and Ahnell, W.H. 1964. Success of pink salmon spawning relative to size of spawning bed materials. Special Scientific Report-Fisheries No. 469. USDI Fish and Wildlife Service, Washington, DC. 15pp.

Moring, J.R. 1975a. The Alsea Watershed Study: effects of logging on the aquatic resources of three headwater streams of the Alsea River, Oregon. Part II. Changes in environmental conditions. Fish. Res. Rep. 9. Oregon Dept. of Fish and Wildlife, Corvallis, OR. 39pp.

Moring, J.R. 1975b. The Alsea Watershed Study: effects of logging on the aquatic resources of three headwater streams of the Alsea River, Oregon. Part III. Discussion and recommendations. Oregon Dept. of Fish and Wildlife, Corvallis, OR.

Moring, J.R. 1981. Changes in populations of reticulate sculpins (*Cottus perplexus*) after clear-cut logging as indicated by downstream migrants. Amer. Midl. Natur. 105:204–207.

Moring, J.R. 1982. Decrease in stream gravel permeability after clear-cut logging: an indication of intragravel conditions for developing salmonid eggs and alevins. Hydrobiologia 88:295–298.

Moring, J.R., and Lantz, R.L. 1975. The Alsea Watershed Study: effects of logging on the aquatic resources of three headwater streams of the Alsea River, Oregon. Part I. Biological studies. Fish. Res. Rep. 9. Oregon Dept. of Fish and Wildlife, Corvallis, OR. 66pp.

Murphy, M.L., Heifetz, J., Johnson, S.W., Koski, K.V., and coauthors. 1986. Effects of clear-cut logging with and without buffer strips on juvenile salmonids in Alaskan streams. Can. J. Fish. Aquat. Sci. 43:1521–1533.

Nickelson, T.E., Rodgers, J.D., Johnson, S.L., and Solazzi, M.F. 1992. Seasonal changes in habitat use by juvenile coho salmon (*Oncorhynchus kisutch*) in Oregon coastal streams. Can. J. Fish. Aquat. Sci 49:783–789.

Northcote, T.G. 1992. Migration and residency in stream salmonids: some ecological considerations and evolutionary consequences. Nordic J. Freshw. Res. 67:5–17.

Northcote, T.G. 1997. Why sea-run? An exploration into the migratory/residency spectrum of coastal cutthroat trout, pp. 20–26. In: J.D. Hall, P.A. Bisson, and R.E. Gresswell, editors. Sea-Run Cutthroat Trout: Biology, Management, and Future Conservation. Oregon Chapter, American Fisheries Society, Corvallis, OR.

Pella, J.J., and Myren, R.T. 1974. Caveats concerning evaluation of effects of logging on salmon production in southeastern Alaska from biological information. Northw. Sci. 48:132–144.

Phillips, R.W., and Koski, K.V. 1969. A fry trap method for estimating salmonid survival from egg deposition to fry emergence. J. Fish. Res. Board Can. 26:133–141.

Phillips, R.W., Lantz, R.L., Claire, E.W., and Moring, J.R. 1975. Some effects of gravel mixtures on emergence of coho salmon and steelhead trout fry. Trans. Amer. Fish. Soc. 104:461–466.

Platts, W.S., and Nelson, R.L. 1988. Fluctuations in trout populations and their implications for land-use evaluation. N. Amer. J. Fish. Manage. 8:333–345.

Platts, W.S., Shirazi, M.A., and Lewis, D.H. 1979. Sediment Particle Sizes Used by Salmon for Spawning with Methods for Evaluation. EPA-600/3-79-043. U.S. Environmental Protection Agency, Corvallis, OR. 39pp.

Ringler, N.H. 1970. Effects of Logging on the Spawning Bed Environment in Two Oregon Coastal Streams. M.S. Thesis. Oregon State Univ., Corvallis, OR. 96pp.

Ringler, N.H., and Hall, J.D. 1975. Effects of logging on water temperature and dissolved oxygen in spawning beds. Trans. Amer. Fish. Soc. 104:111–121.

Ringler, N.H., and Hall, J.D. 1988. Vertical distribution of sediment and organic debris in coho salmon (*Oncorhynchus kisutch*) redds in three small Oregon streams. Can. J.Fish. Aquat. Sci. 45:742–747.

Rodgers, J.D., Ewing, R.D., and Hall, J.D. 1987. Physiological changes during seaward migration of wild juvenile coho salmon (*Oncorhynchus kisutch*). Can. J. Fish. Aquat. Sci. 44:452–457.

Scarnecchia, D.L. 1981. Effects of streamflow and upwelling on yield of wild coho salmon (*Oncorhynchus kisutch*) in Oregon. Can. J. Fish. Aquat. Sci. 38:471–475.

Smoker, W.A. 1955. Effects of Streamflow on Silver Salmon Production in Western Washington. Ph.D. Thesis. Univ. of Washington, Seattle, WA. 175pp.

Systat. 1996. Systat 6.0 for Windows. SPSS, Inc., Chicago, IL.

Terhune, L.B. 1958. The Mark VI groundwater standpipe for measuring seepage through salmon spawning gravel. J. Fish. Res. Board Can. 15:1027–1063.

Wilzbach, M.A., Cummins, K.W., and Hall, J.D. 1986. Influence of habitat manipulations on interactions between cutthroat trout and invertebrate drift. Ecology 67:898–911.

Wolf, P. 1951. A trap for the capture of fish and other organisms moving downstream. Trans. Amer. Fish. Soc. 80:41–45.

Young, M.K., Hubert, W.A., and Wesche, T.A. 1990. Fines in redds of large salmonids. Trans. Amer. Fish. Soc. 119:156–162.

Young, M.K., Hubert, W.A., and Wesche, T.A. 1991a. Biases associated with four stream substrate samplers. Can. J. Fish. Aquat. Sci. 48:1882–1886.

Young, M.K., Hubert, W.A., and Wesche, T.A. 1991b. Selection of measures of substrate composition to estimate survival to emergence of salmonids and to detect changes in stream substrates. N. Amer. J. Fish. Manage. 11:339–346.

Chapter 6
The Oregon Forest Practices Act and Forest Research

Anne B. Hairston-Strang, Paul W. Adams, and George G. Ice

Origins of Oregon Forest Practice Regulation

The Alsea Watershed Study (AWS) was the first to combine assessment of water quality, stream channel habitat, and fish response to forest management on a watershed basis. It is therefore not surprising that it influenced the initial development of regulations in Oregon to address water quality and fishery concerns in forest operations. The Oregon Forest Practices Act (FPA) was passed in 1971, not long after release of many of the AWS results.

The FPA was developed from a long history of forest regulation in Oregon; the earliest laws were enacted in 1864 for fire control on forest operations. In 1941 the state Conservation Act expanded regulation to include sustained-yield forest management concepts, a profound change in an era when most timberland was abandoned after logging. The act emphasized reforestation and fire control. The basis for the 1941 Act was forest practice standards developed by the forest industry to preclude potentially less workable agency regulation (Pacific Northwest Loggers Association 1937). Forestry research results available then apparently had little direct influence on the formation of the State Conservation Act.

Although Oregon had been regulating certain aspects of forest operations for years, the late 1960s were a time of new ideas about forestry, ecology, and environmental protection. These new ideas challenged the adequacy of existing forest practice regulation. Nationwide, attention and concern were growing about the quality and safety of air and water, stimulated by books like Rachel Carson's *Silent Spring* that described environmental hazards of pesticides (Buck 1991). Expanding urban populations were also resulting in greater demands for forest recreation and aesthetic considerations.

A new breed of federal legislation was emerging at this time. The 1960 Multiple-Use Sustained-Yield Act, 1969 National Environmental Policy Act

Anne B. Hairston-Strang
Maryland Department of Natural Resources, Annapolis, MD 21401
astrang@dnr.state.md.us

J. D. Stednick (ed.), *Hydrological and Biological Responses to Forest Practices.*   95
© Springer 2008

(NEPA), 1970 amendments to the Clean Air Act and 1972 Federal Water Pollution Control Act reflected the increasing breadth of forest resources and environmental considerations in which the public and its policymakers were interested.

In Oregon, the broader concerns for public health and the environment were evidenced in the transformation of the State Sanitary Authority into the Department of Environmental Quality (DEQ) in 1969. Severe weather events had also left impressions on public perception of environmental hazards. High winds in the 1962 Columbus Day storm that pummeled the Northwest produced extensive blowndown timber and debris, and the 1964 Christmas week storm produced severe flooding events estimated to have a return interval of 100 years or more for many parts of the Northwest. Damage from the flooding was exacerbated by the load of woody debris from upstream, blocking culverts and battering bridges and buildings in floodplains (Froehlich 1971). The Columbus Day windstorm had provided a large debris source, and the extensive salvage logging it had triggered also was noticeable to the public. Logging debris left in stream channels was blamed for intensifying the woody debris load in the river and subsequent property damage. The 1964 flood shaped both the streambeds and public perceptions of rivers and woody debris management for years afterward.

The 1972 Federal Water Pollution Control Act (FWPCA) was probably the piece of federal legislation most directly related to the original 1971 Oregon Forest Practices Act. The Forest Practices Act was developed before the FWPCA was passed, anticipating the call for nonpoint source pollution control. The FWPCA was federally enacted, but state agencies were given responsibility for developing programs to meet the requirements. Section 208 of the Act required the development and implementation of area-wide waste treatment management plans. According to Rey (1980), "...EPA initially interpreted Section 208 to: (1) emphasize the control of point rather than nonpoint sources; and (2) concentrate on development of pollution control programs for primarily urban rather than rural areas." Nonpoint source water pollution is typically defined as "pollution that is not discharged through pipes" (Puget Sound Water Quality Authority 1986). Nonpoint pollution sources such as forest or agricultural operations are diffuse, difficult to isolate and identify, and severity is often determined by storms and other natural events. Court decisions and a reversal in the EPA's initial position resulted in extension of Section 208 planning to include rural areas and nonpoint sources.

Best Management Practices (BMPs) were identified as appropriate tools for controlling nonpoint pollution sources. Best Management Practices are "... a practice or combination of practices that are determined ... to be the most effective, practicable (including technical, economic, and institutional considerations) means of preventing or reducing the amount of pollution generated by nonpoint sources to a level compatible with water quality goals" (U.S. Forest Service 1980). BMPs are designed to prevent rather than fix problems. When the Oregon DEQ developed its nonpoint source pollution control program, the

FPA was used to meet requirements to control nonpoint pollution from silvicultural activities such as road and landing construction, harvesting, and reforestation (Brown et al. 1978).

Creation of the Oregon Forest Practices Act

Within this context of expanding environmental concern and perceived threats to water and air quality, the 1941 State Conservation Act was reviewed for a major overhaul in 1968. Both regional public concern about hazards from logging, and knowledge that federal law was aiming at broader attention to environmental impacts, provided motivation to update the Act. Public concerns about logging included human health and safety from pesticide use, especially herbicides in reforestation, as well as erosion and instream woody debris during large storms and floods.

The Oregon Board of Forestry (BOF), a committee appointed by the governor and approved by the state legislature (Adams 1996), was responsible for the review and revisions of the 1941 Conservation Act. The BOF makes forest policy decisions and approves administrative rules and regulations for state and private forest lands, while the Oregon Department of Forestry (ODF) has primary responsibility for implementing these rules and other policies. In the late 1960s the BOF had six representatives from the forest industry and a representative from each of: (1) county, (2) agricultural, (3) range, and (4) environmental or labor interests. At that time, the BOF was chaired by the dean of the Oregon State University School of Forestry, completing a total of 11 voting members, the majority representing the forest industry.

In 1968 the BOF established a subcommittee to review and update the 1941 State Conservation Act in light of changed forests and public interests. Forest practices as well as public environmental concern had evolved since the Conservation Act. New methods in nurseries and greater availability of high-quality seedlings had made planting more common than seeding or use of seed-tree cuts for reforestation. Harvesting and road-building equipment and techniques had changed. The ODF had surveyed private forest land for reforestation and watershed problems in 1967 (Schroeder 1971), providing an information base with which to evaluate performance of the existing regulations.

The goal of the 1968 update was to set minimum standards for forest operations, a baseline from which to prevent clearly detrimental practices. The 1941 Act had reflected the concepts and practices of good forest management developed by forest industry in that era. The updated FPA was intended to reflect changes in accepted good forest management that had occurred over the previous three decades, practices that most responsible landowners met without regulation. The 1941 State Conservation Act was viewed as "out of step with the times" and minimum acceptable standards in forest practices were supported by public and private foresters to prevent abuses by problem landowners (Oregon

Dept. of Forestry 1968). It was hoped that removing the poorest practices would counter public criticism and enhance credibility of forest management.

The BOF had a history of cooperating closely with the forest industry, from the industrial origins of the Conservation Act to strong industry representation in BOF membership. Keeping with the tradition of close cooperation, the BOF convened representatives of forest industry and natural resource agencies to discuss problems, identify successful forest practices, and recommend practical ways of implementing requirements. Agreement was reached that a range of forest resources were of interest to the public, including timber, forage, soil, water, wildlife, fish, recreation, and scenery; specific rules or practices were more contentious (Schroeder 1971). The approach emphasized BMPs to prevent damage to soil or water resources. Reforestation standards were to be performance-based, focusing on goals like successful seedling stocking rather than procedures like planting density. This combated criticism of the 1941 Act that the requirement of a minimum reforestation investment allowed inferior work and poor seedling survival to count for compliance without achieving reforestation.

The BOF directed the development of draft legislation and the state legislature passed the Forest Practices Act (FPA) in 1971. The Act acknowledged the importance of maintaining forest tree species; soil, air, and water resources; and habitat for wildlife and aquatic species, but did not mention recreational and scenic considerations (ORS 527.010 Sec. 4). The Act did not specify forest practice standards; instead, it created three regional committees to draft detailed rules tailored to the conditions for northwest Oregon, southwest Oregon, and eastern Oregon. The BOF retained the responsibility to enact the proposed requirements into enforceable administrative rules. The initial set of specific rules was approved in June 1972 and became effective 1 July 1972.

Oregon's Forest Practice Rules (FPR) were intended to change with new information and experience in implementation. The regional committees were created to facilitate local input and participation, while the requirement of BOF approval provided a level of consistent state oversight. Administrative approval through the appointed committees and BOF, even with the public review and hearing requirements, was more flexible than legislative action, which could occur only during legislative sessions held every two years. The BOF also had more direct experience with forestry issues and the forest industry than the legislature, and forestry interests did not have to compete for attention.

The FPR legally applied to state and private forest lands. However, substantial areas of Oregon forest land are in federal ownership that is not subject to state regulation. To assure uniform minimum standards, the ODF, U.S. Forest Service, and U.S. Bureau of Land Management developed a memorandum of understanding that the forest practice rules would be met or exceeded on federal forest land in Oregon. Because the federal agencies had sufficient expertise on their staffs, enforcement was left to the federal agencies.

Use of Research Results in Developing Forest Practice Rules

The general objective of the Oregon FPA was to mandate widely used good management practices. However, unlike the 1941 Conservation Act, forest research results were instrumental in justifying the need for and nature of some of the regulations. The AWS was one of the earliest and most comprehensive watershed studies in the Pacific Northwest, and was especially notable in its consideration of long-term effects of logging practices on fish resources (Hall et al. 1987). Initial results were available at the time the FPA was being considered (e.g., Hall and Lantz 1969; Brown and Krygier 1970; Krygier and Hall 1971), so the Alsea findings were in a unique position to influence policy. Other long-term studies of logging impacts on fisheries were available, but they did not compare effects on the scale of entire watersheds (Hall et al. 1987). In the policy debates over the FPA, conventional wisdom and personal observations on the effect of logging on fisheries abounded. Opinions that could be supported by scientific data gained substantial credibility.

The AWS results were relevant to several areas of concern for revised forest practices, particularly stream buffers and logging debris in streams. The intensive logging and slash burn on Needle Branch demonstrated dramatic stream effects. Results for stream temperature and sediment production supported the concept of buffers for shading and sediment filtration. The low dissolved oxygen found in areas ponded by accumulated logging debris (Hall et al. 1987) added to existing fisheries concerns about woody debris and fish passage.

As mentioned above, the process of creating the Oregon FPR involved two major steps: (1) the state legislature passing the FPA authorizing the regulations and (2) the Board of Forestry and Regional Forest Practice Committees creating specific administrative rules for desired forest practices. At both of these steps, research results were included in the discussions among the scientific community, natural resource agencies, members of the BOF and regional committees.

Oregon Board of Forestry members who directed the development and review of draft rules often were not trained as natural resource specialists, but most had familiarity or experience with forest industry and natural resource management. The regional committee members also were chosen for familiarity with forestry issues in their region. Relevant technical information was available from staff of the ODF and other state agencies such as the Fish and Game Commissions and DEQ. Despite the general expertise available within the state agencies, researchers had an important role in relating the latest findings, particularly those linking water quality, fish habitat, and forest operations.

The Forest Land Uses and Stream Environment symposium in October 1970 at Oregon State University gathered scientists, foresters, and policy-makers to discuss forest practices, sediment, logging debris, chemicals and stream temperature (Krygier and Hall 1971). Results from the AWS were highlighted, along with other Pacific Northwest research and experiences such as those from the H. J. Andrews Experimental Forest in Oregon. This symposium was the most formal and complete compilation of regional forest watershed research, and management knowledge and experience, at the time, occurring while the FPA was being debated and before specific rules were created.

Both the draft of the FPA and the regional rules received public circulation and comment before enactment. Occasionally, the AWS results were specifically referenced in support of a position. For example, a water resources analyst from the Oregon Fish Commission quoted the Alsea results on stream temperature increases when recommending requirements for stream shading to a state legislator (Haas 1971). The State Fish and Game Commission made similar recommendations to the BOF and Regional Forest Practice Committees during development of specific rules. Advocacy groups such as the Oregon Environmental Council and the National Wildlife Federation sometimes used scientific studies to support positions and recommendations, although these groups were considerably more active during later rule changes.

Rules for Watershed Protection

Riparian Buffer Strips

Water temperature data from the Needle Branch clearcut showed large temperature increases whereas few effects were seen on the patch cut area with stream buffers (Brown and Krygier 1970). These results, combined with laboratory studies in fish physiology on temperature tolerance and lethal effects, provided a scientific argument for stream shade requirements (Lantz 1971). The fivefold increase in sediment after intensive logging and slash burning without buffers on Needle Branch further supported the concept of riparian buffers (Brown and Krygier 1971). Patch cutting of the Deer Creek watershed with stream buffer strips did not yield these large sediment increases, although pulses of sediment from roadfill failures were later noted (Hall et al. 1987).

Studies of dissolved nutrients showed increases after clearcutting and burning on the H. J. Andrews Experimental Forest, but they seldom exceeded standards and rapidly decreased (Fredriksen 1971). Later data from the AWS showed similar patterns of short-term increases in nitrate-nitrogen and potassium that remained within water quality limits (Brown et al. 1973). Although interest in effects of clearcutting on nutrients had been raised by high nitrate exports after experimental clearcutting and total vegetation removal through repeated herbicide applications in a Northeastern deciduous forest basin (Likens et al. 1970), the small response in Northwest forests under much more conventional treatments appeared to preclude much concern about the need for large buffers in nutrient control.

The 1972 FPR ultimately required that 75% of the original shade be left along fish-bearing streams to prevent unacceptable stream temperature rise. At the Forest Land Uses Symposium, the concept of buffers was presented as allowing both timber and fisheries values to be met (Lantz 1971). Directional felling and careful yarding would be needed to preserve a vegetated strip around the stream, usually raising logging costs. However, any of the marketable conifers could be removed as long as sufficient shade was provided by remaining shrubs and hardwoods.

Under the 1972 FPR, a fringe of merchantable trees could be necessary if shrubs and hardwoods did not meet the 75% requirement, but this might be waived if other means such as staggered cuttings avoided unacceptable temperature rise and loss of wildlife cover. On nonfish-bearing streams, the FPR required leaving a strip of understory vegetation in widths sufficient to prevent sediment from entering the stream. This approach deliberately left substantial flexibility in application because of variable site conditions for different logging operations. State forest practices officers could make more specific requirements based on a site visit, but not every site could be visited. In the first four years of implementing the FPA, less than 10% of the operations were visited prior to cutting, although these included most of the high-priority sites with fish-bearing streams (Brown et al. 1978).

Dissolved Oxygen

The Alsea Watershed Study also identified low concentrations of dissolved oxygen (DO), both in surface stream water and within streambed gravels, at levels potentially detrimental to fish. As with temperature, laboratory studies had shown the importance of high DO in the survival and development of fish, especially salmonids. The AWS documented that large amounts of fine organic debris deposited in Needle Branch lowered surface and intergravel DO concentrations to levels below those needed to support healthy fish (Moring 1975). Factors contributing to the low DO concentrations observed in Needle Branch included extremely low discharge, removal of shade near the stream, introduction of fresh fine slash with high oxygen demand, and possibly a decrease in the rate of reaeration caused by heavy clogging of the channel with slash and sediment (Hall et al. 1987). These results led to FPR that were designed to maintain shade over and keep organic debris and sediment out of streams (Ice et al. 1989). A minimum amount of shade was to be retained near streams, in part because cool water can hold more oxygen than warm water. Equipment and logging slash were to be kept out of streams and above the high-water line of the channel.

Concerns about logging debris impacts on DO, combined with perceived fish-passage problems and property damage associated with wood in streams, contributed to aggressive slash and wood removal programs. These widespread removals of logging slash and natural debris clearly reduced large woody debris (e.g., root wads, logs, and large branches) in many forest streams of the Pacific

Northwest. Recruitment of large wood for aquatic habitat is now an important issue for forest streams in the region, at least in part due to well-intended past practices based on an incomplete understanding of ecological functions of debris in streams (See Chapter 13).

Large Woody Debris

Although the role of natural woody debris for aquatic habitat was partly recognized, the concern in the late 1960s was with impenetrable and suffocating loads of logging debris, blockage of culverts, and related damage. Narver (1971) described a jam of old logging debris that blocked adult fish access to eight miles of upstream spawning area in British Columbia. Awareness of excess woody debris was heightened by the combination of the 1962 Columbus Day windstorm and the 1964 flood (Froehlich 1971).

When logging near any stream, the FPR required that felling and skidding occur away from the stream if possible. Latitude was allowed for situations like severely leaning trees that were difficult to fell in another direction. On fish-bearing streams, felling, bucking, and limbing were to be done so that the tree and any parts would not fall in or across the stream. Debris added by the logging operation was to be removed as an ongoing process and placed above the high-water level. This was intended to prevent the introduction of mobile debris during high flows that could create fish barriers, block culverts, or cause other damage. Although no mention was made of removing woody debris naturally present, clearing the debris generated during logging road construction or maintenance was required.

In practice, clearing debris from logging and road construction was not a delicate task, and heavy machinery such as bulldozers was often used for stream cleaning. With this method, it was difficult to distinguish between natural and added woody debris, and negative perceptions of debris often dominated. There was also a tendency for operators or enforcement personnel to want to leave a stream looking "neat" to clearly show their efforts. Even with good intentions, heavy equipment in streams probably produced impacts far beyond the envisioned benefits of removing debris obstructing fish passage and preventing low DO.

Pesticides

Studies in the Alsea watershed also investigated contamination of streams by pesticides used in forest operations. Preliminary analysis of samples from forest streams revealed detectable but low levels of endrin, a seedcoating chemical, with some evidence of accumulation in stream organisms (Hall 1971). An evaluation of a greater range of chemicals from sites around Oregon concluded that only low levels of chemicals reached streams from properly managed forest applications (Norris and Moore 1971). Although levels exceeding EPA standards were not reported, the inconclusive results and reliance on vigilant on-site application failed to deflect public concern about possible health risks. The unknowns associated with manufactured chemicals and perceptions of significant health risks contributed to continued public concern and inclusion of rules for control of pesticide application to avoid introduction into streams.

Even without FPA requirements, pesticides must be applied according to federally mandated label instructions. Each label has certain requirements and restrictions on how a pesticide is used, and the FPR further required detailed record-keeping and notification to the state. Rules included spray buffers around streams and open water and protection of surface waters during mixing of chemicals. Chemicals could only be applied during conditions where temperature, humidity, wind, and other factors specified by the state forester were appropriate to minimize drift to streams.

Sediment

The Alsea Watershed Study included substantial monitoring of stream suspended sediment that allowed annual yields to be estimated (Harris 1977). Although the observed watershed-scale responses did not clarify relationships with specific forest practices and site conditions, they were compelling enough to suggest the benefits of limiting disturbance along streams. Forest Practice Rules were therefore adopted to require a minimum buffer strip of undergrowth vegetation on permanent streams. This was intended to be a relatively undisturbed area that would allow sediment to deposit before entering streams and avoid disturbance of sediment stored in the stream channel.

The FPR also contained provisions intended to minimize sediment production at the source. Many of the rules specifically designed to reduce sediment were related to road construction and maintenance. Unstable areas and stream crossings were to be avoided whenever physically and economically possible. Debris from road construction was to be kept out of the channel and above the high-water line. Machine activity in streambeds was to be minimized. Channel relocation in fish-bearing streams required written approval of the state forester. Some of the rules for harvesting also were aimed at keeping sediment out of streams, including landing construction, areas appropriate for tractor skidding, and location of skid trails and fire trails.

Rule Changes in the Late 1970s and 1980s

The FPR were expected to change as implementation and new research revealed needed improvements. Initially, the BOF, the regional committees and forest industry provided the impetus for several rounds of changes and additions to the FPR. Later, advocacy groups and the state legislature started to take that initiative. The first substantial rule changes strengthened regulations on pesticide application, prompted by a hatchery fish kill after nearby spraying of

2,4,5-T in May 1977 (Oregon Dept. of Forestry 1977a). Environmental advocacy groups had been dissatisfied with the initial regulations and used the fish kill to focus public concern and involve the state governor. Greater public notification and wider spray buffers were the result, despite studies concluding low probability of 2,4,5-T causing the fish kill and previous studies finding generally low chronic pesticide levels from forest operations(Oregon Dept. of Forestry 1977b). Another study in the Alsea area investigating miscarriage rates and 2,4,5-T was one of the considerations in a review by EPA of the use of 2,4,5-T and Silvex. Although some local scientists formally evaluated and criticized the methods and conclusions of the miscarriage investigation (Wagner et al. 1979), EPA banned the chemical in 1979, taking the 2,4,5-T issue out of state control.

A wet winter in 1981 resulted in increased landsliding in western Oregon, again prompting environmental groups to petition the BOF for additional rules. The Alsea Watershed Study had noted the contribution of roads in triggering mass movements and supplying sediment to streams (Ice 1991). New rules were enacted in 1983 focusing on road and landing layout to avoid high-risk landslide areas. Recommendations from environmental advocacy groups to prohibit timber harvest in unstable areas such as steep ravines were more controversial than adjusting harvest layout. It was another two years before the BOF worked out rules for addressing "in-unit" mass movements, (i.e., those inside the harvest boundary not associated with roads).

The greatest evolution in the rules in this period occurred in the riparian vegetation requirements. The role of natural woody debris in streams gained attention as the problems of excessive debris loading were replaced by "cleaned" streams being too debris-poor for quality fish habitat. The AWS had illustrated the general value of buffer strips but did not provide much insight into more specific objectives for width, composition, and management.

Through the late 1970s and early 1980s, concern developed about sufficiency of buffer strips for aquatic habitat, particularly the role of natural woody debris. Concern extended to wildlife habitat, especially species using decaying snags or mature timber. The conversion of older forests to young plantations brought with them the question of long-term sources of large woody debris for aquatic habitat and snags or mature forest for certain wildlife species.

A 1985 report from a BOF-appointed task force of state and federal technical specialists formed the nucleus of major revisions to the riparian rules (Carleson and Wilson 1985). New rules in 1987 required snags and mature trees, including commercially sized conifers, to be left in riparian areas of fish-bearing streams. In situations where water temperature was a concern, rules for fish-bearing streams were expanded to include nearby areas of important nonfish-bearing tributaries. The initial concept of buffers, described in the 1970 Forest Land Uses Symposium, envisioned fully compatible multiple use, allowing extraction of valuable conifers while nonmerchantable species were left for water quality and habitat (Lantz 1971). The 1987 changes shifted the focus from multiple use towards adequate provision of habitat, with potentially significant costs to landowners (Olsen et al. 1987).

The year 1987 also brought new legislation and a fundamental shift towards more active legislative and public involvement in the FPR. The 1987 Oregon Legislature made major changes in the composition of the BOF and the role of the regional committees, a revision intended to avoid conflict of interest and self-regulation by the forest industry. The forest industry representation on the BOF was changed to a minority, and the Regional Committees lost the role of drafting regulations but retained review and comment responsibilities. The 1987 legislation also required the BOF to develop new rules to protect sensitive resources such as wetlands, threatened and endangered species habitat, and bird nesting, roosting, and watering sites. Both the specificity of the directives and the fundamental nature of procedural changes marked a substantive turning point in legislative action and public involvement regarding the FPR.

Issues of the 1990s

Debates on the adequacy of the FPR continued into the 1990s. Issues expanded to include detailed riparian management and stream habitat questions critical in understanding and addressing the continued declines of wild salmon runs. Cumulative effects and biodiversity considerations added complexity to these questions, once thought to be simply answered by the hardwoods and understory vegetation left primarily to buffer temperature and sediment changes.

The BOF's tasks of periodically assessing and revising the FPR broadened in scope of both the issues and public involvement. In 1990 the ODF organized a public forum to provide a clearer picture of current issues of concern (Oregon Dept. of Forestry 1991). The list was substantial: stream protection, anadromous fish, clearcut size, scenic values, reforestation, land use conversion, wildlife trees, site productivity, landslides, pesticides, biodiversity, cumulative effects, and procedural issues.

The state legislature became even more active in forest practice rules by passing Senate Bill 1125 (SB 1125) in 1991, which for the first time directly placed some specific forest practice requirements into law instead of using the more deliberative and flexible administrative rulemaking process. As in some earlier efforts involving regulatory changes, the forest industry helped initiate SB 1125 to avoid more burdensome legislation from citizen initiatives like those occurring then in California (Davis et al. 1991).

SB 1125 set requirements for clearcut limits and wildlife leave trees in law and mandated review and/or development of revised rules for reforestation, stream reclassification, leave tree requirements for habitat, land use conversion, and scenic highway buffers. It also called for, and directed substantial funding toward, detailed studies of anadromous fish, cumulative effects, regulatory effectiveness, and harvest rates. Faced with this challenge and using public input from the Forestry Forum, the ODF developed its 1991 Strategic Plan to outline and coordinate the various tasks and specific project plans and their timelines for completion.

Many new projects were thus initiated and led by ODF staff, as well as by research contractors and cooperators, such as Oregon State University (OSU). Initial agency reports from these efforts contributed relevant information and helped stimulate discussion among decision-makers and the public at meetings about the FPR. These reports included an evaluation of existing riparian rule effectiveness (Morman 1993) and two reports focused on proposed riparian and stream protection: the Water Classification and Protection Project (Andrus and Lorensen 1992) and the Report on the Analysis of Proposed Water Classification and Protection Rules (Lorensen et al. 1993). The latter reports reviewed relevant research, presented new field data from a wide range of sites, and assessed current and proposed rules and policies.

The focus on riparian and stream protection led by SB 1125 and related activities produced a rapid increase in new information and regulatory proposals for the ODF and BOF to consider. The information and proposals came from many sources, and included both solicited and unsolicited input from technical specialists from outside Salem. In many cases, the input provided extended well beyond research that had been subject to refereed, scientific, peer review. One important issue thus emerged: A significant amount of the technical input appeared subjective or weighted toward value-based opinions vs. validated research findings. Janet McLennan, then BOF chair, described difficulties in using such input and also noted some contrasting input whose "power derives not from passion but from the disinterested presentation of relevant facts" (McLennan 1992). Similar observations of scientists and other technical experts involved in this and other state and federal forest resource policy-making prompted development of some guidance for policy and decision-makers (Adams and Hairston 1994, 1996).

Even with significant scientific and other technical input, the ODF and BOF had much difficulty in publicly resolving a desirable level of riparian and stream protection that also would not greatly impact private property rights. The latter was a key issue, as ODF analyses indicated that proposals for increased riparian protection could as much as double forest landowner costs vs. the current rules, largely due to substantially greater commercial tree retention (Lorensen and Birch 1994). The BOF ultimately named an advisory committee consisting of a spectrum of interest group representatives to help develop a mutually acceptable set of final rule changes, which contributed to the BOF adopting revised riparian and stream classification and protection rules that were implemented September 1994. These final steps in rule-making had some very important socio-political dimensions, and the ODF later released a report to help clarify both science and policy considerations that were key in rule development (Lorensen et al. 1994).

The FPR changes implemented in September 1994 represented perhaps the largest single modification since the rules were first implemented in 1972. Stream classifications expanded to nine, and were based on average flow and primary resource values (Oregon Dept. of Forestry 1993). There were generally wider riparian management areas required, with greater retention of commercial trees and other vegetation, including a 20 foot "no-cut" zone and substantial minimum basal area requirements next to most streams. Design standards for new stream crossings were increased to accommodate larger (i.e., 50 year) flows and both upstream and downstream adult and juvenile fish passage. Some complex options for stream and riparian protection and enhancement also were included to address certain unique situations, and to provide incentive for landowners and operators to make immediate improvements to stream habitat. These many rule changes were summarized in a 60 page document (Oregon Dept. of Forestry 1994).

Although completed shortly after the major rule changes in 1994, two extensive studies mandated by SB 1125 provided important additional technical information and research findings relevant to the issues of forest watershed and stream protection. The first reviewed and discussed an extensive amount of research on the topic of cumulative effects of forest practices (Beschta et al. 1995). This work revealed major complexities and uncertainties about cumulative effects, and thus stressed the need for additional research as well as careful consideration and analysis of such effects to improve management and policy decisions. Clearcut size limits and general reforestation requirements remain the primary means of addressing cumulative effects in the FPR.

The other major report released in 1995 provided an extensive assessment of the status and future of salmon in western Oregon and northern California (Botkin et al. 1995), an issue that remains very timely. Key findings included a widespread lack of data for accurately assessing long-term fish population trends. For two rivers with reliable fish counts, mixed population trends (both increases and decreases) were observed since the implementation of the FPR in 1972. In the authors' opinion, the 1994 rule changes will reduce the negative effects of forest practices, but whether these practices are adequate or optimal is not known and can only be determined through careful monitoring. Although results from such monitoring are yet limited, early research findings showed the 1994 rules provide substantially greater amounts and sizes of riparian trees that potentially can supply large woody debris to stream channels (Hairston-Strang and Adams 1997, 1998).

Oregon's FPR came under even greater public scrutiny during the late 1990s, as wild salmon populations remained very low and major storms in 1996 triggered numerous landslides that were especially visible in cutover areas and along forest roads. The deaths of four rural residents from a landslide originating in an area clearcut 10 years earlier raised the level of public concern and controversy to an especially high level. This prompted special action (i.e., Senate Bill 1211) by the 1997 Oregon legislature to grant temporary authority to the ODF to restrict forest operations on the basis of public safety considerations.

However, many important questions about specific relations between forest practices and landslides also prompted some major research efforts to be initiated. Focusing on many new slides that occurred in 1996, the ODF conducted an extensive ground-based analysis of landslide frequency and other characteristics (Dent et al. 1997). The public and political significance of the forestry and landslide issue was further shown when Governor Kitzhaber asked faculty at OSU to conduct a detailed review of the research literature (Pyles et al. 1998). The ODF and OSU studies pointedly revealed the complexity of both the technical and policy aspects of landslides and forest practices.

Fundamental study of landslides is complicated by the questionable reliability of aerial surveys (Pyles and Froehlich 1987), and although some ground-based studies reveal a general increase in landslide rates in the first decade after logging in very unstable terrain, other areas show no apparent increases or decreases in landslide rates in forests ten to 100 years of age. Moreover, while unstable terrain can be generally identified, the location of specific sites where individual landslides can be expected during large storms remains problematic, which presents major challenges in managing the costs and benefits of policies that limit human activities.

As many wild salmon populations declined in the 1990s, the National Marine Fisheries Service (NMFS) gave formal notice that it was considering listing some coastal coho salmon populations as threatened species under the Endangered Species Act (ESA). The population trends, combined with the prospect of significant new federal regulatory controls, prompted a major state-led program (widely referred to as the "Oregon Plan") to restore salmon habitat largely through coordinated voluntary efforts directed at private lands. Although watershed research was not directly used to develop the Plan, a "Science Team" of primarily state and federal agency personnel provided input on technical issues and university scientists and other experts conducted a review of the initial draft plan (State of Oregon 1996a, 1996b).

One objective of the Oregon Plan was to address concerns raised under the ESA, and thus the state of Oregon and NMFS developed a Memorandum of Agreement (MOA) to collaborate during the implementation of the plan and to pointedly assess the need for additional or modified measures (including changes in Oregon's FPR) to restore habitat for viable salmon populations. The latter assessments involved input from several additional science groups with members primarily from federal agencies and state universities, and included the use of both watershed research and less-formal observations and opinions. One notable characteristic of the science groups was the broad number and scope of disciplines represented, although consideration of the social sciences remained very limited.

The National Marine Fisheries Service's draft proposal for modifying forest practices to conserve salmon was released about a year after the MOA, and it called for significantly greater stream protection on private forest lands (NMFS 1998a). The report indicated that the science groups "...found it uncertain to unlikely that the existing [forest practice] rules would achieve coho habitat objectives...." However, to clarify their role, NMFS noted that the "...scientists are not the authors of the proposal...," and that in submitting it, NMFS was "...not implying that there is complete consensus among the scientists about these matters, nor that the scientists endorse all elements of the proposal."

These and other issues were further revealed in detailed documentation from the science group discussions and other activities (NMFS 1998b), and the BOF subsequently deferred action on the proposal in part due to questions about its technical basis (Adams 2007).

2000 and Beyond

Watershed research findings will continue to be important in helping shape Oregon's FPR and other natural resource policies. For example, questions about the role and adequacy of existing or proposed FPR in protecting and improving fish habitat undoubtedly will persist, given the inherent complexity of the links between fish populations, freshwater habitat, and forest practices. Such questions already have arisen as policy makers have directed more attention to small or intermittent headwater streams, which historically have had few or no unique management requirements. Newer research has shown some important functions and values of headwater areas for habitat and water quality (e.g., Moore et al. 2005), but a persistent lack of clarity about the effectiveness of alternative policy approaches has contributed to some widely variable requirements for these areas in different states and forest ownerships in the region (Adams 2007).

Another important trend that will contribute to both needs and challenges in watershed research and related policies on forest lands is changing practices and technologies. The initial investment in forest watershed research to better understand how alternative forest practices may protect water quality and habitat has not been sustained as forest practices and operations technology have evolved since the mid-1970s. Thus, some substantial gaps exist in research that documents the watershed effects of contemporary forest practices. Recognition of these gaps has been a key stimulus for new or revived paired watershed studies on forest lands in Oregon, including Hinkle Creek, Trask Creek, and the Alsea Watershed Study Revisited. These studies are employing powerful new technologies for data collection and analysis (e.g., WRC 2006) that were unheard of during the previous era of watershed research. Scientists also recognize, however, that new data and understanding often raise as many questions as they answer, especially when put in the context of changing management practices and policies.

Natural processes and forest management concerns involve substantially longer timeframes than are often considered in policy and legislative actions. Stochastic, cyclical and evolving influences and related issues will add some significant but perhaps less anticipated needs and challenges in watershed research and policies. Infrequent natural disturbances such as major storms and wildfires will substantially alter and "reset" watershed conditions, while incremental trends like climate change may produce more subtle but perhaps equal or more significant effects on watershed characteristics and behavior.

Another challenging issue is scale, including persistent questions about how the effects seen in small watersheds relate to the pattern or magnitude of effects observed over much larger landscapes. Responses for large-scale questions will likely rely heavily on computer models, which in turn will require supporting process-based research and ongoing monitoring and validation. Clearly, the evolving nature and scope of watershed concerns will increase the value of long-term investigations and of broad-scale, interdisciplinary research that can reveal complex resource interactions in watersheds with diverse conditions and land uses and practices.

As the range and level of concerns about watershed resources increase, however, it also becomes clearer that more than technically oriented watershed studies will be needed to effectively address policy and management issues. For example, as the scope of forest riparian protection restrictions is expanded they can encompass a significant amount of land area and timber assets (Adams et al. 2002; Ice et al. 2006). Survey and other social science research show that forest owners and managers in the region have substantial concerns about the economic and other social impacts of such current and proposed stream protection policies, which may result in some important unintended consequences (Adams 2007). Thus, a more holistic approach to watershed research in managed areas would pointedly integrate social science data and analyses. Although such an approach remains unusual, there can be little doubt that the 21st century can be expected to further broaden our approaches and thinking in both watershed research and public policies such as Oregon's FPR. Each will continue to evolve in unique and challenging ways, as only rarely does research or policy provide definitive or final answers, especially given ever-changing and diverse resource conditions and human interests.

Literature Cited

Adams, D.M., Schillinger, R., Latta, G., and Van Nalts, A. 2002. Timber harvest projections for private land in western Oregon. Research Contribution 37. Forest Research Laboratory, Oregon State Univ., Corvallis, OR. 44pp.

Adams, P.W. 1996. Oregon's Forest Practice Rules. Extension Circular 1194. Oregon State University Extension Service, Corvallis, OR. 11pp.

Adams, P.W. 2007. Policy and management for headwater streams in the Pacific Northwest: Synthesis and reflection. Forest Science. 53(2):104–118.

Adams, P.W., and Hairston, A.B. 1994. Using scientific input in policy and decision making. Extension Circular 1441. Oregon State University Extension Service, Corvallis, OR. 19pp.

Adams, P.W., and Hairston, A.B. 1996. Calling all experts: Using science to direct policy. J. of For. 94(4):27–30.

Andrus, C.W., and Lorensen, T. 1992. Water classification and protection project. Oregon Dept. of Forestry, Salem, OR.

Beschta, R.L., Boyle, J.R., Chambers, C.C., Gibson, W.P., and coauthors. 1995. Cumulative effects of forest practices in Oregon: literature and synthesis. Prepared for Oregon Dept. of Forestry. Oregon State Univ., Corvallis, OR.

Botkin, D., Cummins, K., Dunne, T., Reiger, H., and coauthors. 1995. Status and future of salmon of western Oregon and northern California. The Center for the Study of the Environment, Santa Barbara, CA. 300pp.

Brown, G.W., Carlson, D., Carter, G., Heckeroth, D., and coauthors. 1978. Meeting water quality objectives on state and private forest lands through the Oregon Forest Practices Act. Report prepared for the State Forester. Oregon State Dept. of Forestry.

Brown, G.W., Gahler, A.R., and Marston, R.B. 1973. Nutrient losses after clear-cut logging and slash burning in the Oregon Coast Range. Water Resour. Res. 9:1450–1453.

Brown, G.W., and Krygier, J.T. 1970. Effects of clear-cutting on stream temperature. Water Resour. Res. 6:1133–1139.

Brown, G.W., and Krygier, J.T. 1971. Clearcut logging and sediment production in the Oregon Coast Range. Water Resour. Res. 7:1189–1198.

Buck, S.J. 1991. Understanding Environmental Administration and Law. Island Press, Washington, DC. 216pp.

Carleson, D., and Wilson, L.W. 1985. Report of the riparian habitat task force. Oregon Dept. of Forestry and Dept of Fish and Wildlife, Salem, OR.

Davis, L.S., Ruth, L.W., Teeguarden, D.E., and Henly, R.K. 1991. Ballot box forestry. J. For. 89(12):10–18.

Dent, L., Robison, G., Mills, K., Skaugset, A., and coauthors. 1997. Oregon Department of Forestry 1996 storm impacts monitoring project –Preliminary report. Oregon Dept. of Forestry, Salem, OR. 53pp.

Fredriksen, R.L. 1971. Comparative chemical water quality –natural and disturbed streams following logging and slash burning, pp. 125–137. In: J.T. Krygier and J.D. Hall, editors. Proceedings of a symposium: Forest Land Uses and Stream Environment. Oregon State Univ., Corvallis, OR.

Froehlich, H.A. 1971. Logging debris: managing a problem, pp.112–117. In: J.T. Krygier and J.D. Hall, editors. Proceedings of a symposium: Forest Land Uses and Stream Environment. Oregon State Univ., Corvallis, OR.

Haas, J.B. 1971. Letter to State Representative William E. Markham from staff of Fish Commission of Oregon.

Hairston-Strang, A.B., and Adams, P.W. 1997. Oregon's streamside rules: achieving public goals on private land. J. For. 95(7):14–18.

Hairston-Strang, A.B., and Adams, P.W. 1998. Potential large woody debris sources in riparian buffers after harvesting in Oregon, U.S.A. Forest Ecol. and Manage. 112:67–77.

Hall, J.D. 1971. Contributions of the Federal Water Quality Administration to the Alsea Watershed Study, pp. 245. In: J.T. Krygier and J.D. Hall editors. Proceedings of a symposium: Forest Land Uses and Stream Environment. Oregon State Univ., Corvallis, OR.

Hall, J.D., Brown, G.W., and Lantz, R.L. 1987. The Alsea Watershed Study: a retrospective, pp. 399–416. In: E.O. Salo and T.W. Cundy, editors. Streamside Management: Forestry and Fishery Interactions. Univ. of Washington Inst. of Forest Resour., Seattle, WA.

Hall, J.D., and Lantz, R.L. 1969. Effects of logging on the habitat of coho salmon and cutthroat trout in coastal streams, pp. 355–375. In: T.G. Northcote, editor. Symposium on Salmon and Trout in Streams. Univ. of British Columbia, H.R. MacMillan Lectures in Fisheries, Vancouver, BC.

Harris, D.D. 1977. Hydrologic changes after logging in two small Oregon coastal watersheds. Water-Supply Paper 2037. U.S. Geological Survey Washington, DC. 31pp.

Ice, G.G. 1991. Significance of the Alsea watershed studies to development of forest practice rules, pp. 22–28. In: The New Alsea Watershed Study. Technical Bull. 602. Natl. Council of the Paper Industry for Air and Stream Improvement, New York.

Ice, G.G., Beschta, R.L., Craig, R.S., and Sedell, J.R. 1989. Riparian protection rules for Oregon forests, pp. 533–536. In: Proceedings of the California Riparian Systems Conference. General Technical Report PSW-110. USDA Forest Service, Berkeley, CA.

Ice, G.G., Skaugset, A., and Simmons, A. 2006. Estimating areas and timber values of riparian management zones on forest lands. J. Amer. Water Resour. Assoc. 42:115–124.

Krygier, J.T., and Hall, J.D. 1971. Forest Land Uses and Stream Environment, Proceedings of a Symposium, October 19–21, 1970. Oregon State University Continuing Education Publications, Corvallis, OR.

Lantz, R.L. 1971. Guidelines for Stream Protection in Logging Operations. Oregon State Game Comm., Portland, OR. 29pp.

Likens, G.E., Bormann, F.H., Johnson, N.M., Fisher, D.W., and coauthors. 1970. Effects of forest cutting and herbicide treatment on nutrient budgets in the Hubbard Brook watershed ecosystem. Ecol. Monogr. 40:23–47.

Lorensen, T., Andrus, C., Mills, K., Runyon, J., and coauthors. 1993. Report on the analysis of proposed water classification and protection rules. Oregon Dept. of Forestry, Salem, OR.

Lorensen, T., Andrus, C., and Runyon, J. 1994. Oregon Forest Practices Act water protection rules: scientific and policy considerations. Oregon Dept. of Forestry, Salem, OR. 38pp.

Lorensen, T., and Birch, K. 1994. Economic analysis of proposed water classification and protection rules (13 December 1993 draft rule proposal). Oregon Dept. of Forestry, Salem, OR.

McLennan, J. 1992. The Endangered Species Act and aquatic natural resources: perspectives for the future. J. Amer. Fisheries Soc. Symp. on the Endangered Species Act, Rapid City, SD.

Moore, R.D., Apittlehouse, D.L., and Story, A. 2005. Small stream channels and their riparian zones in forested catchments in the Pacific Northwest. Special Issue. J. Amer. Water Resour. Assoc 41:813–834.

Moring, J.R. 1975. The Alsea Watershed Study: effects of logging on the aquatic resources of three headwater streams of the Alsea River, Oregon. Part II. Changes in environmental conditions. Fish. Res. Rep. 9. Oregon Dept. of Fish and Wildlife, Corvallis, OR. 39pp.

Morman, D. 1993. Riparian rule effectiveness study report. Oregon Dept. of Forestry, Salem, OR. 198pp.

Narver, D.W. 1971. Effects of logging debris on fish production, pp. 100–111. In: J.T. Krygier and J.D. Hall, editors. Proceedings of a symposium: Forest Land Uses and Stream Environment. Oregon State Univ., Corvallis, OR.

National Marine Fisheries Service (NMFS). 1998a. A draft proposal concerning Oregon forest practices –Submitted by the National Marine Fisheries Service to the Oregon Board of Forestry Memorandum of Agreement Advisory Committee and the Office of the Governor, Portland, OR. 154pp.

National Marine Fisheries Service (NMFS). 1998b. Forest practice science discussion groups – Review materials, notes, evaluations and comments. Vols. I and II. Compiled by Oregon Branch, Habitat Conservation Division, Portland, OR.

Norris, L.A., and Moore, D.G. 1971. The entry and fate of forest chemicals in streams, pp. 45–67. In: J.T. Krygier and J.D. Hall editors. Proceedings of a symposium: Forest Land Uses and Stream Environment Oregon State Univ., Corvallis, OR.

Olsen, E.D., Keough, D.S., and LaCourse, D.K. 1987. Economic impact of proposed Oregon Forest Practice Rules on industrial forest lands in the Oregon Coast Range: a case study. Res. Bull. 61. Forest Research Lab, College of Forestry. Oregon State Univ., Corvallis, OR. 15pp.

Oregon Dept. of Forestry. 1968. Conservation act said out of step with times. Forest Log 38(2):1–4.

Oregon Dept. of Forestry. 1977a. Gnat Creek: We still don't know what caused the fish-kill. Forest Log 47:2–4.

Oregon Dept. of Forestry. 1977b. Minutes of meeting, 8 June 1977. Oregon Dept. of Forestry, Salem, OR.

Oregon Dept. of Forestry. 1991. Final report on Oregon forest practices public forum issues. Forest Log 60:4–70.

Oregon Dept. of Forestry. 1993. Water classification. Tech. Note FP1. Oregon Dept. of Forestry, Salem, OR. 14pp.

Oregon Dept. of Forestry. 1994. Forest practice water protection rules – Divisions 24 and 57. Effective September 1, 1994. Oregon Dept. of Forestry, Salem, OR.

Pacific Northwest Loggers Association. 1937. Joint committee on forest conservation. Forest Practice Handbook: Presenting the Rules of Forest Practice for the Douglas Fir Region. PNW Loggers Assn., Seattle, WA.

Puget Sound Water Quality Authority. 1986. Issue Paper: Nonpoint source pollution. PSWQA, Seattle, WA.

Pyles, M.R., Adams, P.W., Beschta, R.L., and Skaugset, A.E. 1998. Forest practices and landslides – A report prepared for Governor John A. Kitzhaber. Forest Engineering Dept., Oregon State Univ., Corvallis, OR.

Pyles, M.R., and Froehlich, H.A. 1987. Discussion of "Rates of landsliding as impacted by timber management activities in northwestern California," by M. Wolfe and J. Williams. Bull. Assoc. Eng. Geol. 24(3):425–431.

Rey, M. 1980. The effect of the Clean Water Act on forestry practices, pp. 11–30. In: Symposium on U.S. Forestry and Water Quality: What Course in the '80s. Water Pollution Control Federation, Richmond, VA.

Schroeder, J.E. 1971. Programs and policies – Oregon Department of Forestry, pp. 215–218. In: J.T. Krygier and J.D. Hall, editors. Proceedings of a symposium: Forest Land Uses and Stream Environment. Oregon State Univ., Corvallis, OR.

State of Oregon. 1996a. The governor's coastal salmon restoration initiative, Salem, OR.

State of Oregon. 1996b. Peer review 1996—coastal salmon restoration initiative, Salem, OR.

U.S. Forest Service. 1980. An approach to water resource evaluation of nonpoint silviculture sources (A procedural handbook). EPA-600/8-80-012. U.S. Environmental Protection Agency Research Laboratory, Athens, GA.

Wagner, S.L., Witt, J.M., Norris, L.A., Higgins, J.E., and coauthors. 1979. A scientific critique of the EPA Alsea II study and report. Oregon State Univ., Environmental Health Sciences Center, Corvallis, OR. 92pp.

Watershed Research Cooperative (WRC). 2006. Watershed Research Cooperative 2004–2005 Annual Report. Oregon State Univ., Forest Engineering Department, Corvallis, OR.

Chapter 7
The New Alsea Watershed Study

John D. Stednick

The effects of timber harvesting or other forestry practices on water resources are frequently assessed by paired watershed studies (Bethlahmy 1963). Water yields or other hydrograph responses are compared to pretreatment relations to ascertain any departure, which is then attributed to the land use (e.g. Harr 1980, 1983). Water quality changes may be assessed in a similar fashion; changes in nutrient concentrations are compared to the pretreatment period. Water quality-related studies measure constituent concentrations or other water quality parameters as they relate to water quality standards (or criteria) and as a best management practice (BMP) assessment. Water quality effects have not received as much attention as water yield.

Water yield studies under almost all environmental conditions indicate that vegetation removal will result in increased annual water yield (Bosch and Hewlett 1982; Stednick 1996). The annual water yield increases decline over time as vegetation reestablishes. Water utilization is maximized once the site is fully stocked by vegetation. Soil moisture changes after timber harvesting in the California subalpine were estimated to become insignificant after 16 years (Zeimer 1965), streamflow records from Coweeta Hydrologic Laboratory (eastern hardwoods) suggested a return to pretreatment levels after 30 years (Kovner 1956) and streamflow records at Fraser Experimental Forest (Colorado subalpine) indicated 80 years (Goodell 1958), but subsequent data evaluation suggested full hydrologic recovery after 80 to 100 years (Troendle and King 1985). Assessment of the long-term effect of silvicultural treatments on water yields can only be done at study sites with long-term hydrologic records.

Post-logging measurement of annual water yield increases after conifer harvesting in the central Oregon Cascade Mountains decreased in a linear fashion, although the decreasing trend was less apparent the last eight years (Harr 1976, 1979). A predictive equation using years since harvest accounted for 68% of the water yield variability; if annual precipitation was added as a second variable,

John D. Stednick
Department of Forest, Rangeland, and Watershed Stewardship, Colorado State University, Fort Collins, CO 80523

J. D. Stednick (ed.), *Hydrological and Biological Responses to Forest Practices.*   115
© Springer 2008

the variability accounting increased to 87% (Harr 1979). This time function predicted that Needle Branch would return to pretreatment levels of annual water yield in 1991, while the time and precipitation equation predicted that pretreatment conditions would be reached in 1985.

No effort outside the original study (1959 to 1973) was made to measure annual water yields in the three study watersheds at Alsea. No interpretation was made on the seven years of posttreatment data to predict a return to pretreatment (or control) water yields. Water quality monitoring stopped once certain measured constituents returned to pretreatment concentrations, often only months after harvesting. Reactivation of the Alsea water resources monitoring program provided a unique opportunity to assess long-term changes in hydrology and water quality after silvicultural treatments.

The results of the original Alsea Watershed Study showed an increase in annual water yield and three-day peak flows for Needle Branch, the clear-cut watershed, and no significant change for any streamflow metric on Deer Creek, the patch-cut watershed. Because the watersheds "generally appear to be returning to pre-logging conditions" the authors of the original study believed hydrologic recovery had occurred (Harris 1977). The primary work that has been done on the hydrologic effects following timber harvest consists of the short-term paired watershed studies similar to those described in the previous sections. New work on paired watershed studies is addressing the recent finding that some watersheds, deemed to be "recovered", are still showing departures from the pretreatment relations established prior to timber harvest (Hicks et al. 1991; Stednick and Kern 1992). Studies of hydrologic recovery estimate that approximately 30 years are required for recovery in the Cascade region of Oregon and approximately 24 years for the Coastal Range of Oregon (Harr 1983; Stednick and Kern 1992). This long-lasting change indicates that the original, short-term studies were only considering the changes immediately following timber harvest, and not the complex interactions that occur over the long-term, when variations in climatic and terrestrial systems can complicate the issue of hydrologic recovery. There is a need for further research on hydrologic recovery and for a new definition that considers recovery over the long-term.

There were no changes in low flows or lowflow days for either watershed after treatment (Harris 1973). However, periods of no flow were recorded on Needle Branch during the early 1990's, something that was never observed in the original 15 year study (Stednick unpublished data). The emergence of low flows as a critical hydrologic factor in determining salmonid survival especially those located in first and second-order streams is an important area of research that can be addressed by this study. Research again from the central Oregon Cascades suggested that low flows are reduced for 20% of the 70–100 year rotation schedule for timber harvests following timber harvest (Hicks et al. 1991). The length of this reduction could have serious implications to salmonid fisheries. Further study of this phenomenon in salmonid rearing streams is essential and will help land managers schedule harvests to protect salmon-rearing streams.

The effects of timber harvesting and other forest practices on water resources are a concern to land managers, especially when forest watersheds provide fish habitat or public water supplies. Logging, road-building, and slash disposal can upset natural processes that maintain water quality. Water quality changes observed after logging in headwater basins in western Oregon included increased erosion of steep, unstable land, and subsequent sedimentation (Fredriksen 1970; Brown et al. 1973; Harr et al. 1975), nutrient-enriched runoff (Fredriksen 1971; Brown et al. 1973), and increased solar heating of streamwater (Brown and Krygier 1967, 1970; Brown 1972). Changes in chemical concentrations in stream waters are often short-lived (Brown 1972; Brown et al. 1973; Adams and Stack 1989) while increases in water yield may be long-lived (Harr 1976). Many of these findings were based on results from the Alsea Watershed Study. The results of this study were used to help develop BMPs for Oregon and other western states (See Chapter 6).

The Federal Water Pollution Control Act Amendments of 1972 (PL92-500) established a legislative framework from which an assessment of the sources and extent of nonpoint pollution and development of guidelines and procedures to control nonpoint pollution could be made. Land use activities and resultant water quality changes, either positive or negative, are considered nonpoint source pollution. Nonpoint source pollution was addressed in Section 208 of the 1972 amendments. The most effective unit of available prevention and control of nonpoint source pollution is a BMP. This is an administrative creation and initially consists of a combination of practices that are determined after problem assessment. Practices are examined, with appropriate public participation, for practicality and effectiveness in preventing or reducing the amount of pollution generated by diffuse sources, to a level compatible with water quality goals (Rice et al. 1975; Meier 1976; McClimans et al. 1979). Nonpoint water pollution can only be controlled by managing the type of activity that takes place on a watershed. Many nonpoint sources of pollution are present naturally in varying quantities (Stednick 1980, 1991, 2000). It must be assumed that Congress was not intending water in any given stream be of a quality suitable for a specific downstream use; but aiming to ensure that streams maintain or progress toward the ecosystem that existed in its pristine state.

Water quality is an expression of all hydrologic processes occurring in a watershed. Thus, land use activities and their potential environmental impact may be assessed by water quality measurements. Perhaps the easiest to measure, and certainly the most documented, are the effects of land use on water yield. There are few long-term hydrologic records that enable determination of the long-term effects (if any) of harvesting on annual water yield, peak flows, and low flows or other prescribed flow interval recurrences. Water quality changes after harvesting have not been considered long-term, but other biogeochemical studies suggest delayed water quality responses may occur after disturbance (Schlesinger 1991). Understanding the long-term effects of silvicultural treatments on water resources will enable resource managers to better predict the effects of harvesting on water yield and timing. Reactivating the Alsea monitoring program and interpretation of these data should provide new perspectives on long-term effects and cumulative impact assessment.

Assessment of potential water quality changes due to land use activities may be measured by water quality standard compliance and/or definition of cumulative watershed effects. Cumulative impact is the impact that results from the incremental impact of the action when added to other past, present, and reasonably foreseeable future actions regardless of what agency or person undertakes such other actions. Such impacts can result from individually minor, but collectively significant, actions taking place over a period of time (Idaho Dept. of Health and Welfare 1987). Most cumulative impact assessment efforts use physical characteristics of the stream channel (State of California 1992).

Cumulative impacts to a stream transcend ownership boundaries. This is of particular concern in mixed ownership watersheds (checkerboard lands) where no single agency or owner has overall regulatory authority. The federal regulatory authorities have approached cumulative impacts from two different directions to date. The first approach is an attempt to develop alternative regulatory perspectives to better define cumulative watershed effects. The second approach is to determine an allowable impact akin to performance standards. The alternative regulatory perspective is institutional and can only be evaluated after implementation. The allowable impact is often an inventory used to define a threshold of concern, and is not process oriented.

The need for the best possible management of land and water resources is also a motivation for long-term research in the Coast Range. Assessment of long-term effects of timber harvesting on water and water related resources and salmonid habitat and populations is rare. The Carnation Creek study in British Columbia investigated the effects of timber harvesting on rainforest-salmonid stream ecosystems in a single watershed using an intensive pre-post treatment assessment (Hogan et al. 1998). The Alsea Watershed Study provides an excellent opportunity to study the long-term effects of timber harvesting in the Coast Range of Oregon. This widely cited study has contributed important information about the response of watersheds to timber harvest.

The Alsea Watershed Study was reactivated as the New Alsea Watershed Study (NAWS) in 1989 to assess long-term responses of watersheds to timber harvest (Stednick 1991). Preliminary analysis of the data collected during NAWS with the regression equations developed (Harris 1977) for AWS, has shown that the streamflow was still not within the 95% confidence limits, indicating recovery to pre-harvest conditions had not yet occurred (Stednick and Kern 1992). The New Alsea Watershed study provides an opportunity to study the many interrelated effects of timber harvest in an active salmon-rearing area, and in a commercially important timber-reserve.

Flynn Creek was designated a long-term Research Natural Area by the USDA Forest Service in 1977, and used to characterize undisturbed temperate coniferous forest ecosystems in the Oregon Coast Range (See Chapter 8).

Deer Creek had a second timber harvesting entry in 1978 of 20 ha and two units of 14.5 ha and 8.4 ha were logged in 1987 and 1988 (Fig. 7.1). No additional

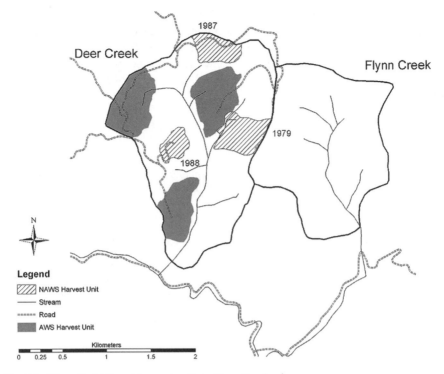

Fig. 7.1 Map of additional harvest units in Deer Creek

harvesting is currently scheduled for Deer Creek in the Forest Service land management plan. A small precommercial thinning (<2 ha) was done in upper Deer Creek, but the size and location of the thinning will have minimal effect. Approximately 39% of the watershed has now been harvested. The multiple entries in Deer Creek provide the opportunity to assess the adequacy of our ability to predict hydrologic recovery and to identify potential cumulative watershed effects on water yield and water quality.

Since timber harvesting, forest management on Needle Branch has included pre- and commercial thinning. Approximately 25% of the middle third of the watershed was precommercially thinned in 1981. In 1997–1998 approximately 40% of this area was commercially thinned with a 30% basal area removal. No additional timber stand management activities have been done in Needle Branch Creek.

The New Alsea Watershed study provides the opportunity to assess the long-term effects of timber harvesting on water and water related resources using the paired watershed approach. The following chapters provide an update on these efforts. The control watershed Flynn Creek was designated a Research Natural Area and serves as an undisturbed control watershed (Chapter 8). Streamflow (Chapter 9) and water quality (Chapter 10) changes after single- and multiple timber harvesting entries were assessed, and new sediment studies were conducted on Flynn Creek (Chapter 12). A risk assessment approach for salmon as related to sediment and nitrate concentrations was developed (Chapter 11). The long term change in salmonid populations and habitat are addressed (Chapters 13, 14, and 15). Finally, a review of watershed management practices (Chapter 16) leads to the lessons learned at Alsea (Chapter 17).

Literature Cited

Adams, P.W., and Stack, W.R. 1989. Streamwater quality after logging in Southwest Oregon, project completion report. Supplement No. PNW-87-400. USDA Forest Service, Pacific Northwest Forest and Range Experiment Station. 19pp.

Bethlahmy, N. 1963. Rapid calibration of watersheds for hydrologic studies. Bull. Int. Assoc. Sci. Hydrol. 8:38–42.

Bosch, J.M., and Hewlett, J.D. 1982. A review of catchment experiments to determine the effect of vegetation changes on water yield and evapotranspiration. J. Hydrol. 55:3–23.

Brown, G.W. 1972. Logging and water quality in the Pacific Northwest, pp.330–335. In: Watersheds in Transition. Amer. Water Res. Assoc. Publ., Proceedings Series No. 14.

Brown, G.W., Gahler, A.R., and Marston, R.B. 1973. Nutrient losses after clear-cut logging and slash burning in the Oregon Coast Range. Water Resour. Res. 9:1450–1453.

Brown, G.W., and Krygier, J.T. 1967. Changing water temperatures in small mountain streams. J. Soil and Water Conserv. 22:242–244.

Brown, G.W., and Krygier, J.T. 1970. Effects of clear-cutting on stream temperature. Water Resour. Res. 6:1133–1139.

Fredriksen, R.L. 1970. Erosion and sedimentation following road construction and timber harvest on unstable soils in three small western Oregon watersheds. Research Paper PNW-104. USDA Forest Service, Pacific Northwest Forest and Range Experiment Station, Portland, OR. 15pp.

Fredriksen, R.L. 1971. Comparative chemical water quality – natural and disturbed streams following logging and slash burning, pp. 125–137. In: J.T. Krygier and J.D. Hall, editors. Proceedings of a symposium: Forest Land Uses and Stream Environment. Oregon State Univ., Corvallis, OR.

Goodell, B.C. 1958. A preliminary report on the first years effect of timber harvest on water yield from a Colorado watershed. Station Paper No. 36. USDA Forest Service, Rocky Mountain Forest and Range Experiment Station, Fort Collins, CO.

Harr, R.D. 1976. Forest practices and streamflow in western Oregon. General Technical Report GTR-PNW-49. USDA Forest Service, Pacific Northwest Forest and Range Experiment Station, Portland, OR. 18pp.

Harr, R.D. 1979. Effects of timber harvest on streamflow in the rain-dominated portion of the Pacific Northwest. In: Proceedings of Workshop on Scheduling Timber Harvest for Hydrologic Concerns. USDA Forest Service, Pacific Northwest Forest and Range Experiment Station, Portland, OR. 45pp.

Harr, R.D. 1980. Streamflow after patch logging in small drainages within Bull Run municipal watershed, Oregon. Research Paper RP-PNW-268. USDA Forest Service, Pacific Northwest Forest and Range Experiment Station, Portland, OR. 16pp.

Harr, R.D. 1983. Potential for augmenting water yield through forest practices in western Washington and western Oregon. Water Resour. Bull. 19:383–393.

Harr, R.D., Harper, W.C., Krygier, J.T., and Hsieh, F.S. 1975. Changes in storm hydrographs after road building and clear-cutting in the Oregon Coast Range. Water Resour. Res. 11:436–444.

Harris, D.D. 1973. Hydrologic changes after clearcut logging in a small Oregon coastal watershed. U.S. Geol. Surv. J. Res. 1:487–491.

Harris, D.D. 1977. Hydrologic changes after logging in two small Oregon coastal watersheds. Water-Supply Paper 2037. U.S. Geological Survey, Washington, DC. 31pp.

Hicks, B.J., Hall, J.D., Bisson, P.A., and Sedell, J.R. 1991. Responses of salmonids to habitat changes, pp. 483–518. In: W.R. Meehan, editor. Influences of Forest and Rangeland Management on Salmonid Fishes and their Habitats. Amer. Fish. Soc. Spec. Publ. 19.

Hogan, D.L., Tschaplinski, P.J., and Chatwin, S. 1998. Carnation Creek and Queen Charlotte Islands Fish/Forestry Workshop: Applying 20 Years of Coastal Research to Management Solutions. Ministry of Forests Research Program, Victoria, BC. 41pp.

Idaho Dept. of Health and Welfare. 1987. State of Idaho forest practices water quality management plan. Water Quality Bureau Report, Boise, ID.

Kovner, J.L. 1956. Evapotranspiration and water yields following forest cutting and natural regrowth, pp. 106–110. Proceedings of Soc. Amer. Foresters, Bethesda, MD.

McClimans, R.J., Gebhardt, J.T., and Roy, S.P. 1979. Perspectives for silvicultural best management practices. Research Report No. 42. Applied Forestry Research Institute, State Univ. of New York, Syracuse, NY.

Meier, M.C. 1976. Research needs in erosion and sediment control, pp. 86–89. In: Soil Erosion: Prediction and Control. Soil Conservation Society of America, Special Publication No. 21.

Rice, R., Thomas, R., and Brown, G.W. 1975. Sampling water quality to determine the impact of land use on small streams. In: Watershed Management, Proceedings of a Symposium by Irrigation and Drainage Division. Amer. Soc. Civil Eng., Logan, UT.

Schlesinger, W.H. 1991. Biogeochemistry: An Analysis of Global Change. Academic Press, San Diego, CA. 351pp.

State of California. 1992. Forest practice cumulative impact assessment process. Tech. Role Addendum No. 2. State Board of Forestry, Sacramento, CA.

Stednick, J.D. 1980. Alaska water quality standards and BMPs, pp. 721–730. In: Proceedings of Watershed Management Symposium. Amer. Soc. of Civil Eng., Boise, ID.

Stednick, J.D. 1991. Purpose and need for reactivating the Alsea Watershed Study, pp. 84–93. In: The New Alsea Watershed Study. Technical Bulletin 602. National Council of the Paper Industry for Air and Stream Improvement, Inc., New York, NY.

Stednick, J.D. 1996. Monitoring the effects of timber harvest on annual water yields. J. Hydrology 176:79–95.

Stednick, J.D. 2000. Effects of vegetation managment on water quality: timber management, pp. 147–167. In: G. Dissmeyer, editor. Drinking Water from Forests and Grasslands. General Technical Report SRS-039. USDA Forest Service, Southern Research Station, Asheville, NC.

Stednick, J.D., and Kern, T.J. 1992. Long term effects of timber harvesting in the Oregon Coast Range: The New Alsea Watershed Study (NAWS), pp. 502–510. In: Interdisciplinary Approaches to Hydrology and Hydrogeology. American Institute of Hydrology, Smyrna, GA.

Troendle, C.A., and King, R.M. 1985. The effect of timber harvest on the Fool Creek wateshed, 30 years later. Water Resour. Bull. 21:1915–1922.

Zeimer, R.R. 1965. Summer evapotranspiration trends as related to time after logging of forests in Sierra Nevada. J. Geophys. Res. 69:615–620.

Chapter 8
Flynn Creek: Research Natural Area

Arthur McKee and Sarah Greene

Research Natural Areas

The Flynn Creek Research Natural Area (RNA) is one of many research natural areas established as part of a national program among federal agencies to identify and protect excellent examples of terrestrial, aquatic, and marine ecosystems for research and education. Most states have similar programs, but these tend to be more oriented to preserving outstanding habitats or rare species. Several not-for-profit organizations such as The Nature Conservancy also establish RNA-like sites to protect and preserve habitats or species. The federal RNA program tends to be somewhat more multifaceted than most state or private programs, but all provide sites where research can be conducted.

The RNA designation assures researchers that the area is protected from disturbances such as logging, mining, or road construction. The management objectives are intended to allow natural processes to occur in a manner analogous to sites with wilderness designation. Research Natural Areas are excellent sites on which to conduct baseline studies of natural processes or conditions, to install permanent sample plots or stream reaches, or to monitor population trends or other variables of interest. Little is required to obtain permission to utilize RNAs, and all field-oriented researchers should seriously examine the possibilities of using them for nonmanipulative kinds of studies.

The Flynn Creek Research Natural Area was established in 1977. It was selected because it contains excellent examples of the highly productive forest and stream ecosystems found in the Oregon Coast Range. It was also the undisturbed control site for the Alsea Watershed Study (AWS), a research program that has been the source for a rich data base and many publications.

The basic objectives of RNA's are to provide sites for research and education, to preserve representative or unusual examples of habitats and species associations found in a region, to protect genetic richness of species, and to

Arthur McKee
Flathead Lake Biological Station, University of Montana, Polson, MT 59860

J. D. Stednick (ed.), *Hydrological and Biological Responses to Forest Practices.*
© Springer 2008

provide a site for collection of baseline data to establish trends in undisturbed systems as well as provide a comparison for nearby managed areas. Flynn Creek RNA is a good example of a site that meets all these objectives.

Location

Flynn Creek RNA lies in the western part of the Oregon Coast Range about 16 km (10 air miles) from the Pacific Ocean, southeast of the town of Toledo (Latitude 44°32′ North; Longitude 123°51′ West) (Fig. 8.1). It includes portions of Sections 1, 2, 11, and 12, T. 12 S., R. 10 W., Willamette Meridian, and lies entirely in lands administered by the Central Coast Ranger District, Siuslaw National Forest. The total area covers about 270 ha (670 acres) and

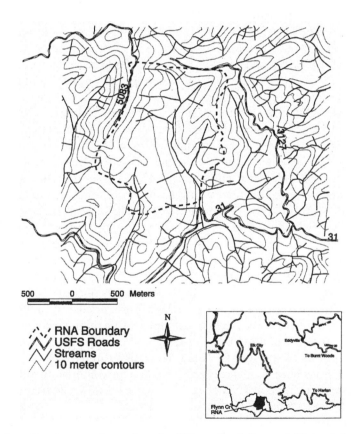

Fig. 8.1 Location, topography, and roads of Flynn Creek Research Natural Area (RNA). Road numbers are U.S. Forest Service numbers. For up-to-date information on road conditions call the Central Coast Ranger Station, Siuslaw National Forest, Waldport, Oregon. Map derived from GIS layers on file at the Siuslaw National Forest, Corvallis, Oregon

contains the complete Flynn Creek watershed. Flynn Creek is a tributary of Drift Creek in the Alsea River basin. The area is easily reached by all-weather roads (Fig. 8.1) that provide access to the stream gauging station as well as several places along the ridge lines.

General Description

The topography of the RNA is rugged and dissected by many small intermittent and perennial streams (Fig. 8.1). Elevations range from about 180 m (600 feet) to 430 m (1400 feet). The relief east of the main stream tends to be steeper than on the west side, although high proportions of both sides of the basin have slopes greater than 35%.

The climate is mild and humid. Corliss (1973) included it in the Tidewater climatic subarea, which he calls the wettest in the Alsea River basin. Precipitation varies from 200–300 cm (80–120 inches) per year. Approximately 90% of this precipitation occurs from October through May, when 6.2 cm (2.5 inches) per day is not uncommon. There is little snowfall and no persistent snow. From June through September, this portion of the Coast Range is generally clear, being too far inland to be affected by coastal fog. Mean daily temperatures range from the mid-60s in summer to the low 40s in the winter. Ocean proximity and high humidity limit diurnal ranges in temperatures to about 8°C (15°F) in winter and 16°C (30°F) in summer.

The RNA is located entirely on sandstones of the Tyee formation, which were deposited under marine conditions in the Eocene. The soils on the site are of the Bohannon-Slickrock association. The soils in the western portion of the area (east aspect) tend to be Slickrock gravelly loams. The soils in the eastern portion (west aspect) tend to be Bohannon gravelly loams. The Bohannon series tends to occur on the sites with the severest relief.

Soil scientists working the Siuslaw National Forest have rated the soils in the area for stability following road construction and timber cutting. The ratings range from moderate to severe risk of slumping associated with road construction over the entire area. The Slickrock series is particularly unstable on steeper slopes (Corliss 1973). Although unstable, these soils are reasonably fertile and contribute to the relatively high plant productivity of the site, which ranges from high III to high II site Douglas-fir.

Mean annual discharge of Flynn Creek is 0.12 m^3 s^{-1} (4.37 ft^3 s^{-1}), with a recorded maximum of 3.94 m^3 s^{-1} (139 ft^3 s^{-1}) and recorded minima of about 0.003 m^3 s^{-1} (0.1 ft^3 s^{-1}). Water temperature varies from about 2°–16°C (35° to 60°F) throughout the year. Total annual sediment yields vary from 21–433 tonnes km^{-2} (959 to 1237 tons mi^{-2}), with values of 28–47 tonnes km^{-2} (120 to 250 tons mi^{-2}) being the most common. There is obviously a great deal of variation in sediment production, which is related to winter storm intensity and frequency.

Terrestrial Vegetation

The terrestrial plant communities are dominated by Douglas-fir (*Pseudotsuga menziesii* [Mirbel] Franco.) and red alder (*Alnus rubra* Bong), which vary in mixture from virtually pure stands of one to virtually pure stands of the other. There is a continuous intergradation of communities dominated by one species or the other, making boundary lines between types difficult to map. In the vegetation type map (Fig. 8.2) this intergradation is obscured. Types are defined by overstory dominance (greater than 60% overstory cover).

The understory is dominated by salmonberry (*Rubus spectabilis Pursh*) on the wettest sites and sword fern (*Polystichum munitum* [Kaulf] Presl) on the upland sites. Vine maple (*Acer circinatum Pursh*) is distributed relatively evenly over the watershed. Shrub cover varies inversely with overstory density, with the densest shrub cover in the most open red alder stands.

The forest appears to have established following a wildfire in the mid-1800s, and the overstory Douglas-fir is between 100 and 150 years of age. Many of the red alder are approaching the maximum age for the species and shifts in dominance between the two species can be expected over the next few decades. The future trend of succession in areas that are dominated by red alder and salmonberry is not clear. Seedlings of tree species are virtually absent under the dense stands of salmonberry, and it is very possible that these areas may go through a period of salmonberry shrub dominance that can last many decades or even centuries.

Beaver (*Castor canadensis*) and Roosevelt elk (*Cervus canadensis* var. roosevelti) activity has disturbed the vegetation in local patches, especially along the main stream and on the small valley floor of the basin. Mountain beaver (*Aplodontia rufa*) appear to play an important role in determining the understory vegetation of the hillslopes. Four Society of American Foresters (SAF) cover types occur in the natural area (Table 8.1)

Because of the intergradation of overstory dominants over much of the area and the extensive shrub cover, the range in basal area within each type is quite great. Basal areas from surveys conducted in 1975 in the Douglas-fir type ranged from 18–78 m^2 ha^{-1} (80–340 ft^2 $acre^{-1}$), and plots in the red alder-dominant type ranged from 9–41 m^2 ha^{-1} (40–180 ft^2 $acre^{-1}$). The pure red alder type was more uniform at 18–30 m^2 ha^{-1} (80–130 ft^2 $acre^{-1}$). There are 1.2 ha to 2 ha (3–5 acre) patches scattered along the tributaries with nearly 100% salmonberry cover. In these patches, the overstory basal areas are very low, about 9 m^2 ha^{-1} (40 ft^2 $acre^{-1}$). The Douglas-fir varied in height in 1975 from 43 m to 61 m (140–200 ft), the red alder from 12 m to 21 m (40–70 feet).

The diameter distribution of the red alder is much more uniform than the Douglas-fir. Overstory red alder diameters generally range from 25–75 cm (10–30 in) whereas Douglas-fir diameters range from 25–165 cm (10–65 in). The largest red alder stems are found near the tributary bottoms; the largest Douglas-fir occur on upland sites, at about mid-slope.

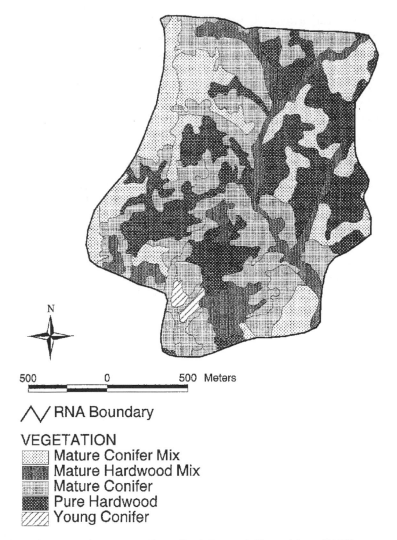

Fig. 8.2 Major vegetation types at Flynn Creek Research Natural Area (RNA) as mapped by Siuslaw National Forest. Map based on air photo interpreted GIS layers. Seral stages derived from information on composition by layer

Table 8.1 Society of American Foresters cover types occurring in the natural area

| SAF type | Dominant species | Mean basal area ||
		$ft^2\ ac^{-1}$	$m^2\ ha^{-1}$
SAF 229 Pacific Douglas-fir	Douglas-fir	190	44
SAF 221 Red Alder	Red Alder	140	32
SAF 221 Red Alder	Red Alder	110	25
Meadow-forest mosaic	–	–	–

Shrub cover varies inversely with overstory density. Shrub species composition also shifts with overstory type. The densest shrub cover occurs in open red alder communities on the lower slopes, where salmonberry and vine maple form a nearly continuous cover. On upland slopes under a denser overstory canopy, salmonberry cover is about 25% and sword fern becomes more important (coverage of 70% to 80%). Salmonberry is absent on the driest sites and under the densest Douglas-fir canopies, but can be found right up to the ridgetops. Red huckleberry (*Vaccinium parvifolium* Smith) and hazel (*Corylus cornuta* var. *californica* Marsh) appear under the upland stands of Douglas-fir and have coverages of 5% to 20%. In general, cover of the understory vegetation is dense throughout the basin and extremely dense in places. Walking is difficult.

The herbaceous layer is well developed and composed of species commonly found in Oregon Coast Range forests. Oregon oxalis (*Oxalis oregona* Nutt.), lady fern (*Athyrium filix-femina* [L.] Roth), and skunk cabbage (*Lysichitum americanum* Hultén & St. John) are common in the bottoms. Major upland herbs are clasping-leaved twisted stalk (*Streptopus amplexifolius* [L.] DC.), beadruby (*Maianthemum dilatatum* [Wood] Nels. and Mach.), western spring-beauty (*Montia siberica* [L.] Howell), sweetscented bedstraw (*Galium triflorum* Michx.), snow queen (*Synthyris reniformis* [Dougl.] Beath.), bracken fern (*Pteridium aquilinum* [L.] Kuhn), and foxglove (*Digitalis purpurea* L. Robust). Moss cover varies considerably due to microsite differences within each of the different types, but seldom exceeds 20% in any area. Lichen and vascular plant species that are expected to occur within the RNA are listed in Table 8.2.

Fish Populations

Three species dominate the fish fauna in the natural area. Coho salmon (*Oncorhynchus kisutch*) and coastal cutthroat trout (*Oncorhynchus clarkii*) are anadromous, while the reticulate sculpin (*Cottus perplexus*) tends to remain within the watershed. Representative values for mean monthly biomass (g m^{-2}) for the three principal species in the Flynn Creek Research Natural Area are:

Reticulate sculpin	3.1
Coho salmon	4.1
Cutthroat trout	4.8
Total	12.0

The reticulate sculpin achieves its maximum biomass in June, its minimum in September. Although known to prey on coho salmon fry, it does

Table 8.2 Lichen and vascular plant species, Flynn Creek Research Natural Area. Lichen list compiled by Linda Geiser, Ecologist, Siuslaw National Forest. Vascular plant list by Katie Grenier, former Botanist, Siuslaw National Forest

Lichen species	
Agyrium rufum	*Cypheliun inquinans*
Alectoria imshaugii	*Diplotomma penichrum*
Alectoria sarmentosai	*Eopyrenula leucoplaca*
Alectoria vancouverensis	*Evernia prunastri*
Arthonia radiata	*Graphis elegans*
Arthothelium macounii	*Hypocenomyce cistaneocenrea*
Arthothelium spectabile	*Hypogymnia appinata*
Bryoria capillaris	*Hypogymnia duplicata*
Bryoria glabra	*Hypogymnia enteromorpha*
Bryoria oregana	*Hypogymnia imshaugii*
Bryoria subcana	*Hypogymnia inactiva*
Buellia disciformis	*Hypogymnia occidentalis*
Calicium glaucellum	*Hypogymnia physodes*
Calicium viride	*Hypogymnia tubulosa*
Caloplaca ceratina	*Hypotrachyna sinuosa*
Caloplaca ferruginea	*Icmadophila ericetorum*
Caloplaca holocarpa	*Japewia tornoensis*
Catillaria endochroma	*Lecanactis megaspora*
Cavernularia hultenii	*Lecanora pacifica*
Cavernularia lophyrea	*Lecanora pulicaris*
Cetraria chlorophylla	*Lecidea botryosa*
Cetraria merrilii	*Lecidella elaeochroma*
Cetraria orbata	*Lecidella euphorea*
Cetraria platyphylla	*Leptogiun bribissonii*
Cetrelia cetrariodes	*Lobaria oregana*
Chaenotheca brachypoda	*Lobaria pulmonaria*
Chaenotheca brunneola	*Lobaria seroviculata*
Chaenotheca chrysocephala	*Lopadiul disciforme*
Chaenotheca furfuracea	*Loxoseporopsis coralifera*
Chaenotheca subroscida	*Melanelia fuliginosa*
Chaenothecopsis pusilla	*Melanelia subaurifera*
Chrysothrix chlorina	*Menegazzia terrebrata*
Cladonia borealis	*Micarea prasina*
Cladonia chlorophaea	*Multiclavula mucida*
Cladonia coniocraea	*Mycobilimbia sabuletorum*
Cladonia cornuta	*Mycoblastus affinis*
Cladonia fimbriaia	*Mycoblastus sanguinarius*
Cladonia ochrochlora	*Nephroma bellum*
Cladonia rei	*Nephroma helveticum*
Cladonia squamosa	*Nephroma laevigatum*
Cladonia subsquamosa	*Nephroma rusupinatum*
Cladonia transcendens	*Normandina pulchella*
Cladonia verruculosa	*Ochrolechia juvenalis*
Cliostomum griffithii	*Leptogium gelatinosum*

(continued)

Table 8.2 (continued)

Lichen species	
Leptogium teretiusculum	*Platismatia glauca*
Ochrolechia laevigata	*Platismatia herreii*
Ochrolechia oregonensis	*Platismatia lacunosa*
Ochrolechia subpallescens	*Platismatia norvegica*
Omphalina sp.	*Platismatia stenophylla*
Opegrapha varia	*Protoparmelia ochirococca*
Ophioparma herrei	*Pseudocyphellaria anomola*
Pannaria leucostictoides	*Pseudocyphellaria anthraspis*
Pannaria lrubiginosa	*Pseudocyphellaria crocata*
Parmelia hygrophila	*Pyrenula occdentallis*
Parmelia pseudosulcata	*Pyrrhospora cinnabarina*
Parmelia saxatilis	*Ramalina dilacerata*
Parmelia squarrosa	*Ramalina farinacea*
Parmelia sulcata	*Ramalina thrausta*
Parmeliopsis hyperopta	*Sphaerophorus globosus*
Parmotrema arnoldii	*Sticta fuliginosa*
Parmotrema chinensis	*Sticta limbata*
Parmotrema crinitum	*Sticta weigelii*
Peltigera collina	*Thelotrema lepadinum*
Peltigera degenii	*Trapelia coarctata*
Peltigera membranacea	*Trapelia corticola*
Peltigera neopolydactyla	*Trapeliopsis flexuosa*
Pertusaria ophthalmiza	*Trapeliopsis pseudogranulosa*
Pertusaria sommerfeltii	*Usnea cornuta*
Pertusaria subambigens	*Usnea filipendula*
Pertussaria amara	*Usnea glabrata*
Pertussaria glaucomela	*Usnea inflata*
Phlyctis agelaea	*Usnea plicata*
Phlyctis argena	*Usnea wirthii*
Pilophorus acicularis	*Xylographa abietina*
Placopsis gelida	*Xylographa hians*

Vascular Plant Species	
Latin name	Common name
Acer circinatum	Vine maple
Acer macrophyllum	Bigleaf maple
Achlys triphylla	Vanilla leaf
Actaea rubra	Baneberry
Adenocaulon bicolor	Trail plant
Adiantum pedatum	Maidenhair fern
Alnus rubra	Red alder
Anemone deltoidea	Threeleaf anenome
Anemone lyallii	Lyall anenome
Asarum caudatum	Wild ginger

Table 8.2 (continued)

Vascular Plant Species	
Athyrium filix-femina	Lady fern
Berberis nervosa	Oregon grape
Blechnum spicant	Deer fern
Bromus vulgaris	Columbia brome
Campanula scouleri	Bluebell
Cardamine angulata	Angled bittercress
Cardamine integrifolia	Toothwort
Corylus cornuta	Hazel
Dicentra formosa	Bleeding heart
Digitalis purpurea	Foxglove
Disporum hookeri	Fairybells
Dryopteris austriaca	Wood fern
Festuca californica	California fescue
Festuca occidentalis	Western fescue
Galium aparine	Catchweed bedstraw
Galium oreganum	Bedstraw
Galium triflorum	Sweetscented bedstraw
Gaultheria shallon	Salal
Hieracium albiflorum	White hawkweed
Holodiscus discolor	Oceanspray
Luzula campestris	Woodrush
Lysichitum americanum	Skunk cabbage
Maianthemum dilatatum	Beadruby
Marah oreganus	Oregon bigroot
Menziesia ferruginea	Rusty menziesia
Mimulus guttatus	Common monkeyflower
Montia sibirica	Western spring-beauty
Oplopanax horridum	Devil's club
Osmorhiza chilensis	Sweetroot
Oxalis oregana	Oregon oxalis
Poa laxiflora	Bluegrass
Polypodium glycyrrhiza	Licorice fern
Polystichum munitum	Sword fern
Pseudotsuga menziesii	Douglas-fir
Pteridium aquilinum	Bracken fern
Ranunculus muricatus	Buttercup
Rhamnus purshiana	Cascara buckthorn
Romanzoffia californica	Mist maiden
Rosa gymnocarpa	Baldhip roses
Rubus parviflorus	Thimbleberry
Rubus spectabilis	Salmonberry
Rubus ursinus	Trailing blackberry
Sambucus racemosa	Elderberry
Senecio jacobaea	Tansy ragwort

(continued)

Table 8.2 (continued)

Vascular Plant Species	
Smilacina racemosa	Feather solomonplume
Smilacina stellata	Starry solomonplume
Stachys mexicana	Hedge nettle
Streptopus amplexifolius	Claspleaf twisted stalk
Streptopus roseus	Twisted stalk
Syntheris reniformis	Snow queen
Tellima grandiflora	Alaska fringecup
Thuja plicata	Western redcedar
Tolmiea menziesii	Menzies tolmeia
Trientalis latifolia	Western starflower
Trillium ovatum	Trillium
Tsuga heterophylla	Western hemlock
Vaccinium ovatum	Box blueberry
Vaccinium parvifolium	Red whortleberry
Vancouveria hexandra	Inside-out flower
Viola glabella	Pioneer violet
Viola sempervirens	Redwoods violet

not appear to exert a major influence on the salmon population. Unlike other sculpins, it does not appear to migrate downstream to spawn, but rather tends to remain in a relatively short section of the stream.

The cutthroat trout grows to much larger size than the reticulate sculpin. The population includes both resident and anadromous forms. Gravid females only 15 cm long (6 in) that never left the study area have been observed over redds in the Flynn Creek tributaries. Sea-run females return to spawn between November and January, with the maximum upstream migration generally occurring in December. The cutthroat migrate downstream after one to two years, during the period from February to May. The peak downstream migration occurs during April and May. It appears that the cutthroat redds tend to be in the tributaries of the main stream. No coho salmon juveniles have been observed in those tributaries having the greatest concentrations of cutthroat trout fry.

The spawning populations of coho salmon are quite variable. The ratio of males to females remains about 2 to 1, however. The outmigration of smolts is somewhat less variable, but there was some indication that smolt numbers declined in the later years of the AWS. The timing of upstream migration coincides with, or lags slightly behind, the cutthroat trout. See Table 8.3 for population statistics for coho salmon in Flynn Creek for the period 1959–1973.

For the year that the juvenile coho remain in Flynn Creek, the population decline follows a reversed J-shaped curve. It appears that the mortality is density dependent, for the smolt output is less variable than the other

Table 8.3 Some population statistics for coho salmon in Flynn Creek are given below for the period 1959–1973 (Moring and Lantz 1975; Hall et al. 1987)

Category	Number of fish	
	Mean	Range
Spawning females	17	2–55
Spawning males (including jacks)	36	3–80
Smolts	610	138–1284
Outmigrant fry	7780	24–29,877

categories. The coho salmon fry are aggressive and exhibit territorial behavior that would enhance other density-dependent mechanisms.

Terrestrial Animals

The list of vertebrate species (excluding fish) presented in Table 8.4 is compiled from observations and from lists made for similar areas in the Oregon Coast Range. For that reason it may be overly inclusive or include species that are extremely rare visitors to the RNA.

There are three mammals that are quite abundant in the RNA and affect the vegetation: Roosevelt elk, black-tailed deer *(Odocoileus hemionus* var. *columbianus),* and mountain beaver. The grazing pressure from these three species can be locally quite heavy. The mosaic of meadow and forest on the benchy western portion of the watershed seems particularly heavily grazed by deer and elk. Because the two species selectively avoid bracken and foxglove, the plant species composition of the meadow communities could be shifted in favor of those two species in the future. Mountain beaver are extremely numerous in portions of the natural area. Their burrows are most dense in lush sword fern sites, and are obviously contributing significantly to down-slope soil movement on the steeper slopes. A survey in 1980 found no beaver. A 1994 survey noted the presence of numerous beaver dams. Amphibians are present in more species than the reptiles, which appear to be represented only by three snake species.

Mineral Deposits

No mineral explorations are known to have occurred in the natural area. No mineralized bodies are known to exist there. The RNA was withdrawn from mineral entry after establishment.

Table 8.4 Terrestrial vertebrate species expected to be found in Flynn Creek Research Natural Area. Compiled from lists for similar habitats in the Oregon Coast Range, with help from Carl Frounfelker, Wildlife Biologist, Siuslaw National Forest

Mammals

Latin name	Common name
Aplodontia rufa	Mountain beaver
Canis latrans	Coyote
Castor canadensis	Beaver
Cervus canadensis	Roosevelt elk
Clethrionomys californicus	Western red-backed vole
Didelphis virginianus	Opossum
Erethizon dorsatum	Porcupine
Eutamias townsendi	Townsend chipmunk
Felis concolor	Mountain lion
Glaucomys sabrinus	Northern flying squirrel
Lasionycteris noctivagans	Silver-haired bat
Lasiurus cinereus	Hoary bat
Lepus americanus	Snowshoe hare
Lutra canadensis	River otter
Lynx rufus	Bobcat
Martes americans	Pine marten
Mephitis mephitis	Striped skunk
Microtis oregoni	Creeping vole
Mustela erminea	Short-tailed weasel (ermine)
Mustela frenata	Long-tailed weasel
Mustela vison	Mink
Myotis lucifugus	Little brown bat
Myotis volans	Long-legged bat
Neotoma cinerea	Bushy-tailed woodrat
Neurotrichus gibbsi	Shrew-mole
Odocoileus hemionus	Black-tailed deer
Peromyscus maniculatus	Deer mouse
Phenacomys albipes	White-footed vole
Phenacomys longicaudus	Red tree vole
Procyon lotor	Raccoon
Scapanus orarius	Coast mole
Sorex pacificus	Pacific shrew
Sorex trowbridgei	Trowbridge's shrew
Spilogale putorius	Spotted skunk
Sylvilagus bachmani	Brush rabbit
Tamiascirius douglasi	Douglas' squirrel
Ursus americanus	Black bear
Zapus trinotatus	Pacific jumping mouse

Birds (includes resident and migratory birds)

Accipiter gentilis	Northern goshawk
Accipiter striatus	Sharp-shinned hawk
Aegolius acadicus	Northern saw-whet owl
Agelains phoeniceus	Red-winged blackbird

Table 8.4 (continued)

Birds (includes resident and migratory birds)

Aix sponsa	Wood duck
Ardea herodias	Great blue heron
Asio otus	Long-eared owl
Bombycilla cedrorum	Cedar waxwing
Bombycilla garrulus	Bohemian waxwing
Bonasa umbellus	Ruffed grouse
Brachyramphus marmoratus	Marbled murrelet
Bubo virginiamo	Great-horned owl
Buteo jamaicensis	Red-tailed hawk
Butorides striatus	Green-backed heron
Callipepla californica	California quail
Carduelis pinus	Pine siskin
Carpodacus cassinii	Cassin's finch
Carpodacus mexicanus	House finch
Carpodacus purpureus	Purple finch
Cathartes aura	Turkey vulture
Catharus guttatus	Hermit thrush
Catharus ustulatus	Swainson's thrush
Certhia americana	Brown creeper
Ceryle alcyon	Belted kingfisher
Chaetura vauxi	Vaux's swift
Chamaea fascinata	Wrentit
Charadrius vociferns	Killdeer
Chordeiles minor	Common nighthawk
Cinclus mexicanus	American dipper
Coccothraustes vespertinea	Evening grosbeak
Colaptes auratus	Northern flicker
Columba fasciata	Band-tailed pigeon
Contopus sordidulus	Western wood-pewee
Corvis brachyryhnchos	American crow
Corvus corax	Common raven
Cyanocitta stelleri	Steller's jay
Dendragapus obscurus	Blue grouse
Dendroica auduborii	Audubon's warbler
Dendroica coronata	Yellow-rumped warbler
Dendroica nigrescens	Black-throated gray warbler
Dendroica occidentalis	Hermit warbler
Dendroica petechia	Yellow warbler
Dendroica townsendi	Townsend's warbler
Dryocopus pileatus	Pileated woodpecker
Empidonax difficilis	Pacific-slope flycatcher
Empidonax hammondii	Hammond's flycatcher
Empidonax traillii	Willow flycatcher
Euphagus cyanocephalus	Brewer's blackbird
Geothlypis trichas	Common yellowthroat

(continued)

Table 8.4 (continued)

Birds (includes resident and migratory birds)

Glaucidium gnoma	Northern pygmy owl
Haliaeetus leucocephalus	Northern bald eagle
Hirundo pyrrhonto	Cliff swallow
Icteria virens	Yellow-breasted chat
Ixoreus naevius	Varied thrush
Junco hyemalis	Dark-eyed junco
Loxia curvirostra	Red crossbill
Melospiza lincolnii	Lincoln's sparrow
Melospiza melodia	Song sparrow
Mergus serrator	Red-breasted merganser
Molothrus ater	Brown-headed cowbird
Myadestes townsendi	Townsend's solitaire
Nuttallornis borealis	Olive-sided flycatcher
Oporornis tolmiei	MacGillivray's warbler
Oreortyx pictus	Mountain quail
Otis kennicottii	Western screech-owl
Parus atricapillus	Black-capped chickadee
Parus gambeli	Mountain chickadee
Parus rufescens	Chestnut-backed chickadee
Passerella iliaca	Fox sparrow
Passerina amoena	Lazuli bunting
Perisoreus canadensis	Gray jay
Pheucticus melanocephalus	Black-headed grosbeak
Picoides pubescens	Downy woodpecker
Picoides villosus	Hairy woodpecker
Pipilo erythrophtalmus	Rufous-sided towhee
Piranga ludoviciana	Western tanager
Psaltriparus minimus	Common bushtit
Regulus calendula	Ruby-crowned kinglet
Regulus satrapa	Golden-crowned kinglet
Sayorsis nigricans	Black phoebe
Selasphorus rufus	Rufous hummingbird
Sialia mexicana	Western bluebird
Sitta canadensis	Red-breasted nuthatch
Sitta carolinensis	White-breasted nuthatch
Sphyrapicus ruber	Red-breasted sapsucker
Spinus tristis	American goldfinch
Spizella arborea	American tree sparrow
Spizella passerina	Chipping sparrow
Stelgidopteryx serripennis	Northern rough-winged swallow
Stellula calliope	Calliope hummingbird
Strix occidentalis caurina	Northern spotted owl
Strix varia	Barred owl
Sturnus vulgaris	European starling
Tachycineta bicolor	Tree swallow
Tachycineta thalassina	Violet-green swallow

Table 8.4 (continued)

Birds (includes resident and migratory birds)	
Thryomanes bewickii	Bewick's wren
Troglodytes aedon	House wren
Troglodytes troglodytes	Winter wren
Turdus migratorius	American robin
Vermivora celata	Orange-crowned warbler
Vermivora ruficapilla	Nashville warbler
Vireo gilvus	Warbling vireo
Vireo huttoni	Hutton's vireo
Vireo olivaceus	Red-eyed vireo
Vireo solitarius	Solitary vireo
Wilsonia pusilla	Wilson's warbler
Zenaida macroura	Mourning dove
Zonotrichia atricapilla	Golden-crowned sparrow
Zonotrichia leucophrys	White-crowned sparrow

Reptiles and amphibians	
Latin name	Common name
Abystoma gracile	Northwestern salamander
Aneides ferreus	Clouded salamander
Ascaphus truei	Tailed frog
Charina bottae	Rubber boa
Dicamptodon ensatus	Pacific giant salamander
Elgaria coerulea	Northern alligator lizard
Ensatina eschscholtzi	Ensatina
Plethodon dunni	Dunn's salamander
Plethodon vehiculum	Western redback salamander
Pseudacris regilla	Pacific treefrog
Rana aurora	Red-legged frog
Rhyacotriton variegatus	Southern torrent salamander
Taricha granulosa	Rough-skinned newt
Thamnophis ordinoides	Northwestern garter snake
Thamnophis sirtalis	Common garter snake

Summary

The Flynn Creek Research Natural Area provides the unique opportunity to revisit the Alsea Watershed Study because the loss of a control watershed is often the biggest limitation to long-term watershed studies.

Literature Cited

Corliss, J.F. 1973. Soil Survey of Alsea Area, Oregon. USDA Soil Conservation Service and Forest Service, USDI Bureau of Land Management, in cooperation with Oregon Board of

Natural Resources and Oregon Agricultural Experiment Station, Superintendent of Documents, Washington, DC. 82pp.

Hall, J.D., Brown, G.W., and Lantz, R.L. 1987. The Alsea Watershed Study: a retrospective, pp. 399–416. In: E.O. Salo and T.W. Cundy, editors. Streamside Management: Forestry and Fishery Interactions. Univ. of Washington Inst. of Forest Resour., Seattle, WA.

Moring, J.R., and Lantz, R.L. 1975. The Alsea Watershed Study: effects of logging on the aquatic resources of three headwater streams of the Alsea River, Oregon. Part I. Biological studies. Fish. Res. Rep. 9. Oregon Dept. of Fish and Wildlife, Corvallis, OR. 66pp.

Chapter 9
Long-term Streamflow Changes Following Timber Harvesting

John D. Stednick

Studies under virtually every environmental condition indicate that vegetation removal results in increased annual water yield (Rothacher 1970; Harr 1976, 1979, 1983; Bosch and Hewlett 1982; Stednick 1996). However, treatment responses are variable and depend on the vegetation complex, landform, and climate of the particular watershed system studied (Hewlett and Hibbert 1967; Bosch and Hewlett 1982; Stednick 1996). Vegetative recovery, following harvest, leads to return of water yield to pretreatment levels, though at different rates for each climatic and geographic zone studied (Kovner 1956; Ziemer 1964; Harr 1979, 1983; Troendle and King 1985; Keppeler and Ziemer 1990; Stednick 1996). Thus the treatment response is time-dependent.

There have been many papers written regarding short-term hydrologic changes resulting from both timber harvest and road-building in the United States (Rothacher 1970; Harr et al. 1975; Harris 1977; Harr 1980; Ziemer 1981; Harr et al. 1982, among others). Generally speaking, these studies have been short-term due to the expense and commitment required for long-term monitoring or the loss of a control watershed, i.e. land use activities (Stednick and Kern 1992).

The monitoring techniques used at Alsea typify methods employed in most paired watershed studies. The effects of timber harvesting are measured by differences between pre- and post-logging hydrologic relations or hydrologic events of interest on Needle Branch and Deer Creek, compared to the control Flynn Creek. Regression equations developed to describe the prelogging relationships predict expected values for the dependent variables (Needle Branch or Deer Creek flow parameters) from observed values of the independent variable (Flynn Creek) (Rothacher and Miner 1967; Harris 1977).

The results of the Alsea Watershed Study showed an increase in annual water yield and the three-day peak flow for Needle Branch, the clearcut watershed, and no significant change for any streamflow characteristic on Deer Creek, the

John D. Stednick
Department of Forest, Rangeland, and Watershed Stewardship, Colorado State University, Fort Collins, CO 80523

J. D. Stednick (ed.), *Hydrological and Biological Responses to Forest Practices.*
© Springer 2008

patchcut watershed. After 7 years of post-harvest monitoring, conclusions drawn on the original analysis were that the watersheds "generally appear to be returning to pre-logging conditions" that hydrologic recovery had occurred (Harris and Williams 1971). Hydrologic recovery is a broad term used to describe the re-establishment of the individual systemic processes that are altered following timber harvest. The primary evidence used to characterize hydrologic recovery following timber harvest was derived from short-term paired watershed studies similar to those described earlier. Additional streamflow data collection indicates that some watersheds, deemed "recovered", are still showing departures from the pretreatment relations (Hicks et al. 1991; Stednick and Kern 1992). For example, extrapolation of streamflow records from the Oregon Cascades suggested that annual water yields in Douglas-fir (*Pseudotsuga menziesii*) watersheds would return to pretreatment conditions 27 years after timber harvesting (Harr 1983) and approximately 24 years are required for similar recovery in the Coastal Range of Oregon (Harr 1983; Stednick and Kern 1992). This long-lasting change indicates that the original, short-term studies were only considering the changes immediately following timber harvest, and did not reflect the complex interactions that occur over the long term. Natural variability in climatic and terrestrial system response can complicate the issue of hydrologic recovery, particularly in the short term. There is a need for continued research on hydrologic recovery and for a new definition that considers recovery over the long term. Most paired watershed studies are in smaller first- and second-order streams that occur in the headwaters regions of larger river systems. The first- and second-order streams are those primarily affected by timber harvest and the road-building activity associated with forest management (Chamberlin et al. 1991). Since watersheds of this size dominate the landscape, and because timber harvests can include most, if not all, of the watershed area, timber harvest and road building effects may be observed more easily on-site or at the scale of the first order watershed than off-site or larger order watersheds.

The recovery of vegetation and soil are the primary terrestrial processes that control the recovery of the runoff characteristics to those observed prior to timber harvest. Vegetation recovery is defined as growth sufficient to restore evapotranspiration and interception processes such that their effect on runoff approximates that observed prior to timber harvest. This does not mean that these systems must exhibit the exact characteristics as those observed prior to timber harvest, rather that the hydrologic processes that utilize energy and water have been re-established to have similar results to those observed prior to timber harvest. Determining when hydrologic recovery occurs is not a simple task. A vast array of potential system responses, and a wide array of tests, are available to determine if a watershed has completely "recovered" from a disturbance. Recovery of any hydrologic parameter or metric may be defined as the return to pretreatment behavior or to some predictable stationary or meta-stationary state (Thomas 1990).

Annual Water Yield

Change in annual water yield following timber harvesting is probably the most commonly analyzed hydrologic metric. Annual water yield is estimated as the accumulated total runoff from a watershed over the water year (Calder 1993). Annual water yield is the precipitation entering the watershed minus the evapotranspiration losses, changes in soil water storage, and losses to deep seepage. The term evapotranspiration includes interception, evaporation, and transpiration losses. Water yield is thus susceptible to change if any of these components are changed by land use practices. Annual water yield generally increases following timber harvest, due to decreased evapotranspiration and interception on the harvested site, and potentially any physical disturbance caused by harvesting that may alter infiltration characteristics or intercept soil water. This increase is generally observed immediately following timber harvest, and decreases as vegetation recovers and soil disturbances are stabilized (Harr 1976; Hewlett and Helvey 1976; Chamberlin et al. 1991). An analysis of paired watershed studies suggests that harvesting (or basal area removal) of at least 20% of vegetation on the watershed is needed in order to generate a detectable change using the standard streamflow measurement methods. A key factor governing changes in annual water yield is the proximity of harvest to streamflow source areas (Stednick 1996).

The impact that timber harvest has on the water balance is variable but predictable from a process standpoint. Following harvest, the soil mantle is generally wetter during the growing season as a result of less vegetation to transpire moisture from the soil. This can lead to higher runoff during the beginning of the wet season, or whenever precipitation occurs, because there is less soil moisture deficit to make up. As vegetation recovers following timber harvest, more soil moisture is transpired during the summer months. This leads to an increasingly greater soil water deficit, which in turn leads to lower levels of runoff following precipitation events. This period of lower runoff continues until the soil moisture deficit has been satisfied. Ultimately the recovery of the soil moisture deficit in the summer contributes significantly to decreasing annual water yield, and a return to pre-harvest conditions. However, reductions in interception losses also contribute to the change in water yield and the recovery of the interception process may not occur at the same rate as for transpiration.

Site revegetation, sometimes including undesirable as well as pioneer species, is usually rapid after site preparation following timber harvest, reaching nearly 100% cover in as few as 4 years on some sites (Adams et al. 1991). The hydrologic impact of the successional stages that occur prior to the climax stage are of significance when considering hydrologic recovery. Often, hardwood species are the first trees to be re-established on disturbed sites. There is evidence that the increase in water yield is short-lived, lasting only a few years following the initial disturbance, depending upon the type of vegetation that regrows in the riparian area of the watershed (Hicks et al. 1991). Hardwoods have been shown to transpire at a higher rate than conifer species in some environments (Hicks et al. 1991). Conversion of conifer cover to hardwood cover, especially in the riparian zone, can result in higher water use by the pioneer vegetation that occupies the site following timber harvest, initially this may reduce water yield below pre-harvest levels (Adams et al. 1991). A soil moisture study in Oregon showed that increases in soil moisture are very short-lived following the re-establishment of phreatophytic vegetation species, causing soil moisture to drop below pretreatment levels for at least 14 years following the initial period of increase (Adams et al. 1991). This suggests that hydrologic recovery is a complex relationship between the physical and vegetative processes.

The effects of timber harvest on runoff are also dependent on the degree of soil disturbance as influenced by the method of harvest and site preparation used. Although high-line cable yarding techniques, for example, can disturb the soil surface, and have been shown to leave from 30% to 60% of soils exposed following a timber harvest, they are considered to be less impacting than tractor yarding (Smith and Wass 1980). Exposure of the soil surface may reduce infiltration rates, alter flow paths, and thus cause greater surface runoff (Chamberlin et al. 1991). Site preparation through the burning of the slash residue from timber harvest can also expose the soil surface, depending on the duration and severity of the burn. High intensity burns can remove the organic layer that protects the soil surface, leaving the surface exposed to the same impact mentioned above.

Only a few paired watershed studies have sufficiently long periods of observation to verify annual water yield returns to pretreatment levels after timber harvesting. Reactivation of the original Alsea Watershed Study provides a unique opportunity to assess the long term effects of timber harvesting on water resources in the Pacific Northwest.

Peak Flows

The controlling hydrologic factor in temperate coniferous forest environments is rainfall, the seasonality of precipitation leads to peak flows in the winter months when the majority of rainfall occurs, and low flows in the summer when little rain occurs. The highest annual peak flows generally occur during the winter months when precipitation is highest, the soil is generally the wettest and at or near saturation, and the vegetation is not transpiring at peak levels (Harr 1976).

Seasonal precipitation effects can also cause variable source areas for streamflow. During the wet season, hollows and other near-stream depressions may saturate and become part of the active stream network while perennial segments of the stream network may widen. This process of enlarging the stream network to allow greater quantities of water to be effectively moved through the watershed, results in higher, quicker peak flows observed during the wet season (Harr 1976).

Following harvest, decreased evapotranspiration during the growing season leaves the soil wetter than during pre-harvest conditions, resulting in earlier saturation of the soil mantle when precipitation does occur, and results in potentially higher early season peak flows (Harr 1976). This effect is not generally observed in the winter, when the soil moisture is fully recharged in both harvested and unharvested watersheds, and peak flows are generally highest. The timing of peak flows is dependent on the site-specific impacts of the particular timber harvest. The increase in fall soil moisture associated with decreased evapotranspiration has the greatest effect on peak flows with a 1 to 5-year recurrence interval. Larger peak flows are not as susceptible to change by timber harvest, since the amount of precipitation during these storms will exceed increased soil moisture due to timber harvest. The effect of timber harvest on the largest peaks primarily reflects the effect of interception differences between cut and uncut areas (Harr 1976).

Many studies have shown that few changes in peak flows occur as a result of timber harvest, even clearcutting (Harris 1977; Harr 1980; Harr et al. 1982). This evidence suggests that changes in peak flows are not as important as were once thought, especially since the small to average peak flows (channel maintenance flows), not the larger channel forming flows, are the most affected by timber harvest.

Previous studies in the Pacific Northwest, including the Alsea Watershed Study, suggested that there was no appreciable increase in peak flows after timber harvesting and road building (Harr et al. 1975). In a reanalysis of paired peak discharges over 34 years from 2 small experimental watersheds and over 50 years in larger and adjacent watersheds in the Oregon Cascade Mountains, forest harvesting increased peak discharges by as much as 50% in the small basins and 100% in the larger basins (Jones and Grant 1996). An alternative interpretation of these results suggested that only the smallest peak flows on the clearcut watershed could be measured and the percent treatment effects decreased as flow event size increased and were not detectable for flows with 2-year return intervals or greater on the treated watersheds (Thomas and Megahan 1998). The later paper also offered alternatives on the peak flow matching algorithm, the classification of large events, and the statistical analysis of percent change.

Yet another review of both of these papers addressed the concern of using estimated peak flows, especially when the streamflow measurement error with a weir or flume is approximately 3–5%, and detecting a treatment effect of less than 5% is extremely difficult (Beschta et al. 2000). Additionally, the percent area harvested is a non-unique and weak variable for streamflow difference models. The difference variable may be good for statistics, but is disassociated with peak flow magnitudes. A better comparison is to use unit area peak flow rates. Given the complex nature of the effects of timber harvesting and roads on streamflow, it is not surprising that the literature provides mixed messages about peak flow responses (Thomas and Megahan 1998). It seems that much of the interpretation of peak flow responses is a function of the quantiles into which the peak flows are grouped.

Low Flows

Low flow is defined as the minimum discharge observed during the year, associated with drainage of the subsurface of the watershed (Calder 1993). Summer and fall, when precipitation is low, the soil is dry, and vegetation is actively transpiring, usually coincide with the annual low flows in these systems. These low flows are generally dominated by baseflow, but can be augmented by larger summer precipitation events (Harr 1976). Interception and evapotranspiration of the majority of total volume of precipitation during the summer months leads to a drying of the soil mantle (Harr 1976). Generally, precipitation is used to fill soil pore spaces depleted by atmospheric withdrawal via evapotranspiration, leaving little moisture to generate stormflow during the summer. The streamflow that is observed during the summer is generally due to the slow, lateral draining of the soil storage (Harr 1976). Stormflow that does occur during this period is generally caused by interception of precipitation by the active stream channel (Harr 1976). Decreases in vegetation result in decreased evapotranspiration rates on the harvested site, ultimately resulting in increased soil moisture during the dry months. The increased soil moisture results in increased drainage to the stream, resulting in higher flow during the dry period (Harr 1976). Low flows were shown to have tripled following timber harvest in the Oregon Coast Range (Rothacher 1970; Harr and Krygier 1972). These increases may be fairly short-lived as vegetation begins to re-establish on the harvested sites (Hicks et al. 1991). Most studies indicate soil moisture is as wet or wetter after harvest and this would tend to result in the same or higher base flows. This should be universal. Vegetative recovery could cause greater soil moisture depletion and therefore lower base flows after even a few years.

Periods of low flow are stressful to fish because they mark periods of increased stream temperature and decreased dissolved oxygen. Increases in stream temperature are generally caused because there is less water to be warmed in the stream during the low flow period (Chamberlin et al. 1991). Decreases in the length of the low flow period will translate directly into decreased time that aquatic organisms are susceptible to stress from increased stream temperature. Research from the Oregon Cascades suggested that low flows after timber harvesting are reduced for 20% of the 70–100 year timber rotation schedule (Hicks et al. 1991). The length of this reduction could have serious implications to salmonid fisheries, especially those located in first and second-order streams. Further study of this phenomenon in active salmonid rearing streams is essential to understanding this potential impact, and will help land managers adjust harvest schedules to protect salmon-rearing streams.

The Alsea Watershed Study was reactivated as the New Alsea Watershed Study (NAWS) in 1989 to assess long-term responses of watersheds to timber harvest (Stednick 1991). Interpretation of the early streamflow data suggested that streamflow recovery to pre-harvest conditions had not yet occurred (Stednick and Kern 1992). Other studies showed that long-term effects of timber harvest may persist for 40 to 60 years following the initial timber harvest (Hicks et al. 1991). The New Alsea Watershed Study provides an opportunity to study the many interrelated effects of timber harvest in an active salmon-rearing area, and in a commercially important timber-reserve.

This study focused on the response of four hydrologic metrics: annual water yield, peak flow, low flow, and low flow days, in an attempt to determine whether the study watersheds have recovered hydrologically from timber harvest. These metrics were selected because they follow the original study, and represent the major aspects of the flow regime of the temperate forest that can be affected by timber harvesting and road building. In addition they deal only with streamflow data and are not dependent on associated precipitation to make an accurate determination of whether data falls into a specific category or not.

Methodology

Few abiotic research efforts outside the original study were made in the Alsea area until 1989 (Stednick and Kern 1991). Flynn Creek was designated a long-term Research Natural Area by the USDA Forest Service, and used to characterize undisturbed temperate coniferous forest ecosystems in the Oregon Coast Range (Fig. 9.1) (See Chapter 8). Since being clearcut in 1966, forest management on Needle Branch has included pre-commercial and commercial thinning. Approximately 25% of the middle-third of the watershed was pre-commercially thinned in 1981. In 1997–1998 approximately 40% of this thinned area was commercially thinned with a 30% basal area removal (Plum Creek Timber Company, unpublished data). Deer Creek had a second timber harvesting entry in 1978 when one 20 ha unit was harvested with a streamside buffer. Two additional units of 11 ha each were logged in 1987 and 1988 (Figs. 9.2 and 9.3). In total, approximately 39% of the watershed area is now harvested. The multiple entries in Deer Creek provide the opportunity to assess the adequacy of our ability to detect incremental response and assess potential cumulative watershed effects on water yield and water quality. Because of the timeframe over which treatments were implemented, hydrologic recovery may complicate the assessment process.

Fig. 9.1 Undisturbed Douglas-fir forest on Flynn Creek

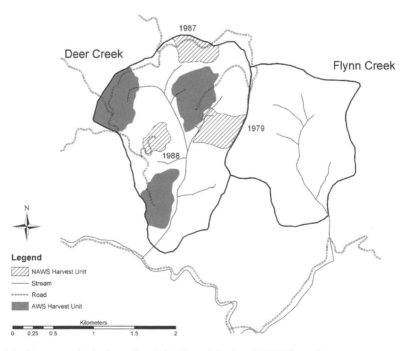

Fig. 9.2 Harvest units in Deer Creek for the original and NAWS studies

Fig. 9.3 View in Deer Creek looking to the northeast. Foreground is unit harvested in 1988, mid ground is original harvest unit of 1966, and background is unit harvested in 1987 (left) and undisturbed forest (right). Photo taken in 1991

The water resources monitoring program was reactivated with the rebuilding of the stream gauging houses in September 1989 and streamflow measurements began October 1, 1989 (Stednick 1991; Stednick and Kern 1991, 1992). Each stream gauging station has a broad-crested compound concrete weir and is instrumented with a mechanical stream level recorder (Leupold-Stevens® A-35) in 1989, (as done in the original study). The original study had recorders geared at 1:0.1, that is a stage increase of 1.0 was recorded with a line length of 0.1. The NAWS gauges record at 1:1. Programmable electronic data loggers with a float system independent of the Leupold-Stevens® were installed in 1991. Streamflow records are reduced per standard U.S. Geological Survey techniques (Rantz et al. 1982) which are similar to or better than earlier techniques. Stage-discharge relationships are continually recalibrated for each gauging station and stage shifts incorporated in streamflow record reduction. Streamflow data are considered excellent for accuracy and precision, with measurement error less than 2% during low flow, and less than 5% at high flow conditions (Stednick and Kern 1991).

Climate is a key issue in any paired watershed study. Various attempts were made to measure precipitation in the watersheds. Lack of on-site personnel, infrequent site visits, and an over-reliance on electronic data loggers resulted in poor data records. The three study watersheds have similar, though not identical, precipitation records (Harris 1977; Stednick 1991). Analysis of covariance (Snedecor and Cochran 1980) of cumulative annual precipitation indicated no significant differences between any of the study watersheds (Stednick unpublished). Although there are no significant differences in annual precipitation between watersheds, the areal distribution of individual storm event precipitation may account for observed streamflow differences (Harris 1977; Belt 1997; Stednick unpublished). Streamflow metrics were used to assess the effects of timber harvesting practices on water resources.

Results and Discussion

Precipitation data from nearby Tidewater and Newport, Oregon were used to assess the stability of the long term streamflow data from Flynn Creek. Annual water yields from Flynn Creek were compared to annual precipitation values at both sites. Comparisons were made for the pretreatment period (1959–1965), posttreatment (1967–1973) and NAWS period (1990–2002). There were no significant differences within periods to indicate that the relationship between precipitation at Tidewater and Newport and streamflow from Flynn Creek had changed, suggesting streamflow data stationarity and that Flynn Creek remains a suitable control watershed (Belt 1997; Stednick unpublished).

Annual Water Yield

Annual water yield is perhaps the most common hydrologic metric used to assess land use activities on a watershed basis. Annual precipitation influences the magnitude of water yield increases following timber harvest operations. Greater annual water yield increases are observed in wetter years (Harr 1979; Ponce and Meiman 1983). For the AWS data, the three wettest years (1969, 1971, and 1972) had a mean annual precipitation of 3216 mm and the mean difference between observed annual water yield from Needle Branch and the predicted water yield was 618 mm or a 19% increase. Conversely, the three driest years (1990, 1991, and 1992) had a mean annual precipitation of 1998 mm and a mean yield difference of 200 mm for a 10% increase. This supports the suggestions that mean annual precipitation most influences streamflow responses after vegetation manipulation (Bosch and Hewlett 1982).

Following the techniques used in the original study, differences in measured annual water yield minus predicted streamflows were plotted over time (Fig. 9.4). A linear-line fit through the decreasing water yield increases suggests hydrologic recovery after 31 years. Although additional timber harvesting occurred in Deer Creek, the annual water yield increases are not detectable. The Needle Branch model for hydrologic recovery results in an equivalent clearcut area of 5.4%, below the threshold of 20% needed for a detectable response in streamflow measurements at the watershed scale.

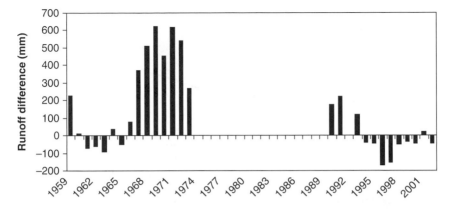

Fig. 9.4 Runoff difference in mm (observed minus predicted) for Needle Branch as predicted by Flynn Greek

Peak Flows

The AWS selection criterion for peak flows, approximately bankfull discharge, was $0.55 \text{ m}^3 \text{ s}^{-1} \text{ km}^{-2}$ (or $50 \text{ ft}^3 \text{ s}^{-1} \text{ mi}^{-2}$) or greater (Harris 1977). For water years 1990–2002, 11 events met the peak flow selection criterion. All but one of the peak flow events were within the confidence intervals identified in the original work (Fig. 9.5). Peak flow events on Needle Branch are not significantly different than Flynn Creek (ANCOVA, $p = 0.74$).

Similarly, using analysis of covariance, Deer Creek peak flows were not significantly different ($p = 0.45$) than predicted (Fig. 9.6). The hydrologic recovery of peak flows in Needle Branch is suggested by these data. However, future study objectives include analysis of the time to peak and recession limb characteristics.

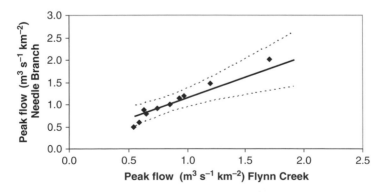

Fig. 9.5 Peak flow comparison for Needle Branch and Flynn Creek

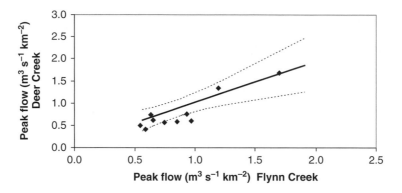

Fig. 9.6 Peak flow comparison for Deer Creek and Flynn Creek

The original streamflow measurement on Needle Branch showed increased peak flows. Given that channel debris was removed by hand and a crawler tractor, channel conveyance efficiency may increase, and decrease the time to peak.

The drainage density could be increased by the road system. The roads may act as stream channels and divert water through the basin. Road construction was found to not significantly alter the hydrologic response at Caspar Creek following logging (Ziemer 1981; Wright et al. 1990) but other studies suggest otherwise (Wemple and Jones 2003). Field observations in Needle Branch indicate a stream network and channel density greater than that suggested by the original maps. A detailed mapping of channel size and location outside the main channel was not done in the original study. Photo documentation of the posttreatment period in Needle Branch (1967–1973) show several areas of significant soil erosion and rill formation. An increase in first order streams and increased stream channel density will decrease the time to peak and increase the peak discharge. Several sections of Needle Branch have been subject to tree blowdown and coarse woody material recruitment from the riparian alder forest. The woody material increase could decrease peak flow timing. Channel mapping and channel evolution monitoring could prove valuable in future research.

Low Flows

In the western temperate coniferous forest, the greatest relative increases in streamflow following timber harvesting have been observed during the summer season low flows, although in absolute terms, larger increases occurred during the rainy season (Keppeler and Ziemer 1990). Low flows and the number of low flow days, where streamflow fell below a preset threshold, were used to evaluate flow changes in AWS; fewer low flow days were found after logging (Harr and Krygier 1972). These summer increases are short-lived, lasting only 2 to 3 years (Harr 1979). Timber harvesting in the California redwoods significantly increased annual water yields and summer low flows, although the response was variable because of variable precipitation. The increased summer flows disappeared within 5 years of harvesting and suggest a possible decline in summer low flows relative to prelogging levels (Keppeler and Ziemer 1990). Re-evaluation of streamflow records in the Oregon Cascades also suggested a decrease in summer low flows (Hicks et al. 1991).

Low flows in Alsea were defined as flows less than $0.01 m^3 s^{-1} km^{-2}$ (Harris 1977). Summer streamflow is a function of precipitation and soil water storage depletion by evapotranspiration. Periods of no flow were measured for Needle Branch, a phenomenon not reported in the original study. Low flows on Needle Branch were not significantly different from those predicted by Flynn Creek using analysis of covariance ($p = 0.75$) (Fig. 9.7).

Multiple timber harvesting entries on Deer Creek had no effect ($p = 0.79$) on mean low flows for any of the treatment period comparisons but higher low flows and lower low flows were observed (Fig. 9.8). If low flow decreases occur sometime after timber harvesting, the harvest schedule in Deer Creek is effective in mitigating that effect.

Continuous stage (streamflow) measurements show pronounced diel fluctuations. Indeed, streamflow responses were almost instantaneous when clouds drifted over the watershed and reduced incoming radiation. Summer diel fluctuations were most pronounced on Needle Branch, resulting in daily streamflow changes up to 50 %. The rapid response shows an intimate linkage between the streamflow generating area (variable source area) and streamflow (and not just a barometric pressure response), and suggest that the riparian vegetation is more influential than hillslope vegetation. These observations suggest that streamflows might be modified by riparian management. However,

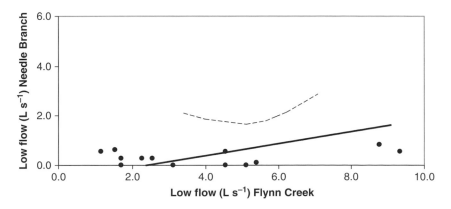

Fig. 9.7 Low flow comparison for Needle Branch and Flynn Creek

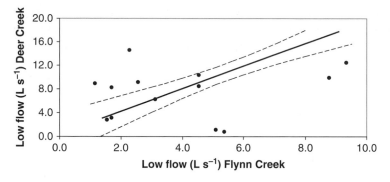

Fig. 9.8 Low flow comparison for Deer Creek and Flynn Creek

Phreatophyte control in a riparian corridor failed to increase water yield (Hicks et al. 1991).

Low Flow Days

Low flow days were defined as days with low flow or less (Harr and Krygier 1972). The period of streamflow record used for this analysis, 1990–2002, included some relatively dry years and low flow conditions. For Needle Branch, there were 3 years with low flow days outside the upper confidence interval and 1 year below the lower confidence interval (Fig. 9.9) but were not significantly different ($p = 0.79$). Similarly for Deer Creek, 2 years were above the upper confidence interval and 2 years were below the lower confidence interval, but all years were not significantly different ($p = 0.99$) than pretreatment conditions (Fig 9.10).

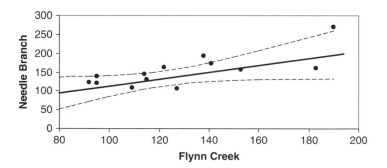

Fig. 9.9 Low flow day comparison for Needle Branch and Flynn Creek

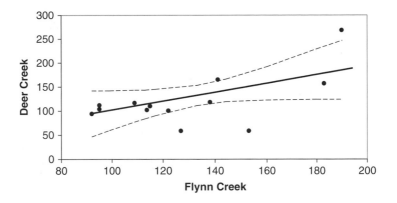

Fig. 9.10 Low flow day comparison for Deer Creek and Flynn Creek

Conclusions

With additional streamflow monitoring in the Alsea Watershed Study, it was determined that the hydrology, as assessed by annual water yield, unit peak flows, low flows, and low flow days, of Flynn Creek is not significantly different than the pretreatment period condition. Thus Flynn Creek remains a suitable control watershed. The principal method used to assess the effects of land use practices on hydrology is to develop pretreatment regressions between the control watershed and the treatment watersheds and conduct an analysis of variance.

Needle Branch hydrology suggested annual water yield increases were not detectable after 31 years and that the peak flow and low flow metrics were also not statistically different than pretreatment. Deer Creek has had additional timber harvesting in the watershed (39%) and no flow metrics are significantly different from the pretreatment. The application of best management practices (BMPs) and the timber harvesting schedule resulted in no measurable change in the streamflow metrics.

Continued research in the New Alsea Watershed Study will be used to identify the hydrologic processes affected by harvesting in Needle Branch and the associated recovery of these processes. Such understanding can be applied to other land use scenarios to better improve resource usage and sustainability.

Acknowledgment A number of my graduate students have reduced and compiled streamflow records from the New Alsea Watershed Study. In particular I would like to acknowledge Rich Belt, Curtis Cooper, and Kelli Jo Rehder. Rich Belt also did the initial statistical analysis on the NAWS streamflow records.

Literature Cited

Adams, P.W., Flint, A.L., and Fredriksen, R.L. 1991. Long-term patterns in soil moisture and revegetation after a clear-cut of a Douglas-fir forest in Oregon. Forest Ecol. and Manage. 41:249–263.

Belt, R.M. 1997. Long Term Hydrologic Recovery Following Timber Harvest in the Alsea River Basin, Oregon. M.S. Thesis. Colorado State Univ., Fort Collins, CO. 76pp.

Beschta, R.L., Pyles, M.R., Skaugset, A.E., and Surfleet, C.G. 2000. Peakflow responses to forest practices in the western Cascades of Oregon, USA. J. Hydrol. 233:102–120.

Bosch, J.M., and Hewlett, J.D. 1982. A review of catchment experiments to determine the effect of vegetation changes on water yield and evapotranspiration. J. Hydrol. 55:3–23.

Calder, I.R. 1993. Hydrologic effects of land-use change, pp. 13.1–13.50. In: D.R. Maidment, editor. Handbook of Hydrology. McGraw Hill, New York, NY.

Chamberlin, T.W., Harr, R.D., and Everest, F.H. 1991. Timber harvesting, silviculture, and watershed processes, pp. 181–205. In: W.R. Meehan, editor. Influences of Forest and Rangeland Management on Salmonid Fishes and their Habitats. Amer. Fish. Soc. Spec. Publ. 19.

Harr, R.D. 1976. Forest practices and streamflow in western Oregon. General Technical Report GTR-PNW-49. USDA Forest Service, Pacific Northwest Forest and Range Experiment Station, Portland, OR. 18pp.

Harr, R.D. 1979. Effects of timber harvest on streamflow in the rain-dominated portion of the Pacific Northwest. In: Proceedings of Workshop on Scheduling Timber Harvest for Hydrologic Concerns. USDA Forest Service, Pacific Northwest Forest and Range Experiment Station, Portland, OR. 45pp.

Harr, R.D. 1980. Streamflow after patch logging in small drainages within Bull Run municipal watershed, Oregon. Research Paper RP-PNW-268. USDA Forest Service, Pacific Northwest Forest and Range Experiment Station, Portland, OR. 16pp.

Harr, R.D. 1983. Potential for augmenting water yield through forest practices in western Washington and western Oregon. Water Resour. Bull. 19:383–393.

Harr, R.D., Harper, W.C., Krygier, J.T., and Hsieh, F.S. 1975. Changes in storm hydrographs after road building and clear-cutting in the Oregon Coast Range. Water Resour. Res. 11:436–444.

Harr, R.D., and Krygier, J.T. 1972. Clearcut logging and low flows in Oregon coastal watersheds. Forest Research Lab. Research Note 54. Oregon State Univ., Corvallis, OR. 3pp.

Harr, R.D., Levno, A., and Mersereau, R. 1982. Streamflow changes after logging 130-year-old Douglas-fir in two small watersheds. Water Resour. Res. 18:637–644.

Harris, D.D. 1977. Hydrologic changes after logging in two small Oregon coastal watersheds. Water-Supply Paper 2037. U.S. Geological Survey Washington, DC. 31pp.

Harris, D.D., and Williams, R.C. 1971. Streamflow, sediment-transport, and water-temperature characteristics of three small watersheds in the Alsea River basin, Oregon. U.S. Geological Survey Circ. 642. 21pp.

Hewlett, J.D., and Helvey, J.D. 1976. Effects of forest clear-felling on the storm hydrograph. Water Resour. Res. 6:768–782.

Hewlett, J.D., and Hibbert, A.R. 1967. Factors affecting the response of small watersheds to precipitation in humid areas, pp. 275–290. In: W.E. Sopper and H.W. Lull, editors. Forest Hydrology, Pergamon, NY.

Hicks, B.J., Beschta, R.L., and Harr, R.D. 1991. Long-term changes in streamflow following logging in western Oregon and associated fisheries implications. Water Resour. Bull. 27:217–226.

Jones, J.A., and Grant, G.E. 1996. Peak flow responses to clear-cutting and roads in small and large basins, western Cascades, Oregon. Water Resour. Res. 34:959–974.

Keppeler, E.T., and Ziemer, R.R. 1990. Logging effects on streamflow: water yield and summer low flows at Caspar Creek in Northern California. Water Resour. Res. 26:1669–1679.

Kovner, J.L. 1956. Evapotranspiration and water yields following forest cutting and natural regrowth, pp. 106–110. Proceedings of Soc. Amer. Foresters, Bethesda, MD.

Ponce, S.L., and Meiman, J.R. 1983. Water yield augmentation through forest and range management – issues for the future. Water Resour. Bull. 19:415–419.

Rantz, S.E., and others. 1982. Measurement and computation of streamflow – Computation of Discharge U.S. Geological Survey Water Supply Paper 2175. 2v. 631pp.

Rothacher, J. 1970. Increases in water yield following clearcut logging in the Pacific Northwest. Water Resour. Res. 6:653–658.

Rothacher, J.S., and Miner, N. 1967. Accuracy of measurement of runoff from experimental watersheds, pp. 705–713. In: W.E. Sopper and H.W. Lull, editors. Forest Hydrology. Proceedings, National Science Foundation Advanced Science Seminar, University Park, Pennsylvania, Sept. 1967. Pergamon Press, New York, NY.

Smith, R.B., and Wass, E.F. 1980. Tree growth on skidroads on steep slopes logged after wildfires in Central and Southeastern British Columbia. Canadian Forestry Service, Pacific Forest Research Centre, Victoria, BC.

Snedecor, G.W., and Cochran, W.G. 1980. Statistical Methods, 7th Ed. Iowa State Univ. Press, Ames, IA. 507pp.

Stednick, J.D. 1991. Purpose and need for reactivating the Alsea Watershed Study, pp. 84–93. In: The New Alsea Watershed Study. Technical Bulletin 602. National Council of the Paper Industry for Air and Stream Improvement, Inc., New York, NY.

Stednick, J.D. 1996. Monitoring the effects of timber harvest on annual water yields. J. Hydrol. 176:79–95.

Stednick, J.D., and Kern, T.J. 1991. Determination of cumulative watershed effects on water resources after multiple timber harvest entries in Coastal Oregon, pp. 351–353. Abstract. American Water Resources Assoc. Meeting, New Orleans, LA.

Stednick, J.D., and Kern, T.J. 1992. Long term effects of timber harvesting in the Oregon Coast Range: The New Alsea Watershed Study (NAWS), pp. 502–510. In: Interdisciplinary Approaches to Hydrology and Hydrogeology. American Institute of Hydrology, Smyrna, GA.

Thomas, R.B. 1990. Problems in determining the return of a watershed to pretreatment conditions: techniques applied to a study at Caspar Creek, California. Water Resour. Res. 26:2079–2087.

Thomas, R.B., and Megahan, W.F. 1998. Peak flow responses to clearcutting and roads in small and large basins, western Cascades, Oregon: a second opinion. Water Resour. Res. 34:3393–3403.

Troendle, C.A., and King, R.M. 1985. The effect of timber harvest on the Fool Creek watashed, 30 years later. Water Resour. Bull. 21:1915–1922.

Wemple, B.C., and Jones, J.A. 2003. Runoff production on forest roads in a steep, mountain catchment. Water Resour. Res. 39:1220.

Wright, K.A., Sendek, K.H., Rice, R.M., and Thomas, R.B. 1990. Logging effects on streamflow: storm runoff at Caspar Creek in northwestern California. Water Resour. Res. 26:1657–1667.

Ziemer, R.R. 1964. Summer evapotranspiration trends as related to time after logging of forests in Sierra Nevada. J. Geophys. Res. 69:615–620.

Ziemer, R.R. 1981. Storm flow response to road building and partial cutting in small streams of northern California. Water Resour. Res. 17:907–917.

Chapter 10
Long-term Water Quality Changes Following Timber Harvesting

John D. Stednick

Chemical input-output budgets have been calculated for many undisturbed ecosystems at the watershed level. Study criteria require relatively small watershed areas so that (1) vegetation, soils, geology, and microclimate are relatively uniform and (2) all liquid water leaving the system can be measured as streamflow. Water and nutrient inputs are measured by precipitation gauges, and subsamples of precipitation are analyzed for nutrient concentrations. Water and nutrient outputs are usually measured by stream-gauging weirs, and water samples are taken for chemical analysis (Likens et al. 1970; Stednick 1989; Schlesinger 1991). Water chemistries are often influenced by season and volume (streamflow), and water quality sampling programs need to be carefully designed (Stednick 1991a; Aber and Melillo 2001).

Most forest ecosystems retain nutrients very efficiently (Gorham et al. 1979). Timber harvesting usually results in only modest increases in leaching losses (Waring and Schlesinger 1985). Chemical concentrations in Alsea stream waters showed limited spatial and temporal increases after timber harvesting (Brown 1972; Brown et al. 1973), but increased water yield continued for many years (Harr 1979; Stednick and Kern 1992; Harr and Nichols 1993). Traditionally, water quality studies measure chemical constituent concentrations in surface waters as pretreatment concentrations; these data are then compared with similar data collected after treatment. Any chemical constituent concentrations that increase generally return to pretreatment levels after a short time, ranging from days to months. However, increased annual water yields do not return to pretreatment levels as rapidly as water quality does, and may affect long-term site productivity (Federer et al. 1989; Hornbeck 1990; Lynch and Corbett 1990).

Variations in the concentration of dissolved ions can often be linked to changes in streamflow generation mechanisms (Abt et al. 1998). As streamflow increases, ion concentrations decline as an increasing proportion of flow is derived from precipitation, surface runoff, and macropore flow with little or no

John D. Stednick
Department of Forest, Rangeland, and Watershed Stewardship, Colorado State University, Fort Collins, CO 80523

J. D. Stednick (ed.), *Hydrological and Biological Responses to Forest Practices.*   157
© Springer 2008

equilibria with solid mineral phases (Schlesinger 1991). The effects of increased runoff should predominate over changes in concentrations.

One of the first watershed-ecosystem studies in the United States began at the Coweeta Hydrologic Laboratory in North Carolina in the 1930s to evaluate land use effects on water resources (Swank and Crossley 1988; Aber and Melillo 2001). Subsequent research at Hubbard Brook Experimental Forest in the White Mountains of New Hampshire indicated that hardwood deforestation resulted in large nutrient losses through leaching to stream water (Likens et al. 1970). These and other studies generated interest in evaluating the effect of timber harvesting on water resources in the Pacific Northwest.

Douglas-fir (*Pseudotsuga menziesii* [Mirb.] Franco) forests of the Olympic Peninsula of Washington and the Oregon Coast Range are among the most productive forests of the world (Peterson and Gessel 1983). Clearcutting is the typical timber harvesting practice, along with slash burning to prepare sites for seeding or planting. Several studies were initiated in the Pacific Northwest to assess effects of timber harvesting on nutrient cycles and input-output budgets (Gessel and Cole 1965; Brown 1972; Fredriksen 1972; Fredriksen et al. 1975; Sollins and McCorison 1981; Feller and Kimmins 1984; Edmonds et al. 1989; Martin and Harr 1989; Edmonds and Thomas 1992). Water budget data from study watersheds suggested that timber harvesting has little effect on nutrient outputs in stream water in the Oregon Cascade Mountains (Martin and Harr 1989) and the Oregon Coast Range (Brown et al. 1973). Natural forest ecosystems usually have nutrient cycles in which inputs approximate outputs, and nutrient losses are minimal (Vitousek and Reiners 1975; Tamm 1979; Vitousek and Melillo 1979; Swank and Waide 1988).

The original Alsea Watershed Study (AWS) in the Oregon Coast Range considered the effects of timber harvesting practices on hydrology (Harris 1977), water quality, stream habitat, and fish populations (Moring and Lantz 1975). This benchmark study utilized three small watersheds: Needle Branch (71 ha), Deer Creek (303 ha), and Flynn Creek (202 ha). Timber harvesting took place from March to October 1966. Needle Branch was clear-cut with no vegetative buffer strip or other streamside protection. Logging debris and slash were broadcast burned. Deer Creek was treated with three clear-cuts of about 25 ha each, with a vegetative buffer strip left along stream channels. Harvest units were burned the following year. Flynn Creek was undisturbed and served as the control watershed.

During the original AWS, water quality samples were collected for two years before and after logging. Samples were obtained once monthly during the first year and twice monthly thereafter. Storm events were sampled occasionally. Dissolved concentrations of nitrate-nitrogen, potassium, and total phosphorus were combined with stream discharge measurements to obtain nutrient fluxes (Brown et al. 1973). Clear-cutting followed by slash burning significantly increased nitrate-nitrogen outputs from Needle Branch from 4 to over 15 kg ha^{-1} yr^{-1}. No significant differences were observed for Deer Creek.

A reconnaissance water quality sampling in early 1973 showed that nitrate-nitrogen concentrations in Flynn Creek were lower than pretreatment concentrations but approximated those observed in fall, 1968. Nitrate-nitrogen concentrations in Needle Branch were the lowest of record, suggesting a rapid return to pretreatment water quality conditions (Brown et al. 1973).

Total phosphorus (P) concentrations remained unchanged before and after treatment during the AWS. Concentrations of P ranged from 0.01 to 0.10 mg L^{-1} in all three watersheds. Potassium concentrations remained unchanged in Flynn Creek and Deer Creek throughout the study and ranged from 0.60 to 1.20 mg L^{-1}. In Needle Branch, potassium concentrations were comparable to Flynn Creek P concentrations prior to logging. During logging, concentrations increased (probably from vegetation foliage dropped in the stream channel) and peaked at about four times the pretreatment concentrations during the first storm after slash-burning, but then returned immediately to prelogging levels (Brown et al. 1973).

Nutrient budget studies are expensive to conduct and long-term data are rare. Also, it is often difficult to maintain watersheds in their original treatment condition; control watersheds are often lost. The New Alsea Watershed Study (NAWS) provides a unique opportunity to evaluate the long-term effects of timber harvest on water resources in the Oregon Coast Range. Flynn Creek remains undisturbed, and watershed comparisons may be made per the original study methodology. Deer Creek had additional harvests in 1979, 1988, and 1989; approximately 40% of the Deer Creek basin has now been harvested (See Chapter 11).

Although water quality changes were short term, water quantity effects from timber harvesting last longer (See Chapter 9). Are there any long-term effects of timber harvesting on water quality? The purpose of this study was to assess water quality 25 years after timber harvest in the Oregon Coast Range.

Methodology

Characteristics of the study watersheds were compiled (Table 10.1). Generally, 100-year-old Douglas-fir (*Pseudotsuga menziesii*) was the principal commercial species harvested. The important hardwood in the area was red alder (*Alnus rubra*). Understory vegetation consisted primarily of salmonberry (*Rubus spectabilis*), sword fern (*Polystichum munitum*), and vine maple (*Acer circinatum*) (Moring and Lantz 1975).

Soils were derived from the Tyee Sandstone Formation. The Slickrock or Bohannon soil series account for approximately 80% of the soils (Brown et al. 1973). The Slickrock soils (medial over loamy, mixed, mesic Typic Dystrandept) were derived from sandstone colluvium (USDA Soil Conservation Service 1990). Organic matter content is classed as medium, with 5–7% organic carbon in surface horizons. The Bohannon series (fine loamy, mixed, mesic Andic

Table 10.1 Characteristics of study watersheds as of 1990 (data from Harr et al. 1975; Moring and Lantz 1975; Stednick unpublished)

	Flynn Creek	Deer Creek	Needle Branch
Treatment	control	patchcut	clearcut
Watershed area (ha)	202	303	71
Median slope (%)	27	39	30
Vegetation (%)			
conifer (mature)	36	33	0
conifer (2nd growth)	0	31	80
hardwood	64	36	20
Area in roads (%)	0	4	5
Area harvested (%)	0	39	82[1]
Area burned (%)	0	22[2]	82

[1] In the early 1950s, 13 ha in the headwaters of Needle Branch were logged.
[2] Only one of the 3 AWS harvest units was successfully burned, vegetation regrowth in the other units made for spotty burns rather than a broadcast burn.

Haplumbrept), a shallow gravelly loam, is derived from the sandstone residuum. Surface horizons contain less than 5% organic carbon. Total nitrogen is approximately 10,600 kg ha^{-1} for the Slickrock soils and 5600 kg ha^{-1} for the Bohannon series (Corliss 1973).

Precipitation measurements and streamflow-gauging methodologies are discussed in Chapters 2 and 9. Precipitation gauges for the study basins were near unimproved roads, and water chemistry samples were potentially contaminated from road dust. Precipitation chemistry from a NADP/NTN site at the Alsea Ranger Station, located approximately 15 kilometers southeast of the study watersheds in Alsea, Oregon (National Atmospheric Deposition Program 1993), was used to calculate inputs.

Stream water quality samples were collected in high-density polyethylene (HDPE) sample bottles. Bottles were washed and then rinsed twice with deionized water. Bottles and caps were rinsed three times with sample water before the sample was collected (Stednick 1991a). The samples were unfiltered and were preserved by freezing. Field measurements included temperature, pH, conductivity, and alkalinity (Stednick 1991a). Other water quality analyses were conducted at Colorado State University; cations were analyzed on a Varian® Inductively Coupled Plasma Emission Spectrophotometer (ICP) and anions by a Dionex® 1200 Ion Chromatograph (IC) (Table 10.2). Quality assurance and quality control formats required 10% duplicate samples, 10% blank samples, and +/–5% error on split and spiked samples. Water quality samples were collected on a random basis for water years 1990 and 1991 (n = 22 and 18, respectively). Water quality data approximated a normal distribution. Water quality parameters were considered normal, and parametric statistics were used in data analysis. Mean concentrations with standard errors of individual parameters were calculated for precipitation and stream waters (Table 10.3).

Table 10.2 Analytical methods for water quality analyses

Variable	Method	Reference
Temperature, water	Thermometer or thermistor	Stednick (1991b)
Dissolved oxygen (DO)	Winkler-azide modification	Hach Chemical Company (1989)
pH, field	Electrode	Stednick (1991b)
pH, lab	Electrode	APHA (1989)
Alkalinity (acid neutralizing capacity)	Titration (with Gran plot)	Stednick (1991b) Gran (1952)
Conductivity	Conductivity cell	U.S. EPA (1979)
SO_4^{2-}, Cl^-, NO_3^-, PO_4^{3-}	Ion chromatography (IC)	O'Dell et al. (1984)
Ca, Mg, Na, K, Si, P	Inductively coupled plasma emission spectrophotometer	U.S. EPA (1979)

Table 10.3 Mean concentrations in mg L^{-1} (and standard error) for precipitation ($n = 50$) (National Atmospheric Deposition Program 1993) and streamflow (n = 40) samples for study watersheds. Means in a row followed by a similar letter are not significantly different ($p > 0.05$)

	Precipitation	Flynn Creek	Deer Creek	Needle Branch
pH	5.47 (0.05)	6.0 (0.1)[a]	6.5 (0.1)[b]	6.2 (0.1)[a]
Ca	0.048 (0.001)	2.24 (0.19)[a]	2.62 (0.16)[a]	1.66 (0.14)[b]
Mg	0.066 (0.005)	1.01 (0.06)[a]	0.99 (0.07)[a]	0.81 (0.08)[a]
Na	0.547 (0.04)	2.97 (0.24)[a]	3.83 (0.31)[a]	3.66 (0.30)[a]
K	0.028 (0.002)	0.85 (0.10)[a]	1.02 (0.16)[a]	0.82 (0.11)[a]
Cl	1.000 (0.077)	3.81 (0.25)[a]	3.75 (0.22)[a]	3.25 (0.25)[a]
NO_3–N	0.050 (0.003)	1.22 (0.32)[a]	0.95 (0.31)[a]	0.40 (0.22)[a]
SO_4	0.361 (0.002)	0.83 (0.07)[a]	0.90 (0.07)[a]	0.80 (0.09)[a]
HCO_3–C	–	1.67 (0.44)[a]	2.50 (0.51)[a]	2.35 (0.63)[a]

Table 10.4 Mean annual input (1990–1991) (precipitation) kg ha^{-1} yr^{-1} (National Atmospheric Deposition Program 1993) and mean annual output (1990–1991) (streamwater) kg ha^{-1} yr^{-1} for dissolved constituents. (Standard error of the flux in parentheses). Means in a row followed by the same letter are not significantly different ($p > 0.05$)

	Precipitation	Flynn Creek	Deer Creek	Needle Branch
Ca	0.67 (0.00)	41.4 (1.5)[a]	44.1 (4.2)[a]	33.6 (1.1)[b]
Mg	1.26 (0.02)	17.6 (0.2)[a]	17.1 (1.6)[a]	15.4 (0.8)[a]
Na	10.66 (0.30)	52.4 (6.3)[a]	66.3 (14.2)[a]	72.4 (1.0)[a]
K	0.44 (0.00)	13.0 (0.2)[a]	27.9 (9.3)[a]	17.3 (6.8)[a]
Cl	19.45 (0.85)	70.6 (14.2)[a]	66.6 (11.1)[a]	77.3 (12.2)[a]
NO_3–N	0.36 (0.07)	21.6 (4.2)[a]	16.8 (3.4)[a,b]	7.9 (1.1)[b]
SO_4	4.98 (0.07)	13.4 (1.9)[a]	14.0 (3.8)[a]	16.4 (6.4)[a]
HCO_3–C	–	28.5 (7.4)[a]	42.2 (14.8)[a]	44.4 (12.2)[a]
Si	–	121.8 (7.0)[a]	141.4 (3.6)[a]	114.0 (13.0)[a]

Annual outputs were calculated using the reverse-flow period weighting technique, where individual concentration values were weighted by the time (and flow) since the previous water quality sampling period (Dann et al. 1986). Stream water outputs were calculated as dissolved load and do not include particulate or organic matter transport. Thus, nutrient exports are considered conservative. Exports were expressed as kg ha^{-1} yr^{-1} to allow for comparisons with data from the original study (Table 10.4).

Results and Discussion

Annual streamflows and water yields for the control watershed (Flynn Creek) were approximately 85% of the long-term mean for the years studied (See Chapter 9). Water yields for water years 1990 and 1991 were similar in storm event size and number and in low flows (Stednick unpublished data). Precipitation was slightly acidic (pH = 5.5) and was dominated by sodium (Na^+) and chloride (Cl^-), reflecting the ocean influence. Precipitation chemistry did not show seasonal differences.

The Alsea streams may be classified as calcium-bicarbonate type waters. Calcium and bicarbonate have the highest concentrations (as expressed as equivalent weights). Formula weights expressed as mg L^{-1} were used in this analysis for easier comparison to the National Atmospheric Deposition Program (NADP) data and the original AWS (Brown et al. 1973).

The average output of the clear-cut watershed (Needle Branch) was 7.9 kg ha^{-1} yr^{-1} nitrate-nitrogen; of the patchcut watershed (Deer Creek), 16.8 kg ha^{-1} yr^{-1} ; and of the control watershed (Flynn Creek), 21.6 kg ha^{-1} yr^{-1} (Table 10.5). The mean nitrate-nitrogen output from Needle Branch was significantly less than the mean nitrate-nitrogen output from Flynn Creek (Table 10.4). The nitrate-nitrogen exports were not significantly different between years (1990–1991) for Flynn Creek or Deer Creek; however, there was a significant difference between years for Needle Branch (6.0 and 9.8 kg ha^{-1} yr^{-1}, respectively). A complete N budget was not calculated. Gaseous transfers were probably small, and stream water concentrations of ammonium were below the analytical detection limit.

Table 10.5 Nitrate-nitrogen flux (kg ha^{-1} yr^{-1}) for the Alsea watersheds (1965–1968; Brown et al., 1973 and 1990–1991)

	Flynn Creek	Deer Creek	Needle Branch
1965	35.0	31.5	2.9
1966	27.4	25.4	4.9
1967	28.5	28.4	15.7
1968	25.0	24.5	15.1
1990	19.4	14.5	6.0
1991	23.8	19.1	9.8

Temperate coniferous forests usually demonstrate nutrient conservation in nitrogen, phosphorus, and sulfur (Henderson et al. 1978; Gorham et al. 1979; Vitousek and Melillo 1979; Gholz et al. 1985; Aber and Melillo 2001). Exceptions to this general pattern are those systems that are so nitrogen rich that microbial immobilization of N cannot compensate for reduced plant uptake (Binkley 1986). Nitrate is often a major portion of the total nitrogen lost from forest ecosystems to surface waters (Driscoll et al. 1989). Processes, such as denitrification within riparian (and wetland) areas and within the stream channels themselves (Hill 1988), are capable of removing nitrate, but the significance of these processes in regulating nitrate flux varies widely. This variation in nitrate-nitrogen fluxes implies that some catchments with increased nitrogen inputs (fertilization and atmospheric fixation) will have increased nitrate-nitrogen outputs, while others have the buffering ability within soils, riparian ecosystems, and the stream channel to mitigate such a response.

There are only minor differences in nitrogen losses (as nitrate-nitrogen) between early and late successional forests (Gorham et al. 1979; Binkley et al. 1982). In general, nitrate-nitrogen losses in stream water are greatest in forests with high nitrogen availability prior to disturbance (Binkley and Brown 1993). Increased nitrification rates will increase the mobile anion pool and, subsequently, cation fluxes in stream water (Likens et al. 1970). The time and vegetation influences on Needle Branch water quality warrant further study.

Each study watershed has various areal extents of red alder (*Alnus rubra* Bong.), especially in the stream corridor and on disturbed sites (Table 10.1). Red alder is a pioneer species that invades recently cleared or disturbed forest areas. Site occupation and rapid juvenile growth have frequently made this species an unwelcome competitor in conifer plantations. Red alder may improve site fertility by increasing organic matter and soil nitrogen. Red alder can be used as an alternative to fertilization in nitrogen-deficient conifer forests. Nitrogen fixation by the red alder increases the nitrogen capital, while reducing the plant's need for nitrogen from the soil (Binkley 1986). Water quality may be influenced by more available nitrogen in the ecosystem.

A subwatershed water quality sampling program helped identify the processes controlling stream water concentrations of nitrate-nitrogen. Stream water in five to seven relatively homogeneous subwatersheds was sampled in each watershed for chemical analyses. Physical and vegetation characteristics of subwatersheds were quantified as landscape elements and used to infer streamflow generation mechanisms and streamflow routing rates (Stednick and Kern 1991). Landscape elements of slope, aspect, vegetation type and biomass, soil type, soil nutrient pools, source areas of streamflow generation, and water-routing mechanisms were related to water quality. Stream water chemistries were influenced most by streamflow generation and routing mechanisms (slope) and near-channel vegetation of differing successional ages (Stednick and Kern 1991). Subwatersheds representing 10–20% of the total watershed area often produced 40–60% of the nitrate-nitrogen measured at the watershed gauging station. Areal distribution of alder played a significant role, particularly as riparian vegetation. NO_3–N concentration at the Flynn Creek gauging station ranged from 0.2 mg L^{-1} to 2.6 mg L^{-1}, and Flynn Creek subbasin had a range of 0.1 mg L^{-1} to 3.3 mg L^{-1}. Thus stream sampling locations on a stream in a "uniform" watershed may show variance due to different landscape elements. Concentrations of nitrate-nitrogen are positively correlated with hydrologic response or streamflow generation mechanisms (Miller and Newton 1983; Stednick and Kern 1991).

During 1990 and 1991, Flynn Creek had beaver activity around the stream-gauging station. Water quality samples collected in, above, and below the beaver ponds showed no differences in nitrate-nitrogen concentrations. Thus, beaver activity did not appear to directly influence nitrate-nitrogen concentrations or dynamics in stream water. The influence of the water inundation on soils along the pond margin was not measured.

Needle Branch had significantly lower nitrate-nitrogen concentrations than did Deer Creek or Flynn Creek (Table 10.3). However, there was no difference in chloride, sulfate, or bicarbonate among the three watersheds. Previous studies in coastal Oregon forests suggested a reciprocity between nitrate-nitrogen and bicarbonate-carbon; nitrate-nitrogen concentrations increased as bicarbonate-carbon concentrations decreased over the growing season into winter (Miller and Newton 1983). In the Alsea watersheds, the bicarbonate-carbon flux increased as nitrate-nitrogen flux decreased (Table 10.5).

The nitrate-nitrogen fluxes calculated for all watersheds for 1990 and 1991 were less than the fluxes calculated from 1965 to 1968 (Table 10.5). Statistical comparisons between the 1965–1968 data and the 1990–1991 data cannot be made since error term or sample variability were not reported for the 1965–1968 data. Presumably, the vegetation uptake of nitrate-nitrogen in Flynn Creek has not changed significantly over time. Differences in water yield between these sampling periods cannot account for such a large difference. Mean nitrate-N concentrations measured in this study were 1.22, 0.95, and 0.40 mg L^{-1} for Flynn Creek, Deer Creek, and Needle Branch, respectively; while the early study averaged 1.15, 1.11, and 0.13 mg L^{-1}. The changes in Needle Branch could be due to fluctuations in vegetation and/or the streamflow generation and routing function as described earlier. Nitrate-N exports from Needle Branch were less than those from Flynn Creek and suggested that vegetation recovery has occurred; however, Needle Branch exports were still greater than during the pretreatment period. Although vegetation recovery/reestablishment is complete, stream generation and routing processes appear to be still affected by timber harvesting.

Is Flynn Creek an appropriate control watershed for water quality? The nitrate-nitrogen output from the control watershed (Flynn Creek) is high compared to the clear-cut watershed (Needle Branch). Flynn Creek chemistry may be different from other control watersheds in the temperate coniferous forest environment. Water quality, expressed as kg ha^{-1} yr^{-1} export, was compiled for other control or undisturbed watersheds (Table 10.6). Nitrate-nitrogen exports from Flynn Creek were considerably higher than any other control watershed, yet other nutrient exports were comparable. This export of

Table 10.6 Mean annual outputs (kg ha^{-1} yr^{-1}) for selected undisturbed forested catchments on the Pacific Northwest Coast. Lack of error terms precludes a statistical comparison

	Alsea[1]	HJ Andrews[2]	Indian River[3]	Jamison Creek[4]	West Twin Creek[5]	Olympic Peninsula[6]
Ca	41.4	34.7	275.0	41.7	223.3	320.0
Mg	17.6	6.9	24.0	8.8	6.1	44.5
Na	52.4	30.5	39.6	25.6	27.0	112.5
K	13.0	5.2	10.3	2.6	59.0	12.0
Cl	70.6	–	90.1	38.1	45.8	–
NO$_3$–N	21.6	<0.1	2.2	0.8	1.3	1.3
SO$_4$–S	7.2	–	22.8	3.0	72.6	83.0
HCO$_3$–C	28.5	47.1	183.5	37.2	144.8	1220.0
Si	121.8	157.9	34.1	48.9	–	102.0
precipitation (mm)	2094	2330	2660	4541	3110	4230
streamflow (mm)	1663	1530	2200	3668	1680	3785

[1] This study
[2] Fredriksen (1972)
[3] Stednick (1981)
[4] Zeman (1975)
[5] Edmonds and Thomas (1992)
[6] Larson (1979)

nitrogen from Flynn Creek probably does not represent a site productivity loss since the soils have relatively high nitrogen pools (USDA Soil Conservation Service 1990) and alder fixation rates range from 50 kg ha^{-1} yr^{-1} to 200 kg ha^{-1} yr^{-1} (Binkley 1986). Approximately 64% of the Flynn Creek watershed area was vegetated with red alder.

Compilation of water quality databases suggests that the absolute losses (kg ha^{-1}) of N in response to cutting are greater on sites that had more N leaching before cutting (Binkley and Brown 1993). The Alsea watersheds do not follow this pattern, as the control watershed nitrate-nitrogen concentrations were usually higher than the treatment watersheds. The nitrogen fixation rates in alder are probably the same as those that regulate overall plant growth; thus, rates decrease with summer drought (Binkley et al. 1994). It would be worthwhile to investigate the relation of nitrate generation and movement to soil water dynamics.

For the Alsea watersheds, there was a greater output of sulfate (SO$_4$$^=$) than input (Table 10.5). No patterns were recognized between precipitation and stream water pH or sulfate, as observed at Coweeta watersheds (Swank and Waide 1988). Approximately 34% of the sulfate output was measured as wet deposition input. Ocean aerosols or, more likely, the weathering of the Tyee Sandstone bedrock was the other source of sulfate.

For all cations measured, outputs were greater than inputs (Table 10.5). This suggests that the soil and parent material weathering processes produced cations sufficient to meet biological requirements and secondary mineral weathering processes, with the excess removed in streamflow.

For all anions measured, outputs were greater than inputs. This is surprising since nitrogen retention is usually the rule rather than the exception in coniferous forests (Vitousek and Melillo 1979; Waring and Schlesinger 1985; Binkley and Brown 1993). Sulfate and chloride (Cl^-) exports may be due to weathering of the marine deposited Tyee Sandstone bedrock or aerosol impaction. Orthophosphate phosphorus (PO_4–P) was below detection levels, and an input-output balance was not calculated.

Geologic weathering processes are rapid in the Coast Range because of heavy precipitation (2500 mm yr^{-1}), mild temperatures (monthly means from 4 to 20°C), and the relative ease of weathering of sedimentary strata (Miller and Newton 1983). The Si export is a conservative estimate of denudation. If Si is assumed to be 20% of the parent material, denudation rates were 27 mm 100 yr^{-1} in Deer Creek, 23 mm 100 yr^{-1} in Flynn Creek, and 22 mm 100 yr^{-1} in Needle Branch, and had a correlation coefficient of 0.91 with median watershed slope. Physical denudation is probably much greater than the chemical denudation and may offer further insight to chemical processes.

Conclusions

Evaluation of the Alsea watersheds 25 years after timber harvesting suggests that water quality, as measured by chemical constituent concentrations (mg L^{-1}) and calculated nutrient fluxes (kg ha^{-1} yr^{-1}), has returned to near pretreatment levels. The Needle Branch watershed was clearcut and slash burned in 1966. Timber harvesting significantly increased nitrate-nitrogen fluxes, and posttreatment water quality sampling showed a quick return to pretreatment conditions. This study showed a mean flux of 7.9 kg ha^{-1} yr^{-1} nitrate-nitrogen from Needle Branch compared to a pretreatment flux of 3.9 kg ha^{-1} yr^{-1}. The control watershed, Flynn Creek, had a nitrate-nitrogen output of 21.6 kg ha^{-1} yr^{-1} for 1990–1991, compared to an output of 29.0 kg ha^{-1} yr^{-1} for 1965–1968.

Although precipitation and streamflows were lower for 1990–1991 than for 1965–1968, not all water quality differences can be attributed to the water balance. Differences in nitrate-nitrogen concentrations and fluxes were related to streamflow generation and routing mechanisms through landscape elements. These landscape elements included physical characteristics, such as slope, soils, and vegetation type.

Multiple timber harvest operations in Deer Creek, following best management practices (BMPs), resulted in no significant water quality changes. The increase in silica export from Deer Creek, compared to the control watershed, is a function of watershed slope rather than land use activities.

The nitrate-nitrogen export from Flynn Creek is the highest export observed from a compilation of undisturbed or control watersheds in the temperate coniferous forest ecosystem. Flynn Creek has 64% of the watershed area vegetated with alder, and the remainder is coniferous. A subwatershed water quality sampling program showed that 10–20% of the total watershed area could produce 40–60% of the nitrate-nitrogen measured at the watershed gauging station.

The increased NO_3–N does not represent a significant degradation of stream quality, nor does it suggest a decrease in site productivity or quality. Surface waters with concentrations that exceed state water quality standards may be considered water quality limited, and a waste load allocation negotiated. The first step is to define background concentrations of water quality constituents. The concept of background water quality may need to be reevaluated.

Literature Cited

Aber, J.D., and Melillo, J.M. 2001. Terrestrial Ecosystems. Academic Press, San Diego, CA. 556pp.

Abt, S.R., Johnson, T.L., Thornton, C.I., and Trabant, S.C. 1998. Riprap sizing at toe of embankment slopes. J. Hydraul. Eng. 124:672–677.

American Public Health Association (APHA). 1989. Standard methods for the examination of water and wastewater, 17th ed. Prepared by the American Public Health Association, American Water Works Association, and Water Pollution Control Federation, Washington, DC.

Binkley, D. 1986. Forest Nutrition Management. John Wiley & Sons, Inc., New York, NY. 290pp.

Binkley, D., and Brown, T.C. 1993. Forest practices as nonpoint sources of pollution in North America. Water Resour. Bull. 29:729–740.

Binkley, D., Cromack Jr., K., and Baker, D. 1994. Nitrogen fixation by red alder: biology, rates and controls, pp.55–72. In: D. E. Hibbs, D.S. DeBell, and R.F. Tarrant, editors. The Biology and Management of Red Alder. Oregon State Univ. Press, Corvallis, OR.

Binkley, D., Kimmins, J.P., and Feller, M.C. 1982. Water chemistry profiles in an early- and a mid-successional forest in coastal British Columbia. Can. J. For. Res. 12:240–248.

Brown, G.W. 1972. Logging and water quality in the Pacific Northwest, pp.330–335. In: Watersheds in Transition. Amer. Water Res. Assoc. Publication, Proceedings Series No. 14.

Brown, G.W., Gahler, A.R., and Marston, R.B. 1973. Nutrient losses after clear-cut logging and slash burning in the Oregon Coast Range. Water Resour. Res. 9:1450–1453.

Corliss, J.F. 1973. Soil Survey of Alsea Area, Oregon. USDA Soil Conservation Service and Forest Service, USDI Bureau of Land Management, in cooperation with Oregon Board of Natural Resources and Oregon Agricultural Experiment Station, Superintendent of Documents, Washington, DC. 82pp.

Dann, M.S., Lynch, J.A., and Corbett, E.S. 1986. Comparison of methods for estimating sulfate export from a forested watershed. J. Environ. Qual. 15:140–145.

Driscoll, C.T., Likens, G.E., Hedin, L.O., Eaton, J.S., and coauthors. 1989. Changes in the chemistry of surface waters. Env. Sci. and Tech. 23:137–143.

Edmonds, R.L., Binkley, D., Feller, M.C., Sollins, P., and coauthors. 1989. Nutrient cycling: effects on productivity of Northwest forests, pp. 17–35. In: D.A. Perry, et al., editors. Maintaining the long term productivity of Pacific Northwest Forest Ecosystems. Timber Press, Portland, OR.

Edmonds, R.L., and Thomas, T.B. 1992. Hydrologic and nutrient cycles in a pristine old growth forested watershed in the Olympic Peninsula, Washington, pp. 461–470. In: Managing Water Resources During Global Change. Amer. Water Resour. Assoc. 28th Annual Conference, Reno, NV.

Federer, C.A., Hornbeck, J.W., Tritton, L.M., Martin, C.W., and coauthors. 1989. Long term depletion of calcium and other nutrients in Eastern U.S. forests. Environ. Manage. 13:593–601.

Feller, M.C., and Kimmins, J.P. 1984. Effects of clear cutting and slash burning on stream water chemistry and watershed nutrient budgets in southwestern British Columbia. Water Resour. Res. 20:29–40.

Fredriksen, R.L. 1972. Nutrient budget of a Douglas-fir forest on an experimental watershed in western Oregon, pp. 115–131. In: J.F. Franklin, L.J. Dempster, and R.H. Waring, editors. Research on Coniferous Forest Ecosystems. USDA Forest Service, Pacific Northwest Forest and Range Experiment Station, Portland, OR.

Fredriksen, R.L., Moore, D.G., and Norris, L.A. 1975. The impact of timber harvest, fertilization, and herbicide treatment on stream water quality in western Oregon and Washington, pp. 283–313. In: Proceedings of the 4th North American Forest Soils Conference on Forest Soils and Forest Land Management. Laval Univ. Press, Quebec.

Gessel, S.P., and Cole, D.W. 1965. Influence of removal of forest cover on movement of water and associated elements through soil. J. Amer. Water Works Assoc. 57:1301–1310.

Gholz, H.L., Hawk, G.M., Campbell, A., and Cromack Jr., K. 1985. Early vegetation recovery and element cycles on a clear cut watershed in western Oregon. Can. J. For. Res. 15:400–409.

Gorham, E., Vitousek, P.M., and Reiners, W.A. 1979. The regulation of chemical budgets over the course of terrestrial succession. Ann. Rev. Ecol. Syst. 10:53–84.

Gran, G. 1952. Determination of the equivalence point in potentiometric titrations. Part 2. Analyst 77:661–671.

Hach Chemical Company. 1989. Water Analysis Handbook, Ames, IA. 829pp.

Harr, R.D. 1979. Effects of timber harvest on streamflow in the rain-dominated portion of the Pacific Northwest. In: Proceedings of Workshop on Scheduling Timber Harvest for Hydrologic Concerns. USDA Forest Service, Pacific Northwest Forest and Range Experiment Station, Portland, OR. 45pp.

Harr, R.D., Harper, W.C., Krygier, J.T., and Hsieh, F.S. 1975. Changes in storm hydrographs after road building and clear-cutting in the Oregon Coast Range. Water Resour. Res. 11:436–444.

Harr, R.D., and Nichols, R.A. 1993. Stabilizing forest roads to help restore fish habitats: a northwest Washington example. Fisheries 18(4):18–22.

Harris, D.D. 1977. Hydrologic changes after logging in two small Oregon coastal watersheds. Water-Supply Paper 2037. U.S. Geological Survey Washington, DC. 31pp.

Henderson, G.S., Swank, W.T., Waide, J.B., and Grier, C.C. 1978. Nutrient budgets of Appalachian and Cascade region watersheds: a comparison. For. Sci. 24:385–397.

Hill, A.R. 1988. Factors influencing nitrate depletion in a rural stream. Hydrobiologia 60:111–112.

Hornbeck, J.W. 1990. Nutrient depletion: a problem for forests in New England and Eastern Canada? pp. 56–67. In: M.K. Mahendrappa, D.M. Simpson, and G.D. van Raalte, editors. Conference on the Impacts of Intensive Harvesting. Forestry Canada-Maritimes Region, Fredricton, NB.

Larson, A.G. 1979. Origin of the Chemical Composition of Undisturbed Forested Streams. Ph.D. Thesis. Washington State Univ., Seattle, WA. 216pp.

Likens, G.E., Bormann, F.H., Johnson, N.M., Fisher, D.W., and coauthors. 1970. Effects of forest cutting and herbicide treatment on nutrient budgets in the Hubbard Brook watershed ecosystem. Ecol. Monogr. 40:23–47.

Lynch, J.A., and Corbett, E.S. 1990. Evaluation of best management practices for controlling nonpoint pollution from silvicultural operations. Water Resour. Bull. 26:41–52.

Martin, C.W., and Harr, R.D. 1989. Logging of mature Douglas fir in western Oregon has little effect on nutrient output budgets. Can. J. For. Res. 19:35–43.

Miller, J.H., and Newton, M. 1983. Nutrient loss from disturbed forest watersheds in Oregon's Coast Range. Agro-Ecosystems 8:153–167.

Moring, J.R., and Lantz, R.L. 1975. The Alsea Watershed Study: effects of logging on the aquatic resources of three headwater streams of the Alsea River, Oregon. Part I. Biological studies. Fish. Res. Rep. 9. Oregon Dept. of Fish and Wildlife, Corvallis, OR. 66pp.

National Atmospheric Deposition Program. 1993. (NRSP-3)/National Trends Network. NADP/NTN Coordination Office, Colorado State Univ., Ft. Collins, CO.

O'Dell, J.W., Pfaff, J.D., Gales, M.E., and McKee, G.D. 1984. Technical Analysis of Water and Wastes, Method 300.0. The determination of inorganic anions in water by ion chromatography. EPA-600/4-85-017. U.S. Environmental Protection Agency, Cincinnati, OH.

Peterson, C.E., Jr., and Gessel, S.P. 1983. Forest fertilization in the Pacific Northwest: results of the regional forest nutrition research project, pp. 365–369. In: Proceedings IUFRO Symposium on Forest Site and Continuous Productivity. General Technical Report GTR-PNW-163. USDA Forest Service, Pacific Northwest Forest and Range Experiment Station, Portland, OR.

Schlesinger, W.H. 1991. Biogeochemistry: An Analysis of Global Change. Academic Press, San Diego, CA. 351pp.

Sollins, P., and McCorison, F.M. 1981. Nitrogen and carbon solution chemistry of an old growth coniferous forest watershed before and after cutting. Water Resour. Res. 17:1409–1418.

Stednick, J.D. 1981. Precipitation and stream water chemistry in an undisturbed watershed in southeast Alaska. Research Paper RP-PNW-291. USDA Forest Service, Pacific Northwest Forest and Range Experiment Station, Portland, OR. 8pp.

Stednick, J.D. 1989. Hydrochemical characterization of alpine and alpine-subalpine waters, Colorado Rocky Mountains. Arctic and Alpine Res. 21:276–282.

Stednick, J.D. 1991a. Purpose and need for reactivating the Alsea Watershed Study, pp. 84–93. In: The New Alsea Watershed Study. Technical Bulletin 602. National Council of the Paper Industry for Air and Stream Improvement, Inc., New York, NY.

Stednick, J.D. 1991b. Wildland Water Quality Sampling and Analysis. Academic Press, San Diego, CA. 217pp.

Stednick, J.D., and Kern, T.J. 1991. Evaluation of landscape elements as water quality determinants in the New Alsea Watershed Study, Abstract. American Geophysical Union 72(43).

Stednick, J.D., and Kern, T.J. 1992. Long term effects of timber harvesting in the Oregon Coast Range: The New Alsea Watershed Study (NAWS), pp. 502–510. In: Interdisciplinary Approaches to Hydrology and Hydrogeology. American Institute of Hydrology, Smyrna, GA.

Swank, W.T., and Crossley, D.A., Jr. 1988. Forest Hydrology and Ecology at Coweeta. Ecological Studies Vol. 66. Springer-Verlag, Inc., New York, NY. 469pp.

Swank, W.T., and Waide, J.B. 1988. Characterization of baseline precipitation and stream chemistry and nutrient budgets for control watersheds, pp. 57–80. In: W.T. Swank, and D.A. Crossley, Jr., editors. Forest hydrology and ecology at Coweeta. Ecological Studies Vol. 66. Springer-Verlag, Inc., New York.

Tamm, C.O. 1979. Nutrient cycling and productivity of forest ecosystems, pp. 2–21. In: A.L. Leaf, editor. Proceedings of a conference on the Impact of Intensive Harvesting on Forest Nutrient Cycling. State Univ. of New York, Syracuse, NY.

U.S. Environmental Protection Agency. 1979. Methods for Chemical Analysis of Water and Wastes. EPA-600/4-79/020. U.S. Environmental Protection Agency, Environmental Monitoring and Support Laboratory, Office of Research and Development, Cincinnati, OH.

USDA Soil Conservation Service. 1990. Soil series of the United States, including Puerto Rico and the U.S. Virgin Islands. Misc. Pub. No. 1483.

Vitousek, P.M., and Melillo, J.M. 1979. Nitrate losses from disturbed forests: patterns and mechanisms. For. Sci. 25:605–619.

Vitousek, P.M., and Reiners, W.A. 1975. Ecosystem succession and nutrient retention: a hypothesis. BioScience 25:376–381.

Waring, R.H., and Schlesinger, W.H. 1985. Forest Ecosystems, Concepts and Management. Academic Press, Orlando, FL. 338pp.

Zeman, T.J. 1975. Hydrochemical balance of a British Columbia mountainous watershed. Catena 2:81–93.

Chapter 11
Risk Assessment for Salmon from Water Quality Changes Following Timber Harvesting

J. D. Stednick and T. J. Kern

Ecological risk assessments evaluate ecological effects caused by human activities. The ecological risk assessment process must be flexible while providing a logical and scientific structure to accommodate a broad array of sensors. The proposed framework consists of three parts: (1) problem formulation (2) analysis and (3) risk characterization (U.S. Environmental Protection Agency 1992).

Problem Formulation

Headwater streams in the Western states may serve as drinking water sources for local communities and as habitat for cold water fisheries, especially salmonid species. Population estimates for juvenile coho salmon (*Oncorhynchus kisutch*) in these reaches range from 20 to 40 fish per 100 m (Ice 1991). Similar numbers are seen for trout populations. These stream ecosystems have coexisted with logging activities for over a century, essentially due to the relatively low-level and dispersed pattern of logging activity in a watershed. Past logging activities not employing BMPs (Best Management Practices) per Federal Water Pollution Control Act Amendments of 1972 (Stednick 1980) to prevent or mitigate water quality impacts (i.e., streamside buffers, proper road drainage systems) may have adversely affected water quality and fisheries resources in some settings. In general, land use practices may reduce salmonid production in streams by decreasing habitat diversity and complexity and increasing stream nutrient concentrations (Bottom et al. 1985; Bjornn and Reiser 1991). Interest

Originally published in Stednick, J.D., and Kern, T.J. 1994. Risk assessment for salmon from water quality changes following timber harvesting, pp. 227–238 In: Environmental Monitoring and Assessment, Volume 32, No. 3, 4 tables, 1994 Kluwer Academic Publishers: With kind permission from Springer Science and Business Media.

J. D. Stednick
Department of Forest, Rangeland, and Watershed Stewardship, Colorado State University, Fort Collins, CO 80523

in cumulative watershed effects (Clean Water Act 1987), water quality, and salmonid populations, is significantly modifying allowable land use practices. Recent legislation preserving old-growth temperate coniferous forests to maintain biotic diversity (often under the Endangered Species Act) decreases the operable land base used for timber production. This reduced land base forces multiple land uses on remaining lands. Quantification of cumulative watershed effects (those effects from multiple management activities in time and/or space) has only recently begun (Stednick and Kern 1992).

The effects of logging on fish and their stream environment are often basin specific (Bormann and Likens 1967; Fredriksen 1971; Harr 1979; Brown 1985; Driscoll et al. 1989; Hicks et al. 1991), but collectively have been used to define Best Management Practices (BMPs) (Ice 1991). Such studies have looked at the effects of silvicultural activities on stream water quality, streamflow, water temperature, sedimentation, loss of nutrients, damage to spawning gravels and reduced migratory fish movement. The use of BMPs is meant to maintain water quality standards, but water quality departures from background can and do occur. Not all issues can be addressed by regulatory procedures; some water quality changes are intrinsic to the land use activity itself. For timber harvesting, this includes possible altered streamflow (due to decreased transpiration and interception), higher stream temperatures, increased primary production, decreased dissolved oxygen, and enhanced nitrification and stream acidification.

Alsea Watershed Study

Concern about the possible influence of logging on water and salmonid resources in the Pacific Northwest led to the initiation of the Alsea Watershed Study in 1958. From 1959 to 1973 a research team from Oregon State University, and the Oregon State Game Commission in cooperation with the U.S. Geological Survey, evaluated the effects of silviculture treatments on water resources and salmon populations in Coastal Oregon. The study looked at the impact of logging on streamflow and water yield, dissolved solids, water temperature, dissolved oxygen, suspended sediments, aquatic nitrogen and phosphorus levels, and fish productivity and populations (Brown and Krygier 1971; Harr and Krygier 1972; Brown et al. 1973; Moring and Lantz 1975; Harr 1976, 1991; Harris 1977).

The original study utilized three small watersheds (Deer Creek, Flynn Creek, and Needle Branch) near the Alsea River in Western Oregon. Following the completion of the original Alsea study in 1973, the study site was essentially abandoned. Flynn Creek is now designated a Research Natural Area by the USDA Forest Service and is undisturbed by human activities. The old-growth Douglas-fir (*Pseudotsuga mensiezii*) stand is approximately 135 years old, with stands of red alder (*Alnus rubra*) dominating disturbed areas and the riparian corridor (Table 11.1).

Needle Branch was clearcut (without streamside buffer strips) and the logging slash broadcast burned in 1966. Some precommercial forest thinning

Table 11.1 Characteristics of study watersheds as of 1990 (adapted from Harr et al., 1975; Moring, 1975; Stednick, unpublished)

	Flynn Creek	Needle Branch
Treatment	control	clearcut
Watershed area (ha)	202	71
Median slope (%)	27	30
Vegetation (%)		
conifer (mature)	36	0
conifer (regrowth)	0	80
hardwood	64	20
Area in roads (%)	0	5
Area harvested (%)	0	82[1]

[1] In the early 1950s, 13 ha in the headwaters of Needle Branch were logged.

occurred in 1981, and is not considered to affect basin hydrology. The forest stand is Douglas-fir, with red alder in the riparian corridor and landslide areas. Deer Creek was not used in this assessment.

The New Alsea Watershed Study, initiated in 1989, will assess the long-term impact of timber harvesting practices on water quantity and quality. This risk assessment utilized water quality and water quantity data collected for water year 1990 (1 October 1989 – 30 September 1990). The basin is considered not to have "recovered", that is, returned to pretreatment conditions with respect to water quality or water quantity (Stednick and Kern 1992).

This assessment used the literature to determine nitrate/nitrite-toxicity relations. Quantification of the exposure probability to critical water quality parameters and concentrations were calculated. The nitrate/nitrite toxicity and exposure probability were used to quantify the risk to salmon populations in an undisturbed catchment and a catchment clearcut 29 years ago.

Analysis

Toxicological Evaluation

Water chemistry results from the Alsea watersheds showed that neither chloride nor nitrate-nitrogen concentrations are at, or near, acute or chronic toxicity levels (Table 11.2). Nitrate-nitrogen in Needle Branch was the only parameter that approached the stream standard within an order of magnitude. Calcium, magnesium, sodium, potassium, chloride, sulfate, and heavy metal concentrations were two or more orders of magnitude lower than stream standards (or criteria) and were not evaluated in this study.

Even though the constituents do not exceed stream standards, this does not imply a risk-free environment. The technique proposed here avoids simplistic assumptions, evaluating all potential parameters of interest. For this case study

Table 11.2 Summary characteristics used for water quality parameters used in risk assessment (n = 22)

	Flynn Creek	Needle Branch
Cl (mg L^{-1})		
Mean	3.08	2.92
Standard deviation	1.19	1.21
Range	0.10–4.21	0.10–4.05
NO$_3$–N (mg L^{-1})		
Mean	1.08	0.31
Standard deviation	0.67	0.21
Range	0.10–3.19	0.10–0.75

we used the model to determine the risk to aquatic life associated with the most likely chemical impact, nitrate/nitrite-sediment toxicity.

The listed EPA criterion is 10 mg L^{-1} nitrate-nitrogen (45 mg L^{-1} as nitrate) for domestic water supply and coldwater fisheries (U.S. Environmental Protection Agency 1976). The EPA based the criterion on health standards for nitrites; nitrates of and by themselves are not toxic. The reduction of nitrate can take place within an organism, allowing nitrites to accumulate in the blood system. Once in the blood plasma, nitrite diffuses into red blood cells. Here it reacts with the iron in hemoglobin to form methemoglobin. This species lacks the capability to bind oxygen reversibly. Higher nitrate levels raise the methemoglobin fraction and reduce the total oxygen carrying capacity of the blood. Salmon blood normally contains up to 11% methemoglobin in the absence of nitrite (Lewis and Morris 1986).

The amount of methemoglobin in the blood necessary to inhibit normal behavior varies with the environment. Levels above 50% inhibit the cough response in salmon, leading to an inability to purge sediment collected in the buccal cavity (Lewis and Morris 1986; Boudou and Ribeyre 1989). When the methemoglobin content of the blood exceeds 70% the fish become torpid (Westin 1974). Since inactive salmon have low oxygen demands, it is only when the fish becomes excited or active that it may die of anoxia. A lengthy exposure is necessary for maximum accumulation of nitrite within a fish (Huey et al. 1980). The earliest LC$_{50}$ for acute exposures is 24 hours unless extreme concentrations are encountered. The 96 hour LC$_{50}$ is significantly lower than the 24 hour LC$_{50}$, implying that longer exposures require a lower nitrite concentration to reach lethal levels (Russo et al. 1974; Westin 1974).

Fish can also experience increased susceptibility to disease and liver dysfunction when high nitrite or ammonia concentrations are coupled with increased sediment loads (Lewis and Morris 1986; Servizi and Gordon 1990). Nitrite and sediment sublethal effects may include gill pathology, hypertrophy, and necrosis (Boudou and Ribeyre 1989). Gill pathology can provide an entry to infectious organisms. Since both sediment and nitrite cause sublethal gill pathology, the combination of enhanced nitrite and sediment levels reduces tolerance to infectious disease. This is especially true with adverse environmental factors such as warm temperatures.

A set of regression equations was developed from previously published nitrate toxicity studies to relate exposure time with reported LC_{50} levels (Russo et al. 1974; Westin 1974). Many studies did not account for confounding factors, so this assessment bases the regressions on the percentage of the 24-hour LC_{50} found for each duration. The regressions were not significantly different at the 90% confidence interval using an analysis of covariance (Snedecor and Cochran 1980). The average regression was:

$$\% \text{ of 24 hour } LC_{50} = 204 - 79 \times \log(\text{exposure duration in hours})$$

Studies show wide ranges of LC_{50} values. This is due to the lack of control of other contributing factors. The ionic strength or anion chemistry of the test solution was rarely mentioned in the earlier literature. Recent work shows more consistency in reporting other chemical parameters. Little work has been done on the influence of other natural factors.

The sediment toxicity mechanism involves accumulation, so there is a time factor involved in the sediment LC_{50} expression. A regression was developed for the relation between suspended sediment (ss) toxicity and LC_{50}:

$$\% \text{ of 24 hour } LC_{50} = 100 - 0.86 \times [\text{ss}] + 0.068 \times [\text{ss}]^2 - 0.0053 \times [\text{ss}]^3$$

The LC_{50} for suspended sediment is quite variable—the particle size fraction of the sediment plays a large role in fish response. Also, the chemistry of the sediment is an issue. Sediments may act as sinks or sources of toxins (heavy metals, organics, etc.) or ions that counter the effects of existing contaminants.

The toxicity of nitrate in fresh water systems is also a function of the chloride concentration. Chloride competes with nitrite for transport across the gills (Perrone and Meade 1977; Russo and Thurston 1977; Meade and Perrone 1980). Salmonid fishes can withstand high nitrite concentrations in waters with high chloride concentrations.

Regressions of chloride (Cl) concentrations and percent of 24 hour LC_{50} were compared by analysis of covariance; there was no significant difference between the regressions at the 90% confidence interval. This gave an average relation of:

$$\% \text{ of 24 hour } LC_{50} = 264 \times (Cl^-)^{0.814}$$

Since nitrite toxicity is a function of time of exposure (t), suspended sediment, and chloride anion concentration, an expression was developed to predict the LC_{50} (renormalized to eliminate the 24 hour LC_{50}):

$$LC_{50}= [0.0049 - 4.2E\text{-}5 \times [ss] + 3.22E\text{-}6 \times [ss]^2 - 2.59E\text{-}7 \times [ss]^3] \times$$

$$[9.95E\text{-}3 - 87E\text{-}4 \times \log(t)] \times [1.30E\text{-}2[Cl^-]^{0.814}$$

$$t = \text{hours of exposure } [24 \leq t \leq 168 \, \text{hours}]$$

To relate the LC_{50} to stream chemistry results, this assessment put the LC_{50} in terms of nitrate-nitrogen concentrations. In surface waters very little of the oxidized nitrogen species exist as nitrite (Lindsay 1979), and the nitrite concentration was approximated by:

$$[NO_2^-] = 0.074 \times [NO_3^-]$$

assuming a constant blood plasma pH+pe $= 14.88$ (Lindsay 1979). A wide range of possible body redox states exist. The pe+pH assumed here is conservative, but can be found in active fish in oxygen depleted situations.

The full expression for the LC_{50} in terms of an allowable nitrate-nitrogen (mg L^{-1}) uptake level is:

$$LC_{50} \text{ for} [NO_3^- - N] = [6.19E\text{-}2 - 5.33E\text{-}4 \times [ss] + 4.08E\text{-}5 \times [ss]^2$$

$$-3.29E\text{-}6 \times [ss]^3] \times [0.126 - 0.049 \times \log(t)] \times [0.164 \times Cl^-]^{0.814}$$

There are a number of assumptions involved in this expression, not the least of which is a multiplicative response to the above factors. The analysis assumes that no other factors influence the relations examined (Table 11.3). No account was taken for other anion effects (bicarbonate or sulfate) or temperature effects on anion uptake. In general, the assumptions are reasonable. Chloride and suspended sediment do not show a functional dependence on each other, and the time factor reflects the cumulative response to the listed chemical parameters. It stands to reason the response (percent of LC_{50} reached before an effect seen) is serially dependent on each of the control variables. If the LC_{50} decreases due to longer exposure duration, the effect of sediment will depress the observed LC_{50} by a specified percent.

Exposure Assessment

To adequately assess the risk stemming from a land use activity, one must be aware of the relations that exist in natural settings. This includes the influence of streamflow generation and routing processes on water quality. Evaluating the

Table 11.3 Predicted NO$_3$–N LC$_{50}$ concentrated based on flow, chloride, and suspended sediment

Exposure duration	Flow probability	Flow (m^3 s^{-1})	Cl$^-$ (mg L^{-1})	Sediment (g L^{-1})	NO$_3$–N (mg L^{-1})
		Flynn Creek			
24	0.1	0.24	2.45	1.84	31.79
24	0.01	0.60	2.16	6.76	27.80
24	0.001	1.53	1.90	24.77	10.12
72	0.1	0.19	2.53	1.32	19.76
72	0.01	0.45	2.25	4.35	17.63
72	0.001	1.04	2.01	14.36	14.14
168	0.1	0.14	2.64	0.86	10.10
168	0.01	0.30	2.38	2.50	9.18
168	0.001	0.64	2.14	7.26	8.18
		Needle Branch			
24	0.1	0.12	2.20	0.54	29.55
24	0.01	0.54	1.91	4.67	25.55
24	0.001	2.40	1.66	40.60[1]	–
72	0.1	0.09	2.27	0.34	18.22
72	0.01	0.36	1.98	2.69	16.06
72	0.001	1.53	1.73	21.24	8.90
168	0.1	0.06	2.35	0.20	9.24
168	0.01	0.23	2.07	1.38	8.26
168	0.001	0.89	1.82	9.70	7.01

[1]Sediment concentration exceeds LC$_{50}$ of 29 g L^{-1}.

dynamics of nitrate-nitrogen in forest stream waters allows one to view a host of related processes that may impact salmonid fishes.

Exposure consists of both the potential source of a contaminant as well as the species uptake mechanism (pathway). Fish ingest nitrate ions through diet, with most also able to accumulate ions through active uptake mechanisms associated with the chloride cells of the gills (Maetz 1971). Lamellar chloride cells excrete ammonium and hydrogen ions in exchange for sodium ions, and replace bicarbonate with chloride. Nitrate and nitrite ions appear to be pumped into the body fluids by the chloride cells, with uptake rates approximating chloride uptake rates. The presence of nitrate/nitrite in fresh water causes enlargement and rapid turnover of chloride cells because the fish maintains a fixed internal chloride concentration even when large amounts of nitrate are present.

Expected Nitrate-nitrogen Concentrations

To estimate an expected exposure level one must be aware of the complex feedbacks in natural systems. The first question to be answered is how nitrate concentrations vary with streamflow.

Nitrate-nitrogen concentrations in both Flynn Creek and Needle Branch were not dependent on streamflow (by comparison of correlation coefficient to t-value) (Snedecor and Cochran 1980). Nitrate-nitrogen concentrations plotted as a probability show the likelihood of exceeding a given nitrate-nitrogen concentration (Rasmussen and Rosbjerg 1989) and were expressed as regressions:

$$\text{Flynn Creek } [NO_3^-] = 0.956e^{0.276E(x)}$$

$$\text{Needle Branch } [NO_3^-] = 0.527e^{0.351E(x)}$$

Since chloride acts antagonistically to nitrite, the next factor to consider is the relation between chloride concentration and streamflow (Q). In both streams the chloride concentration were negatively correlated with stream flow at the 90% confidence level; as streamflow increased, chloride concentrations decreased. Nitrite uptake and toxicity are inversely related to chloride concentration, so the potential of nitrite uptake increased with streamflow (Q).

$$\text{Flynn Creek } [Cl^-] = 3.29Q^{-0.137}$$

$$\text{Needle Branch } [Cl^-] = 2.52Q^{-0.094}$$

Suspended sediment concentrations also vary with streamflow. Assessing sediment concentrations is difficult due to the hysteresis effect with streamflow. Sediment load to a stream is a function of the time of a storm event and the storm season. Instead of looking at the probability distribution of the sediment concentration itself, this assessment used previously derived deterministic relations (Paustian and Beschta 1979; Beschta et al. 1981; Beschta 1991). These regressions put suspended sediment concentrations in terms of a power function of flow (aQ^b). This was the weakest part of the analysis—the regressions explain only 62% of the variation seen ($n = 31$, $r^2 = 0.62$). Any error in these expressions is on the conservative side; extraordinary sediment loads have been reported in these watersheds in response to large storm events (Paustian and Beschta 1979).

$$\text{Flynn Creek } [ss] = 89.3Q^{1.41}$$

$$\text{Needle Branch } [ss] = 67.8Q^{1.44}$$

These sediment relations are the weakest link in this exposure analysis. Sediment concentrations at high flows have not been measured in the study catchments recently. Thus it is unknown if or how the specific suspended sediment transport functions have changed.

Risk Characterization

Risk is a function of both the toxicity and the exposure. The nitrate-nitrogen toxicity was described as a function of suspended sediment, chloride, and time. The exposure is catchment dependent and can only be expressed in terms of a probability of occurrence. Using the exposure relations previously derived, we generated a data base of the probability of a specific stream flow, then related this to concurrent chloride anion and sediment concentrations to calculate the resultant NO_3–N (mg L^{-1}) LC_{50} (Table 11.4). The technique compared the predicted LC_{50} values to the probability of these nitrate- nitrogen concentrations occurring in the catchment as a whole. The importance of subwatershed water quality differences is recognized.

The control watershed, Flynn Creek, had the higher nitrate-nitrogen concentrations. This normally suggests a potential for nitrate impact. But in terms of risk, this reach shows little likelihood of nitrate concentrations adversely impacting aquatic life (approximately 10^{-7}). Likewise, the probability tables generated imply a small risk to aquatic life at Needle Branch, the clearcut watershed. The nitrate-LC_{50}s at the 168 hour exposure duration are quite low, but so is the probability of a high nitrate-nitrogen concentration. Therefore the risk, the probability of exposure times the probability of toxicity (LC_{50}), is fairly low (approximately 10^{-5}).

Table 11.4 Risk assessment calculated from flow probability and LC_{50} exceedance probability

Exposure duration	Flow probability	LC_{50}	LC_{50} Exceedance probability	Risk assessment
		Flynn Creek		
24	0.1	31.79	7.91E-07	7.91E-08
24	0.01	27.80	1.42E-06	1.42E-08
24	0.001	10.12	1.16E-04	1.16E-07
72	0.1	19.76	6.29E-06	6.29E-07
72	0.01	17.63	1.03E-05	1.03E-07
72	0.001	14.14	2.71E-05	2.71E-08
168	0.1	10.10	1.18E-04	1.18E-05
168	0.01	9.18	1.79E-04	1.79E-06
168	0.001	8.18	2.95E-04	2.95E-07
		Needle Branch		
24	0.1	29.44	2.56E-07	2.56E-08
24	0.01	25.55	4.17E-07	4.17E-09
24	0.001	–	1.00E+00	1.00E-03
72	0.1	18.22	1.33E-06	1.33E-07
72	0.01	16.06	2.05E-06	2.05E-08
72	0.001	8.90	1.55E-05	1.55E-08
168	0.1	9.24	1.36E-05	1.36E-06
168	0.01	8.26	2.00E-05	2.00E-07

Summary

The undisturbed catchment, Flynn Creek, had higher nitrate-nitrogen concentrations than the treated catchment. The mature to over-mature red alder stands (alder is a nitrogen-fixing species), coupled with streamflow generation and routing mechanisms, may be the source of this higher nitrate-nitrogen level. The high nitrate-nitrogen concentrations suggest a potential impact, but this catchment shows little likelihood of risk to coho salmon.

The treatment catchment, Needle Branch, had lower nitrate-nitrogen concentrations and higher suspended sediment concentrations resulting in a higher risk to salmonid populations because of the exposure equation sensitivity to suspended sediment concentrations.

The toxicity relations developed for this study were largely from the existing literature. Certain limitations must be recognized. These include the uncertainties in: suspended sediment (nitrite toxicology; nitrate/nitrite-nitrogen equilibria in salmonid fish populations; and current dynamics of sediment transport as related to streamflow for these catchments). This initial approach holds promise for the evaluation of risk, as a function of toxicity and exposure, for water qualities affected by land use activities through nonpoint source pollution. A more complete risk assessment for salmonids would include: dissolved oxygen concentrations, water temperatures, fish habitat as a function of streamflow, channel geomorphology, large woody debris, and other potential control variables.

The goal of this work was to construct an assessment framework that was quantifiable, transferable to land use managers, based on reasonable database requirements, and keyed on long-term forest sustainability. Rather than build a model based entirely on phenomenological data, this study evaluated cumulative watershed effects in terms of the probability of an occurrence, linked to expected basin hydrology and water quality. The effort used the combined assessment to evaluate the risk silvicultural and associated land use activities pose to existing forest resources.

Acknowledgment The ecological risk assessment application developed here was largely the effort of Tim Kern as part of his doctoral research. John Stednick wishes to acknowledge Tim's work here and for his other contributions through data collection and interpretation for several other aspects of the New Alsea Watershed Study.

Literature Cited

Beschta, R.L. 1991. Sediment studies in the Alsea watershed, pp. 64–81. In: The New Alsea Watershed Study. NCASI Technical Bulletin No. 602.

Beschta, R.L., O'Leary, S.J., Edwards, R.E., and Knoop, K.D. 1981. Sediment and organic matter transport in Oregon Coast Range streams. WRRI-70. Water Resources Research Institute, Oregon State Univ., Corvallis, OR. 67pp.

Bjornn, T.C., and Reiser, D.W. 1991. Habitat requirements of salmonids in streams, pp. 83–138. In: W.R. Meehan, editor. Influences of Forest and Rangeland Management on Salmonid Fishes and their Habitats. Amer. Fish. Soc. Spec. Publ. 19.

Bormann, F.H., and Likens, G.E. 1967. Nutrient cycling. Science 155:424–429.

Bottom, D.L., Howell, P.J., and Rodgers, J.D. 1985. The effects of stream habitat alterations on salmon and trout habitat in Oregon. Oregon Dept. of Fish and Wildlife, Portland, OR. 70pp.

Boudou, A., and Ribeyre, F. 1989. Aquatic Ecotoxicology: Fundamental Concepts and Methodologies Vol. II. CRC Press, Inc., Boca Raton, FL.

Brown, G.W. 1985. Forestry and Water Quality, 2nd ed. Oregon State Univ. Book Stores, Inc., Corvallis, OR. 173pp.

Brown, G.W., Gahler, A.R., and Marston, R.B. 1973. Nutrient losses after clear-cut logging and slash burning in the Oregon Coast Range. Water Resour. Res. 9:1450–1453.

Brown, G.W., and Krygier, J.T. 1971. Clearcut logging and sediment production in the Oregon Coast Range. Water Resour. Res. 7:1189–1198.

Driscoll, C.T., Likens, G.E., Hedin, L.O., Eaton, J.S., and coauthors. 1989. Changes in the chemistry of surface waters. Env. Sci. and Tech. 23:137–143.

Fredriksen, R.L. 1971. Comparative chemical water quality – natural and disturbed streams following logging and slash burning, pp. 125–137. In: J.T. Krygier and J.D. Hall, editors. Proceedings of a symposium: Forest Land Uses and Stream Environment. Oregon State Univ., Corvallis, OR.

Harr, R.D. 1976. Forest practices and streamflow in western Oregon. General Technical Report GTR-PNW-49. USDA Forest Service, Pacific Northwest Forest and Range Experiment Station, Portland, OR. 18pp.

Harr, R.D. 1979. Effects of timber harvest on streamflow in the rain-dominated portion of the Pacific Northwest. In: Proceedings of Workshop on Scheduling Timber Harvest for Hydrologic Concerns. USDA Forest Service, Pacific Northwest Forest and Range Experiment Station, Portland, OR. 45pp.

Harr, R.D. 1991. Physical, chemical, and hydrologic responses of the Alsea watersheds to forest management activities, pp. 64–81. In: The New Alsea Watershed Study. NCASI Technical Bulletin No. 602.

Harr, R.D., and Krygier, J.T. 1972. Clearcut logging and low flows in Oregon coastal watersheds. Forest Research Lab. Research Note 54. Oregon State Univ., Corvallis, OR. 3pp.

Harris, D.D. 1977. Hydrologic changes after logging in two small Oregon coastal watersheds. Water-Supply Paper 2037. U.S. Geological Survey Washington, DC. 31pp.

Hicks, B.J., Hall, J.D., Bisson, P.A., and Sedell, J.R. 1991. Responses of salmonids to habitat changes, pp. 483–518. In: W.R. Meehan, editor. Influences of Forest and Rangeland Management on Salmonid Fishes and their Habitats. Amer. Fish. Soc. Spec. Publ. 19.

Huey, D.W., Simco, B.A., and Criswell, D.W. 1980. Nitrite-induced methemoglobin formation. Trans. Amer. Fish. Soc. 109:558–562.

Ice, G.G. 1991. Significance of the Alsea watershed studies to development of forest practice rules, pp. 22–28. In: The New Alsea Watershed Study. Technical Bull. 602. Natl. Council of the Paper Industry for Air and Stream Improvement, New York.

Lewis, W.M., Jr., and Morris, D.P. 1986. Toxicity of nitrite to fish: a review. Trans. Amer. Fish. Soc. 115:183–195.

Lindsay, W.L. 1979. Chemical Equilibria in Soils. John Wiley & Sons, New York, NY. 450pp.

Maetz, J. 1971. Fish gills: mechanisms of salt transfer in fresh water and sea water. Philosophical Transactions of the Royal Society of London, Series B, Biological Sciences 262:209–249.

Meade, T.L., and Perrone, S.J. 1980. Effect of chloride ion concentrations and pH on the transport of nitrite across the gill epithelia of coho salmon. Prog. Fish-Cult. 42:71–72.

Moring, J.R., and Lantz, R.L. 1975. The Alsea Watershed Study: effects of logging on the aquatic resources of three headwater streams of the Alsea River, Oregon. Part I. Biological studies. Fish. Res. Rep. 9. Oregon Dept. of Fish and Wildlife, Corvallis, OR. 66pp.

Paustian, S.J., and Beschta, R.L. 1979. The suspended sediment regime of an Oregon Coast Range stream. Water Resour. Bull. 15:144–154.

Perrone, S.J., and Meade, T.L. 1977. Protective effects of chloride on nitrite toxicity to coho salmon (*Oncorhynchus kisutch*). J. Fish. Res. Board Can. 34:486–492.

Rasmussen, P.F., and Rosbjerg, D. 1989. Risk estimation in partial duration series. Water Resour. Res. 25:2319–2330.

Russo, R.C., Smith, C.E., and Thurston, R.V. 1974. Acute toxicity of nitrite to rainbow trout (*Salmo gairdneri*). J. Fish. Res. Board Can. 31:1653–1655.

Russo, R.C., and Thurston, R.V. 1977. The acute toxicity of nitrite to fishes, pp. 387–393. In: Recent Advances in Fish Toxicology. Ecological Research Series EPA-600/3-77-085. U.S. Environmental Protection Agency.

Servizi, J.A., and Gordon, R.W. 1990. Acute toxicity of ammonia and suspended sediment mixtures to chinook salmon (*Oncorhynchus tshawytscha*). Bull. Environ. Contam. Toxicol. 44:650–656.

Snedecor, G.W., and Cochran, W.G. 1980. Statistical Methods, 7th Ed. Iowa State Univ. Press, Ames, IA. 507pp.

Stednick, J.D. 1980. Alaska water quality standards and BMPs, pp. 721–730. In: Proceedings of Watershed Management Symposium. Amer. Soc. of Civil Eng., Boise, ID.

Stednick, J.D., and Kern, T.J. 1992. Long term effects of timber harvesting in the Oregon Coast Range: The New Alsea Watershed Study (NAWS), pp. 502–510. In: Interdisciplinary Approaches to Hydrology and Hydrogeology. American Institute of Hydrology, Smyrna, GA.

Stednick, J.D., and Kern, T.J. 1994. Risk assessment for salmon from water quality changes following timber harvesting, pp. 227–238 In: Environmental Monitoring and Assessment, Volume 32, No. 3.

U.S. Environmental Protection Agency. 1976. Quality Criteria for Water, Washington, DC. 534pp.

U.S. Environmental Protection Agency. 1992. Framework for Ecological Risk Assessment. Risk Assessment Forum. EPA/630/R-92/001. 41pp.

Westin, D.T. 1974. Nitrate and nitrite toxicity to salmonid fishes. Prog. Fish-Cult. 36:86–89.

Chapter 12
Sedimentation Studies Following the Alsea Watershed Study

Robert L. Beschta and William L. Jackson

One of the major objectives of the 1958–73 Alsea Watershed Study (AWS) was to quantify the effects of forest practices on suspended sediment concentrations and watershed sediment production using a paired watershed approach. However, the AWS study design did not enable direct interpretations of the effects of forest practices on the specific processes that influence erosion rates or sediment transport. Thus, interpreting the results of those studies created additional questions and required development of new hypotheses about processes controlling watershed erosion and stream channel responses to increased sediment delivery. These questions and hypotheses, in turn, formed the basis for additional sedimentation research in the Alsea Watersheds.

Sedimentation research following the AWS focused on documenting, evaluating, and understanding basic instream sedimentation processes. However, the extensive suspended sediment data sets that had been developed as part of the AWS permitted important advances in the modeling of suspended sediment transport dynamics during periods of high stream discharge. There was an additional emphasis of post-AWS sediment studies on the bedload transport dynamics in small streams and on the interactions between instream sedimentation processes, bed material composition, and channel morphology.

Sedimentation research following the AWS was implemented by the Department of Forest Engineering at Oregon State University (OSU) on a project-by-project basis as funding became available. Funding support and cooperators included the Forest Research Laboratory and Water Resources Research Institute at OSU, the National Council of the Paper Industry for Air and Stream Improvement (NCASI), and the USDA Forest Service. The sediment studies reported herein occurred between the mid-1970s and early 1980s.

Robert L. Beschta
Department of Forest Engineering, Oregon State University, Corvallis, OR 97331
robert.beschta@oregonstate.edu

J. D. Stednick (ed.), *Hydrological and Biological Responses to Forest Practices.*   183
© Springer 2008

Sedimentation Research Objectives

Interpreting suspended sediment data during the AWS was complicated by an inability to develop tight statistical relationships between discharge and suspended sediment concentrations. High variability in sediment rating curves was common and thus efforts were undertaken to isolate and understand potential causes of this variability. For example, Brown and Krygier (1971) noted that rising-limb sediment concentrations tended to be substantially higher than falling-limb concentrations at equivalent discharges. Later, Beschta (1978) found the occurrence of seasonal patterns in the relationship between discharge and sediment concentrations. These results suggested that instream sediment supply was interacting with streamflow levels, such that within-storm and seasonal sediment concentrations were being affected. Thus, one of the objectives of the post-AWS research was to develop and incorporate sediment supply relationships into the development of models for suspended sediment transport.

A second question stemming from the original AWS related to the importance of bedload transport as a contributor to watershed sediment yields and the role of bedload transport as a factor affecting channel morphology and fish habitat. Harris (1977) estimated that less than 5% of the total watershed sediment yield occurred as bedload. However, increased rates of mass erosion following road building (Beschta 1978) meant that logging-related activities could potentially deliver substantial amounts of coarse grained sediments to streams that would be available for transport as bedload. Even on the undisturbed (i.e., unharvested) control watershed, substantial amounts of sediment were delivered to streams during relatively large storms. For example, 36% of the total 15 year export of suspended sediment from Flynn Creek (control watershed) occurred over a total of six days from the two largest storms (Beschta 1978). These large storms as well as other smaller events resulted in stream discharges capable of transporting the entire range of sediment sizes comprising the streambed. Thus, a second objective of post-AWS sediment research was to provide better quantification of bedload transport processes during periods of storm discharge and to better describe interrelationships between suspended sediment, bedload sediment, and organic matter transport processes.

A third question following the AWS focused on the response of stream channels and aquatic habitats to sediment increases associated with roading and harvesting practices. There was increasing concern in the fisheries literature that accelerated sediment delivery to streams could alter instream habitats or increase the percentage of fine sediments in the interstitial spaces of spawning gravels, with corresponding adverse effects on spawning success (see literature reviews by Cordone and Kelley 1961; Gibbons and Salo 1973; Sorenson et al. 1977; Iwamoto et al. 1978). Furthermore, with the observed increases in mass erosion processes following road-building and the associated increases in delivery of bed-sized sediments to streams, there were questions concerning the response of stream channel morphology (e.g., the filling of pools) to these additional sediment loads. Thus, a third objective of the post-AWS sediment research was to better document interactions between instream sediment transport, channel morphology, and substrate composition.

Other sedimentation-related questions and issues also originated from the AWS, especially those concerning the influence of road-building, logging, and fire on upland soil properties (e.g., soil shear strength, bulk density, and infiltration capacity) and hillslope erosion processes (both surface erosion and mass soil movements). Subsequent to the AWS, sedimentation research on these experimental watersheds largely focused on instream processes whereas studies of upland erosion processes generally were pursued in other forested watersheds of the Pacific Northwest.

Suspended Sediment Transport

Since high soil infiltration capacities are characteristic of forested watersheds, little surface runoff occurs even during large storms. For Flynn Creek, the undisturbed control watershed of the AWS, inspection of historical aerial photographs of watershed did not show recent evidence of mass failures, although field inspection indicated that small failures (not visible through the forest canopy) had occurred on steep slopes near channels and inner gorges. Thus, it appears that stream channels and the near-stream environment comprised the primary source of sediments available for suspended load transport during storm events on this watershed.

The first post-AWS sedimentation study utilized frequent sampling (either manually or with pumping samplers) at Flynn Creek and Oak Creek (another Coast Range stream but outside the Alsea Watersheds) to obtain information regarding instream turbidities and suspended sediment transport. Sediment sources and changes in channel morphology were also evaluated (Paustian and Beschta 1979). Temporal patterns of turbidity and suspended sediment concentration during storms were identified that were more intensively evaluated in later studies. Results indicated major differences in the amount of fine sediments stored in the stream gravels of these two streams. Furthermore, while local changes in channel morphology (measured at channel cross-sections) were common during individual storms, little net aggradation or degradation occurred along specific stream reaches. These efforts provided an important basis for follow-up studies that continued to address questions related to prediction and modeling of suspended sediment transport in mountain streams.

Although suspended sediment records from the AWS had previously been summarized by Harris (1973, 1977) and Beschta (1978), additional analyses of these data were undertaken by Van Sickle (1981). Results indicated that annual sediment yields from small mountainous watersheds, such as those of the Oregon Coast Range, approximated a log-normal distribution. The log-normal property appears to be a consequence of several empirical characterizations of sediment transport in small streams: (1) most of the total load is carried during high runoff events, (2) frequency curves of streamflow have a pronounced positive skew, and (3) sediment yields are typically related to streamflow by a power function.

Perhaps the most common approach for characterizing suspended sediment transport in streams is to establish an empirical relationship between suspended sediment concentrations (C, mg L^{-1}) and the corresponding stream discharge (Q, in m^3 s^{-1}). Rating curves of stream discharge and suspended sediment generally take the form of a power function:

$$C = aQ^b$$

where "a" and "b" are coefficients determined by the linear regression between "log C" and "log Q."

In the case of data from the AWS, log-log plots of C vs. Q demonstrated an overall positive correlation, but exhibited considerable scatter about the best-fit regression lines. As a consequence of this variability, several patterns of suspended sediment concentration were apparent when the data were examined in a temporal context (i.e., as part of a time series). First, suspended sediment concentrations during a given storm event were typically higher on the rising limb of a hydrograph than those measured at equivalent flows on the falling limb (Brown and Krygier 1971; Paustian and Beschta 1979). As a result, consecutive data points during a given storm typically resulted in a characteristic hysteresis (Fig. 12.1). Secondly, suspended sediment concentrations during the rising limb of the hydrograph generally were positively correlated with the rate of discharge increase (Beschta 1987). Thus, at a given discharge steeply rising limbs of a storm hydrograph tended to have higher suspended sediment concentrations than rising limbs exhibiting a slower increase in flows (Fig. 12.2). Finally, seasonal patterns in the relationship between suspended sediment concentration and discharge were observed (Beschta 1978). Fall and early-winter storms in the Oregon Coast Range prior to the occurrence of the annual peak flow tended to produce higher suspended sediment concentrations for a given discharge than what occurred for similar streamflow levels during late-winter or springtime storms (Fig. 12.3). Within these seasonal trends, it was further noted that sediment concentrations were typically lower for runoff events that had been preceded by a relatively large peak discharge.

All of the observed temporal patterns in suspended sediment concentration indicated sediment supply was an important controlling factor affecting the concentration of suspended sediment observed at a given discharge (Beschta 1978, 1987; Van Sickle and Beschta 1983). Furthermore, these patterns were apparent in the data for Flynn Creek (the "control" watershed) as well as for the two treated watersheds.

Fig. 12.1 Suspended
sediment rating curves from
six consecutive storm events
at Flynn Creek during water
year 1979. Numbers indicate
chronological sequence of
storms (1 = 1st storm, etc.);
broken line with arrows
highlights a hysteresis loop
for data from the largest
storm (Beschta et al. 1981b)

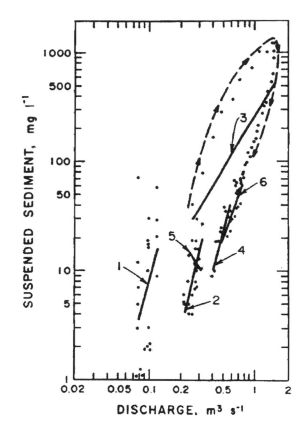

    The large amount of scatter about a traditional suspended sediment rating
curve indicates that the use of instream suspended sediment sampling as a
basis for statistically detecting relatively small changes in sediment transport
represents a difficult monitoring challenge. If instream suspended sediment
sampling is to form the basis for monitoring the effects of forest practices on
watershed erosion and water quality, it is clear that more precise suspended
sediment transport models are needed. Reassessment of suspended sediment
data from the AWS strongly pointed toward the need to incorporate a
sediment supply function in traditional discharge-driven sediment rating
curves. If supply-based models could successfully be developed, it might be
possible to better understand and evaluate the effects of forest practices
(roading, logging, fire, etc.) on instream sediment levels and to focus
future sediment monitoring research on understanding the dynamics of
sediment supplies.
    Van Sickle and Beschta (1983) proposed a structure for supply-based models
of suspended sediment transport in streams based upon a reanalysis of the
Flynn Creek data sets that had been developed during and after the AWS.
Their initial model incorporated a lumped sediment storage variable (S) that

Fig. 12.2 (a) Hydrographs for two large runoff events at Flynn Creek and (b) corresponding suspended sediment concentrations measured during their rising limbs. Over a 15-year monitoring period, these were the largest peak discharges of record (Beschta 1987). Reproduced from Beschta 1987, with permission from John Wiley & Sons, Inc

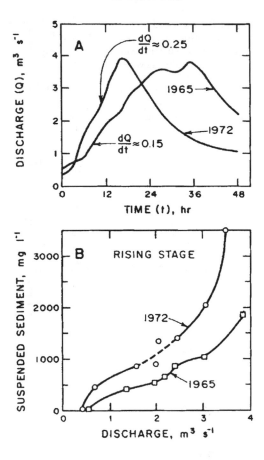

varied with time (t). During a storm event, suspended sediment concentration was modeled with a sediment rating curve modified to include a supply depletion function $g[S(t)]$:

$$C(t) = aQ(t)^b \cdot g[S(t)]$$

Van Sickle and Beschta (1983) found $g[S(t)]$ to be of exponential form and chose:

$$g[S(t)] = p \cdot \exp[rS(t)/S_0]$$

where S_0 represents the total available sediment. The complete model is:

$$dS(t) = -Q(t) \cdot C(t)$$

so that:

$$C(t) = aQ(t)^b \cdot p \cdot \exp[rS(t)/S_0]$$

Fig. 12.3 Regression relationships between monthly stream discharge and monthly sediment concentration for the Alsea Watersheds during the pretreatment period (Water Years 1959–1965). Solid lines show fall and winter months before and including the annual peak discharge whereas dashed lines show winter and spring months following the occurrence of the annual peak discharge (Beschta 1978). Reproduced from Beschta 1978, with permission from the American Geophysical Union

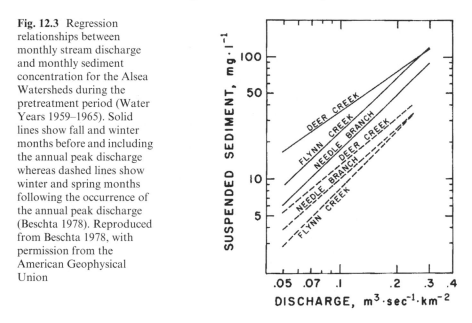

where a, b, p, and r are empirically derived parameters. For the suspended sediment data at Flynn Creek, this approach was shown to model observed suspended sediment concentrations much more closely than a sediment rating curve.

In an attempt to more realistically represent sediment storage and availability to a stream system during high flows, a "distributed" model was also developed whereby the total supply of stored sediment was distributed among several storage compartments (Fig. 12.4). In the distributed model, flow magnitude determines which compartments are accessible for the removal of sediment by the stream. During periods of rising streamflow, an increasing number of storage compartments are progressively accessed by the stream and the sediment stored within them is depleted over time. While both models indicate sediment concentrations increase more rapidly for storm hydrographs with relatively steep rising limbs, modeled sediment concentrations increase more rapidly with the distributed model than for the single compartment model. After streamflow has peaked and is into recession, the accessed storage compartments will have been partially depleted. The decreasing rate of sediment depletion from each compartment in combination with fewer and fewer storage compartments being available during receding flows indicates that modeled suspended sediment concentrations decrease rapidly following a hydrograph peak. Such patterns of suspended sediment concentrations are readily observed during storm hydrographs.

Both the lumped and distributed models were able to replicate single-storm hysteresis patterns in suspended sediment concentration, as well as seasonal

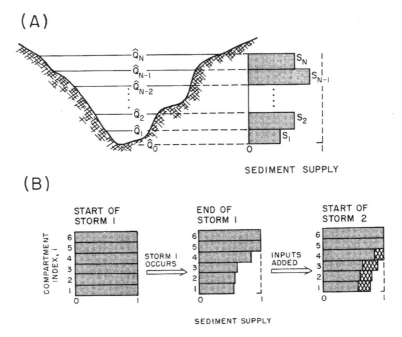

Fig. 12.4 Schematic of distributed supply model: (a) sediment supply (S) as defined by stream discharge (Q) and (b) modeled sediment supply before and after storm 1, 1977, at Flynn Creek (Van Sickle and Beschta 1983). Reproduced from Beschta 1978, with permission from the American Geophysical Union

patterns (Van Sickle and Beschta 1983). If supply-based sediment models can be calibrated to streams during a pretreatment period, it may be possible to more accurately determine the effects of forest practices on suspended sediment transport in the posttreatment period. However, the sensitivity of the sediment storage and depletion coefficients to changes in sediment availability from natural and anthropogenic factors needs additional study if such models are to be used for developing effective monitoring programs.

Bedload Sediment Transport

The absence of widely accepted bedload sampling protocols and types of samplers, as well as the overall difficulty in measuring bedload transport in natural streams, has resulted in relatively few data sets of bedload sediment transport from mountain watersheds. While some data sets are available (e.g., Klingeman 1971; Hayward and Sutherland 1974; Emmett 1976; Emmett et al. 1980), most estimates of bedload transport are derived by applying one of many available bedload transport prediction equations (Vanoni 1975; Stelczer 1981).

Bedload transport equations typically are structured around basic hydraulic theory and empirically calibrated with transport data generated in controlled flume experiments. Most equations have been developed utilizing channels of uniform dimensions and over limited ranges of sediment particle sizes and flows. Thus, it is not surprising that there can be several orders of magnitude variation in estimated bedload transport rates generated by the application of alternative transport equations to a given natural channel.

Results from the AWS indicated that bedload transport was a relatively insignificant contributor to overall sediment yields from Coast Range streams. This conclusion was largely based upon observations of bed sediment accumulation in weir pools at each of the streamflow gauging stations. Harris (1973, 1977) estimated that less than 5% of the total sediment yield in the AWS occurred as bedload transport. However, there was no way of knowing the sediment-trapping efficiency of the weir pools, especially during large storms when relatively high rates of bedload transport occurred. Thus, considerable effort was expended during the post-AWS period characterizing bedload transport phenomena. The emphasis of this research was on Flynn Creek (the undisturbed control watershed) because it was believed that an improved understanding of basic bedload transport processes in undisturbed watersheds was required prior to investigating the potentially more complicated effects of forest practices on bedload transport processes.

Bedload Transport Sampling Methods

Two primary methods for sampling bedload sediment in transport were used in Flynn Creek and detailed descriptions of these methods are provided in Beschta et al. (1981b) and O'Leary and Beschta (1981). A vortex-tube bedload sampler, designed after those utilized by Milhous and Klingeman (1973) and Hayward and Sutherland (1974), was installed across the bottom of a pre-existing fish trap that had been constructed in Flynn Creek more than a decade previously (Fig. 12.5). The fish trap, which was used during the AWS to monitor upstream and downstream movements of anadromous fish (Moring and Lantz 1975), provided a rectangular-shaped channel cross-section. Installation of the vortex sampler on the bottom of the fish trap permitted bedload sampling across the entire width of flow.

Based upon bedload transport measurements immediately upstream and downstream of the vortex tube with a Helley-Smith bedload sampler (Helley and Smith 1971), the sampling efficiency of the vortex sampler for particles <10 mm in diameter was found to be poor (Beschta et al. 1981b). Since much of the bedload transport in Flynn Creek is comprised of particles in the sand-size range (i.e., <2 mm), sampling efficiency represented an obvious limitation of the vortex sampler in this Coast Range stream.

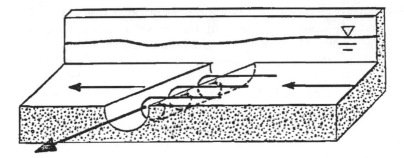

Fig. 12.5 Schematic of vortex-tube bedload sampler at Flynn Creek that extended across the entire width of channel. The water and sediment mixture captured by the vortex tube was discharged into an off-channel depression where bedload sediments were separated from the flow (O'Leary and Beschta 1981)

During WYs 1977–78, bedload measurements with the Helley-Smith sampler (7.6 cm square) orifice used a triangular-shaped, 0.2 mm mesh catch bag with a surface area of 1,950 cm^2. Use of the sampler (Fig. 12.6) required obtaining several subsamples at equally spaced locations across the width of a channel. Thus, a single sample of bedload sediment transport was obtained by combining these subsamples. Because the sampler at any given location only measures bedload transport across a small portion of the total channel width, several potential sources of sampling error were inherent in its use (e.g., spatial and temporal variations in bedload transport along the streambed). Furthermore, sampler efficiency was found to vary significantly over time as the mesh began to plug with sediment (Beschta et al. 1981b). Thus, Helley-Smith samplers used at Flynn Creek were modified by attaching a larger catch bag. This larger bag considerably increased the surface area of porous mesh thus allowing water to flow from the bag without causing the mesh to become plugged. Starting in WY 1979, the sampler was equipped with a longer, cylindrically-shaped catch bag with a bag surface area of 6,000 cm^2 (Beschta et al. 1981b).

Because of the Helley-Smith sampler's ease of use, it quickly became the preferred method for quantifying bedload transport in Flynn Creek. The sampler was well adapted for use on the concrete bed of the fish trap where it was possible to avoid problems of bed-scooping that can occur when used in natural channels. Furthermore, since the sampler was portable it could easily be used at various locations along a stream and it provided a unique means of sampling coarse particulate organic matter that was being transported near the streambed.

Bedload samples obtained during field studies (using either the vortex or Helley-Smith sampler) were subsequently analyzed at OSU's Forest Research Laboratory. Samples were oven-dried at 105°C for 24 hours and weighed. They were then placed in a high-temperature oven at 550°C for 24 hours to oxidize organic matter and re-weighed. The particle-size distribution of individual samples was determined by dry sieving.

Fig. 12.6 Helley-Smith
bedload sampler comparing
the "standard" sized catch
bag and an enlarged catch
bag. Because of sampler
inefficiencies associated
with the standard bag
(Beschta 1981), the larger
bag permitted more
accurate sampling of
bedload transport

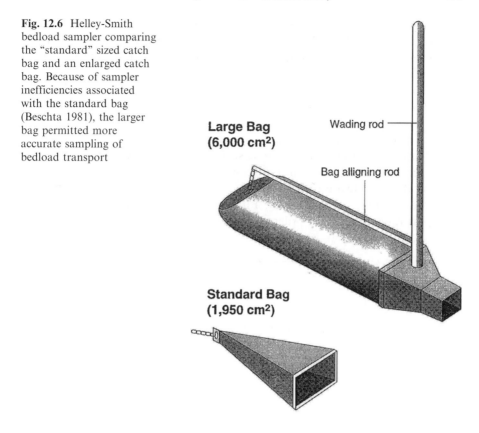

Flynn Creek Bedload Transport Rates

Bedload transport in mountain streams normally occurs during periods of storm discharge. For the 2.02 km^2 watershed Flynn Creek, a discharge of 0.7 m^3 s^{-1} (0.3 m^3 s^{-1} km^{-2}) seemed to represent an approximate "threshold" for measurable bedload transport. This flow represents roughly one-half the average annual peak flow. Bedload transport rates dramatically increased for discharges approaching 1.3 m^3 s^{-1} (0.6 m^3 s^{-1} km^{-2}). This flow corresponds to approximately a 1.3-year recurrence-interval peak discharge and generally represents a "bankfull" condition for Flynn Creek. At these larger discharges, most sediment particle sizes that occur in the streambed are being transported. Since Flynn Creek is underlain by sedimentary rocks comprised mainly of loosely cemented siltstones and sandstones, bedload samples were typically comprised of predominantly sand-sized particles.

During the WYs 1977–1980 when bedload transport studies were conducted at Flynn Creek, significant periods of transport (i.e., >100 kg hr^{-1}) occurred approximately 1% of the time, or about 3.5 days per year. Most of this transport occurred during three storms in WY 1978 although smaller storms in the WYs 1979 and 1980 also resulted in measurable bedload transport. There was essentially no bedload transport during WY 1977, which was one of the driest years on record (Fig. 12.7).

Bedload sediment transport patterns for the December 13–15, 1977, period (using vortex-tube samples) and for the storm of February 6–8, 1979 (using Helley-Smith samples), are illustrated in Figs. 12.8 and 12.9, respectively, for Flynn Creek. Figure 12.9 also illustrates particulate organic matter transport. These data indicate a substantial increase in bedload transport occurs as stream discharge approaches bankfull conditions.

Several other features of bedload transport in mountain streams are illustrated in Figs. 12.8 and 12.9. First, short-term fluctuations in bedload transport rates of up to an order of magnitude were common during periods

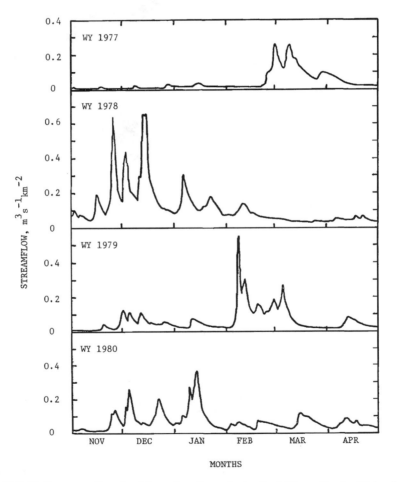

Fig. 12.7 Daily streamflow for November through April at Flynn Creek, Water Years 1977–1980 (Beschta et al. 1981b)

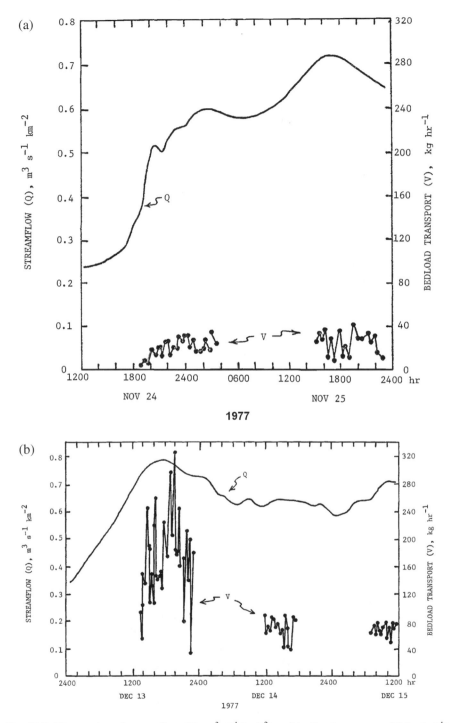

Fig. 12.8 Time series of streamflow (Q, m³ s⁻¹ km⁻²) and bedload transport (V, kg hr⁻¹) measured with the vortex-tube sampler for two storms at Flynn Creek (Beschta et al. 1981b)

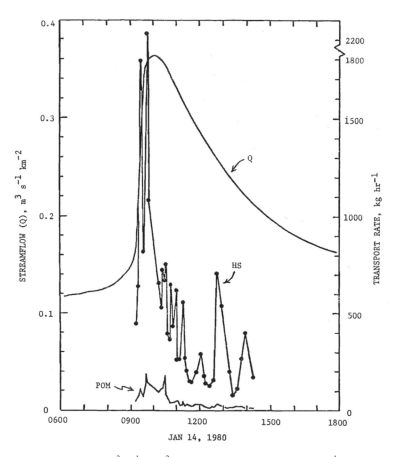

Fig. 12.9 Streamflow (Q, m^3 s^{-1} km^{-2}), bedload transport (HS, kg hr^{-1}) and particulate organic matter transport (POM, kg hr^{-1}) for the February 6–8, 1979, storm at Flynn Creek (Beschta et al. 1981b). Both bedload and particulate organic matter transport samples were obtained using a Helley-Smith bedload sampler with an enlarged sample bag; see Fig. 12.6)

of high sediment transport. While some of this temporal variability may be associated with sampling error, O'Leary and Beschta (1981) concluded that the data were indicative of real-time fluctuations in transport rates. Studies in other small mountain streams have similarly indicated that high-magnitude, short-term fluctuations are a common feature of bedload transport (e.g., Hayward and Sutherland 1974; Hayward 1980; Beschta 1983; Hubbell 1987).

Secondly, the largest bedload transport rates often occurred after the stream discharge peak. Even though bedload transport rates ultimately declined during the recession limb of the storm hydrograph, for a given flow they tended to be higher on the recession limb than on the rising limb of the hydrograph. This "reverse hysteresis" is in sharp contrast to suspended sediment transport patterns but is consistent with bedload transport theory which suggests that greater forces are required to initiate bed sediment motion than are required to keep sediment in transport following incipient motion. Since storm hydrographs for Coast Range streams typically decline much more slowly than they rise, a generally common feature for streams in the Pacific Northwest and elsewhere, most bedload transport appears to occur during periods of stormflow recession.

Bedload transport rating curves for Flynn Creek, representing two types of bedload samplers and a range of runoff events, are summarized in Table 12.1 (Beschta et al. 1981b). Except for WY 1980, the exponent of the annual rating curves ranged from 3.4 to 4.5 and averaged 3.8 for all samples. These data indicate that a doubling in streamflow results in approximately a 14-fold increase in bedload transport. Such a rapid change in bedload transport with increasing flow emphasizes the importance of large, infrequent runoff events in the routing of coarse sediments down mountain streams.

An upwards shift in bedload rating curves occurred between WYs 1978 and 1979. This shift is thought to be a result of increasing the size of the catch bag attached to the Helley-Smith sampler, hence increasing its sampling efficiency (Beschta et al. 1981b). The negative slope of the relationship between streamflow and bedload transport for the WY 1980 bedload rating curve is an anomaly and is thought to stem from the fact that samples were only obtained over a small range of stream discharges during the recession limb of a hydrograph.

Table 12.1 Summary of regression equations relating bedload transport (kg hr^{-1}) and total suspended solids concentrations (mg L^{-1}) to streamflow (m^3 s^{-1} km^{-2}), Flynn Creek Watershed (Beschta et al. 1981b).

Years	Range of Water discharges (Q, m^3 s^{-1}km^{-2})	n	Regression equations[#]	r^2
Bedload				
1978	0.28–0.80	187	$VT = 442\ Q^{4.51}$	0.56
1978	0.28–0.80	24	$HS_1 = 446\ Q^{3.87}$	0.36
1979	0.45–0.75	86	$VT = 376\ Q^{3.41}$	0.72
1979[##]	0.45–0.75	157	$HS_2 = 800\ Q^{3.41}$	0.92
1980[##]	0.12–0.32	114	$HS_2 = 1.2\ Q^{-1.32}$	0.62
Suspended Solids				
1978	0.28–0.80	82	$TSS = 188\ Q^{1.33}$	0.16
1979	0.45–0.75	118	$TSS = 424\ Q^{1.56}$	0.65
1980	0.12–0.32	145	$TSS = 28\ Q^{-0.11}$	0.01

[#] Q = stream discharge, m^3 s^{-1} km^{-2}

VT = bedload transport using vortex-tube sampler, kg hr^{-1}

HS^1 = bedload transport using Helley-Smith bedload sampler with 0.2 mm mesh and 1,950 cm^2 surface-area catch bag, kg hr^{-1}

HS^2 = bedload transport using Helley-Smith bedload sampler with 0.2 mm mesh and 6,000 cm^2 surface-area catch bag, kg hr^{-1}

TSS = total suspended solids (primarily inorganics) based on samples obtained with pumping samplers, mg L^{-1}

[##] The equations for 1979 and 1980, originally reported in Beschta et al. (1981b), were in error. These are the correct equations.

The median particle size for inorganic sediments collected with the Helley-Smith sampler in Flynn Creek ranged from 0.3 to 1.8 mm and averaged 0.5 mm, with maximum sizes approaching 10 mm in diameter. Median particle sizes tended to be relatively insensitive to discharge.

It was not the intent of the post-AWS bedload studies to determine the percentage of annual sediment yield contributed by bedload. However, the application of sediment rating curves (both suspended sediment and bedload) to the February peak flow event (Fig. 12.9) indicated that bedload and suspended load comprised approximately 25% and 75%, respectively, of the total inorganic export during that storm. For larger storms, the percentage attributed to bedload would likely increase. Thus, the proportion of the total inorganic load in transport as bedload during storm hydrographs can be considerably greater than the annual percentages (i.e., 1–4%) indicated by Harris (1977).

Particulate organic matter (POM) transport is rarely reported in the scientific literature. However, use of the Helley-Smith bedload sampler for inorganic sediments also provided an opportunity to characterize POM transport rates. For Flynn Creek, which drains a forested watershed dominated by Douglas-fir (*Pseudotsuga menziesii*), western hemlock (*Tsuga heterophylla*), and red alder (*Alnus rubra*) forests, POM samples consisted mostly of partially decomposed needles, leaves, cones, barks, twigs, and other small pieces of woody debris and represents an important component of the carbon export budget.

POM transport during the runoff event of February 6–8, 1979 (Fig. 12.9) suggests a pattern of fluctuation that is relatively similar to those associated with suspended sediment concentrations whereby transport rates reach their maximum prior to or during the peak in stream discharge and then rapidly decline (Beschta et al. 1981b). In general, POM transport dynamics tended to exhibit additional patterns similar to that of suspended sediment: (1) transport rates on the ascending limb of the hydrograph being greater than at similar discharges during the falling limb (hysteresis) and (2) seasonal flushing whereby early fall storms have higher amounts of POM in transport than do storms later in the winter or spring. It is not known if, or to what extent, POM transport rates might influence suspended sediment or bedload transport rates.

Sediment Transport – Channel Morphology Interactions

The patterns of bedload sediment transport obtained from frequent sampling during storm hydrographs at Flynn Creek during WYs 1977–80 were significant for several reasons. The highly unsteady nature of bedload sediment transport suggested that there may be strong interactions between bedload sediment transport and processes influencing channel structure and morphology. For example, if bedload sediment transport is occurring at radically different rates along a stream, there must be corresponding changes in storage whereby some reaches may be undergoing scour or an unloading of bedload sediments while other reaches might accumulate sediments in depositional features. Since significant bedload sediment transport appeared to occur less than 1% of the time, these may be periods during which the overall morphology of instream habitats is largely determined.

By the mid-1970s, changes in the physical conditions of stream channels and their associated substrates became the focus of many studies in the Pacific Northwest on the interactions between sedimentation processes and fish habitat (Everest et al. 1987; Sullivan et al. 1987). Several projects were initiated as part of the ongoing sediment studies to develop improved understanding of the interactions between sediment transport processes and channel morphology.

Streambed Scour and Fill

Since order-of-magnitude fluctuations in bedload transport had been found to occur at a given discharge and other field observations/measurements began to indicate bedload sediment transport was non-uniform along the length of Flynn Creek, it soon became clear that a "conveyor belt" model of bedload transport, whereby bedload movement occurred as a thin, uniform veneer along the length of the stream, was inapplicable to Flynn Creek. While such a model might appropriately describe bedload transport in rectangular flumes, it clearly was not applicable to streams with non-uniform channel geometries, sediment-size distributions, and variable hydraulics. Instead, features of streams such as riffles, pools, and bars, which are maintained by scour and fill processes, were thought to be intrinsically linked with non-uniform bedload transport rates such as those measured at Flynn Creek.

Based on field observations and measurements, Jackson and Beschta (1982) developed a descriptive model of bedload transport that accounted for riffle scour and deposition and the non-uniformity of bedload transport rates along a stream (Fig. 12.10). This model originated from concurrent measurements of bedload transport and changes in channel geometry associated with a riffle-pool sequence approximately 100 m upstream from the fish trap on Flynn Creek.

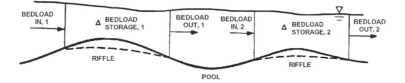

Fig. 12.10 Definition sketch for non-uniform bedload transport along a stream reach, depicting interactions between bedload transport and changes in sediment storage on riffle features (Jackson and Beschta 1982). Reproduced from Jackson and Bescheta 1982, with permission from John Wiley & Sons, Inc

As previously indicated, flows of approximately 0.7 m³ s⁻¹ in Flynn Creek were found to initiate the transport of sands that had been previously deposited on the surface of the streambed in such locations as pools, channel margins, interstices of larger particles that comprise the surface armor layer, and back-water portions of the channel associated with organic debris obstructions. These sand deposits were relatively uniform in size (approximate median particle diameter of 1 mm) and less common in the coarse bed material matrix that forms the channel bed. Rather, they rested on top of this coarser bed matrix, presumably having been deposited during the recession of a previous storm hydrograph after coarser materials were no longer capable of movement. This initial transport of sands over a stable gravel matrix was designated as "Phase I" bedload transport by Jackson and Beschta (1982).

As flows increased and approached ~1.3 m³ s⁻¹ for the storm of February 6–8, 1979 (Fig. 12.11), the gravel armor of the riffle was disrupted. The initiation of gravel transport that occurred as the armor layer began to break up was designated as "Phase II" bedload transport (Jackson and Beschta 1982). During the February 6–8 storm, streamflow increased to a peak discharge of 1.53 m³ s⁻¹ (1.8-yr. recurrence-interval flow). The initiation of Phase II transport was evidenced by a dramatic increase in particle sizes in bedload samples. Particles nearly 13 mm in diameter were sampled during this runoff event. Also, as flows greater than 1.3 m³ s⁻¹ persisted, two dramatic spikes in transport rate occurred

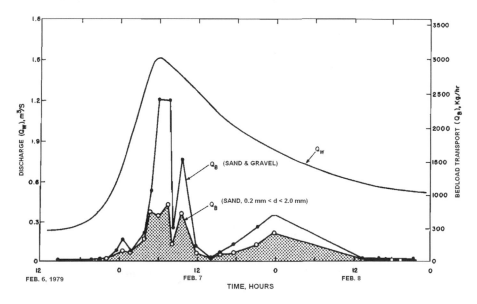

Fig. 12.11 Hydrograph for February 6–8, 1979, storm at Flynn Creek and bedload transport (Q_B) of (1) sand (0.2–2.0 mm) and (2) sand and gravel. High periods of gravel transport (i.e., the differences between the two lines) correspond to periods of scour of a riffle immediately upstream from the sampling station (Jackson and Beschta 1982). Reproduced from Jackson and Bescheta 1982, with permission from John Wiley & Sons, Inc

(the largest spike ranging from 400 kg hr^{-1} to 2,400 kg hr^{-1}). Both of these spikes involved considerable gravel-sized material and both corresponded with scour of the riffle immediately upstream from the sampling station (Fig. 12.11). The maximum measured riffle scour was 22 cm.

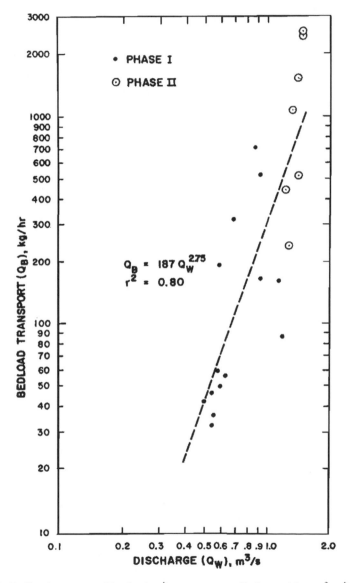

Fig. 12.12 Bedload transport (Q_B, kg hr^{-1}) versus water discharge (Q_W, m^3 s^{-1}) at Flynn Creek for a channel cross-section immediately downstream from a riffle (Jackson and Beschta 1982). Reproduced from Jackson and Bescheta 1982, with permission from John Wiley & Sons, Inc

Based on cross-section measurements that were continued through the storm of February 6–8, 1979, the rapid release of coarse bed material from the upstream riffle was quickly and efficiently routed into a pool situated immediately downstream of the riffle. Located on a bend in the channel, this pool experienced plunging flows during high stream discharges that apparently served to maintain the scoured condition. Thus, sediments scoured from the upstream riffle were rapidly routed through the pool to a downstream location where local hydraulics were incapable of transporting the amount of sediment being delivered from upstream. While there is some indication that the initiation of gravel transport in Phase II also marked a steepening of the bedload sediment rating curve (Fig. 12.12), the evidence is inconclusive due to the narrow discharge range over which Phase II transport was sampled.

Overall, Jackson and Beschta (1982) demonstrated that large bedload transport spikes are also associated with the non-uniformity of bedload transport rates along the stream length. They further showed that this "episodic" nature of bedload sediment transport is part of a larger process of scouring and redepositing sediments that maintains riffle features in streams.

Substrate Composition

The particle-size composition of gravel substrates in small Coast Range streams is primarily of interest because of its potential to influence the quality of spawning habitat. Furthermore, there was concern that if logging practices increased the delivery of fine sediments to a channel system this would increase the proportion of fine sediments in gravels. An increase in the amount of fine sediments occupying interstitial spaces of spawning gravels had been generally shown to reduce egg-to-fry survival (e.g., Cordone and Kelley 1961).

Flume experiments of the process by which fine sediments intrude into existing gravel substrates suggested that deposition (filtering) of fine sediments occurs at or near the surface of a gravel substrate (Beschta and Jackson 1979). While the intrusion of fine sediment into the underlying gravel matrix represents an important concern from a biological perspective, it is not the only process affecting the proportion of fine particles below the streambed surface. A significant consequence of the results from flume experiments was that the particle size composition of riffles is perhaps determined during brief periods of bedload transport, when local coarse-grained sediments are scoured and then redeposited. This also would be the period when fine sediments, which had either intruded into the surface veneer of the gravels or had been previously deposited within the gravel matrix during the deposition of bedload sediments, could be "flushed" from the bed.

As part of the post-AWS sediment studies, freeze-core technologies were used to investigate factors affecting the particle-size composition of streambed substrates (Adams and Beschta 1980). Frozen-core samples (Fig. 12.13) of the streambed for several reaches in Flynn Creek, Deer Creek, Needle Branch and two additional Oregon Coast Range streams were used to analyze the potential effects of watershed factors on the substrate sediment size composition. A total of 21 bed material sampling locations were established within these five streams that were sampled monthly and following large flow events.

Adams and Beschta (1980) found considerable variability, both within streams and between streams, in the percentage of fine sediments (<0.1 mm) comprising streambeds. The proportion of fines (weight basis) averaged 19% across all 21 sampling locations and ranged from 11 to 29% for streambeds associated with watersheds undisturbed by logging. For comparisons between watersheds, the average slope, total area, total relief, and land use of the watersheds accounted for significant differences ($p<0.05$) in the percentage of fine sediments comprising a streambed. Within watersheds, channel sinuosity and bankfull stage were also significantly ($p<0.05$) associated with differences in substrate fine sediment composition. Decreases in percent fines in the channel substrates in the WYs 1978 and 1979 were generally detected only after periods of high streamflow. Overall, changes in fine sediments did not appear to be strong diagnostic variable for detecting potential changes that might be associated with various forest practices (Adams and Beschta 1980). The results of Adams and Beschta (1980) and Jackson and Beschta (1982) suggested that major changes in fine sediment composition of streambed substrates are likely to occur during the brief periods that a stream is experiencing Phase II bedload transport.

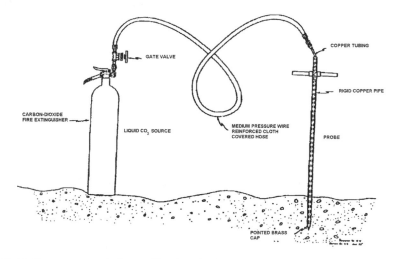

Fig. 12.13 Freeze-core technique utilized to obtain a streambed substrate sample for particle size analysis (Walkotten 1973)

Channel Response to Increases in Sediment Delivery

Based upon field observations and measurements of bed composition by Adams and Beschta (1980) and Jackson and Beschta (1982), a flume study was undertaken to establish whether increased levels of fine sediment in transport during the redeposition of gravel substrates would affect the fine sediment composition of that substrate. The deposition experiments of Jackson and Beschta (1984) showed that increasing the composition of fine sand in transport relative to that of gravel, from 1:1 to 5:1 by weight during riffle deposition, resulted in only a small increase in the fine sand composition of the resulting depositional feature (riffle). Furthermore, as this ratio increased, gravels in transport were less likely to become deposited and previously stable riffles became destabilized. These results were somewhat unexpected and led Jackson and Beschta (1984) to speculate as to the implications of the flume research to processes and conditions in the Alsea Watershed streams. They hypothesized that if land uses result in increased availability of sand-sized sediments to a channel system, the stream's longitudinal structure (i.e., its riffle-pool structure) would potentially be affected more than the fine sediment composition of gravel substrate. For example, increases in the delivery of sand- sized sediments to streams might diminish riffle amplitude and cause the filling of pools. This situation would suggest a decrease in hydraulic roughness of the bed and an increase in the capacity of a channel to transport bedload. If sand increases were great enough, they could result in the "smothering" of the entire streambed, including riffles, with sands. Under such conditions, average channel depths would decrease and there would be a tendency for widths to increase via streambank erosion.

To further investigate possible factors influencing the longitudinal morphology of streams, Stack and Beschta (1989) evaluated the longitudinal profiles of channel thalwegs (i.e., deepest part of stream for a given channel cross-section) for 14 stream reaches in the Oregon Coast Range, including three on Flynn Creek. The length of each stream reach represented approximately 70 bankfull channel widths. This study focused on streams relatively unimpacted by logging activities. As a basis for defining pools from a longitudinal sequence of thalweg depth measurements, Stack and Beschta (1989) developed a "rapid bed profile" (RBP) technique. "Residual pool volumes" from this methodology represented the longitudinal cross-section defined by where water would remain standing within the channel if stream discharge ceased. The crests of shallows or riffles thus provided control points that defined the downstream boundary of each pool (Fig 12.14).

Average residual pool volumes were directly correlated with watershed area whereas the number of pools per 100 meters of channel length was inversely correlated with watershed area; both correlations were significant at $p<0.10$ (Adams and Stack 1989). Stream gradient and pool frequency also appeared to affect processes related to pool formation (e.g., plunging flows,

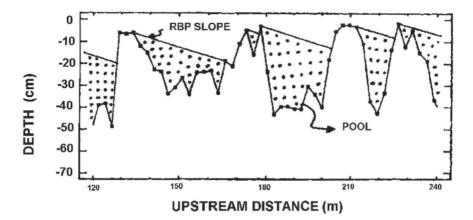

Fig. 12.14 Illustration of rapid bed profile (RBP) technique used to define residual pools from a longitudinal profile of thalweg depths. "RPB slope" is obtained from measurement of channel gradient (Stack and Beschta 1989). Reproduced from Stack/Beschta 1989, with permission from the American Water Resources Association

flow deflection). Large wood debris was a relatively important factor affecting pool formation in lower-gradient streams whereas large boulders appeared to be more influential in steeper-gradient streams. The presence of beavers (*Castor canadensis*) often caused relatively large pools within a given reach.

While Stack and Beschta (1989) did not investigate how various timber harvest and watershed management practices might influence pool morphology, they developed a framework for stream-reach classification and a simple method for characterizing pool morphology. Such approaches may be especially useful if, as Jackson and Beschta (1984) hypothesized, pool-riffle streams respond to increased sediment loads by smoothing their longitudinal profiles and reducing form roughness.

Other Research Results

Although turbidity is often utilized as a water quality standard by various states in the western United States, the relationship of turbidity to suspended sediment loads for mountain watersheds in Oregon had not been previously quantified. Data obtained over a wide range of flows from three Coast Range streams, including Flynn Creek, indicated that turbidity could be used as a general surrogate for suspended sediment levels. However, the relationships between turbidity and suspended sediment concentration for various watersheds were not the same, indicating a need to establish specific relationships for any watershed of interest (Beschta 1980b).

Bed material samples collected from Flynn Creek, Deer Creek, and Needle Branch (Adams and Beschta 1980), in conjunction with data from other Pacific Northwest streams, were used to address questions regarding the particle size distribution of streambed gravels. In the early 1980s the geometric mean diameter (d_g) was being proposed as a promising statistic for characterizing the textural composition of spawning gravels (Platts et al. 1979; Shirazi and Seim 1981). However, analyses of bed material samples from Coast Range streams in Oregon and Washington indicated that particle size distributions of streambed gravels were not log-normally distributed, but instead had a pronounced negative skew (Beschta 1982). As the d_g of a sample increased, the skew coefficient became increasingly negative. Furthermore, d_g was found to be a less sensitive indicator of land use impacts to stream gravels than another commonly used measure of gravel texture (i.e., percent fines).

Several technological spin-offs occurred as a result of the various sedimentation research projects that followed completion of the original AWS. These adaptations contributed to methodologies that improved the accuracy or efficiency of collecting field data and included:

(1) The vortex-tube bedload sampler utilized an off-channel work area and mechanism for controlling bypass flows that allowed rapid sampling of bedload transport during periods of high flow (Beschta et al. 1981b).

(2) The increased size and cylindrical shape of the Helley-Smith catch bag made the sampler a more practical and useful device for bedload sampling in streams with high proportions of sand-sized particles and particulate organic matter in transport (Beschta 1981). Without this larger bag, the sampling efficiency of the standard Helley-Smith sampler rapidly decreases over time, particularly at high rates of bedload transport, thus limiting its utility for obtaining accurate samples.

(3) Where repeated bedload sampling across a natural stream channel was undertaken with the Helley-Smith bedload sampler, there were concerns about sampling crews continually wading in a cross-section, disturbing the bed material during high flow conditions, and potentially influencing the resultant bedload measurements. Furthermore, safety can be a concern when wading streams at high flows. Thus, a bridge design was developed from which bedload sampling could be conducted (Hawks et al. 1987). Bridges based on this design were used to span streams up to 15 m in width for sediment transport studies in Utah (Beschta et al. 1981a), Oregon (Jackson and Beschta 1982), and Alaska (Estep and Beschta 1985).

(4) In several studies, pumping samplers were utilized to obtain discrete samples for suspended sediment and turbidity analysis (continuous sampling was sometimes used for turbidity measurements). Frequent samples required an intake nozzle that would not plug or accumulate organic debris over time. Numerous types of nozzle configurations were tried in streams before one was found whereby the nozzle would not plug. Attaching the nozzle to the end of a metal rod that in turn was hinged from a bridge or cable across

the channel (Beschta 1980a) essentially eliminated the occurrence of "lost" suspended sediment and turbidity samples; plugging of intake nozzles no longer occurred.

(5) A means of consistently and efficiently characterizing pools (Adams and Stack 1989) provided a means of quantitatively evaluating factors affecting pool formation and occurrence.

Summary

Sedimentation research during the original AWS attempted to evaluate the effects of forest practices (roading, logging, and site preparation) on suspended sediment concentrations and yields from small mountain watersheds of the Oregon Coast Range. However, subsequent sediment studies, as reported in this chapter, focused more on instream processes. This process-oriented research reflected the need for improved understanding of basic sediment transport mechanisms and dynamics, and the interaction of transport rates with other factors.

While more basic research is needed to fully understand and predict instream responses to watershed management practices, results of post-AWS studies contributed substantially to the body of literature on small stream functioning in forested watersheds. Reanalysis of the suspended sediment data sets for Flynn Creek, Needle Branch, and Deer Creek that had been collected during the AWS, in conjunction with more detailed measurements of sediment concentrations during the post-AWS period, provided an improved understanding of suspended sediment dynamics and variability for supply-limited streams. This information was essential to the development of supply-based models of suspended sediment dynamics. Important insights and scientific contributions were also gained regarding bedload transport dynamics in small streams draining steep-sloped forested watersheds and the interactions of these processes with stream channel morphology and function. As the fields of forest hydrology, fluvial geomorphology, and watershed management continue to evolve, it is hoped that this sedimentation research has provided important building blocks in the understanding of small stream responses to natural and anthropogenic factors.

Acknowledgment In the years following the 1958–73 Alsea Watershed Study, research was conducted on a variety of topics associated with sediment transport and channel morphology of mountain streams in forested environments. These research projects were undertaken by graduate students in the Forest Hydrology Program at Oregon State University and utilized streams in the Alsea Watersheds for all or part of their field research. They included: Paustian (1977), Adams (1979), Edwards (1979), O'Leary (1980), Jackson (1981), and Stack (1989). Much of what has been summarized in this chapter represents the acquisition of difficult-to-obtain field data due to the persistence of these individuals and others who assisted. We would like to express our appreciation to all of them for their invaluable contributions.

Literature Cited

Adams, J.N. 1979. Variations in Gravel Bed Composition of Small Streams in the Oregon Coast Range. M.S. Thesis. Oregon State Univ., Corvallis, OR. 160pp.

Adams, J.N., and Beschta, R.L. 1980. Gravel bed composition in Oregon coastal streams. Can. J. Fish. Aquat. Sci. 37:1514–1521.

Adams, P.W., and Stack, W.R. 1989. Streamwater quality after logging in Southwest Oregon, project completion report. Supplement No. PNW-87-400. USDA Forest Service, Pacific Northwest Forest and Range Experiment Station. 19pp.

Beschta, R.L. 1978. Long-term patterns of sediment production following road construction and logging in the Oregon Coast Range. Water Resour. Res. 14:1011–1016.

Beschta, R.L. 1980a. Modifying automated pumping samplers for use in small mountain streams. Water Resour. Res. 16:137–138.

Beschta, R.L. 1980b. Turbidity and suspended sediment relationships, pp. 271–282. In: Proceedings of Watershed Management Symposium, Irrigation and Drainage Division. Amer. Soc. Civil Eng., Boise, ID.

Beschta, R.L. 1981. Increased bag size improves Helley-Smith bedload sampler for use in streams with high sand and organic matter transport, pp. 17–25. In: Erosion and Sediment Transport Measurement. Int. Assoc. Hydrol. Sci. Publ. 132.

Beschta, R.L. 1982. Comment on "Stream system evaluation with emphasis on spawning habitat for salmonids" by M. A. Shirazi and W. K. Seim. Water Resour. Res. 18:1292–1295.

Beschta, R.L. 1983. Sediment and organic matter transport in mountain streams of the Pacific Northwest, pp. 1–69 to 1–89. In: Proceedings of D.B. Simons Symposium on Erosion and Sedimentation. Colorado State Univ., Ft. Collins, CO.

Beschta, R.L. 1987. Conceptual models of sediment transport in streams, pp. 387–419. In: C.R. Thorne, J.C. Bathurst, and R.D. Hey, editors. Sediment Transport in Gravel-bed Rivers. John Wiley & Sons, Ltd., Chichester, UK.

Beschta, R.L., and Jackson, W.L. 1979. The intrusion of fine sediments into a stable gravel bed. J. Fish. Res. Board Can. 36:204–210.

Beschta, R.L., Jackson, W.L., and Knoop, K.D. 1981a. Sediment transport during a controlled reservoir release. Water Resour. Bull. 17:636–641.

Beschta, R.L., O'Leary, S.J., Edwards, R.E., and Knoop, K.D. 1981b. Sediment and organic matter transport in Oregon Coast Range streams. WRRI-70. Water Resources Research Institute, Oregon State Univ., Corvallis, OR. 67pp.

Brown, G.W., and Krygier, J.T. 1971. Clear-cut logging and sediment production in the Oregon Coast Range. Water Resour. Res. 7:1189–1198.

Cordone, A.J., and Kelley, D.E. 1961. The influence of inorganic sediment on the aquatic life of streams. Calif. Fish and Game 47:189–228.

Edwards, R.E. 1979. Sediment Transport and Channel Morphology in a Small Mountain Stream in Western Oregon. M.S. Thesis. Oregon State Univ., Corvallis, OR. 114pp.

Emmett, W.W. 1976. Bedload transport in two large gravel-bed rivers, Idaho and Washington. Proceedings, Third Fed. Interagency Sedimentation Conf. Denver, CO. 3:101–114.

Emmett, W.W., Myrick, R.M., and Meade, R.H. 1980. Field data describing the movement and storage of sediment in the East Fork River, Wyoming. Part I. River hydraulics and sediment transport, 1979. USDI Geological Survey, Open File Report 80–1189. 43pp.

Estep, M.A., and Beschta, R.L. 1985. Transport of bedload sediment and channel morphology of a southeast Alaska stream. Research Note PNW-430. USDA Forest Service, Pacific Northwest Forest and Range Experiment Station. 15pp.

Everest, F.H., Beschta, R.L., Scrivener, J.C., Koski, K.V., and coauthors. 1987. Fine sediment and salmonid production: a paradox, pp. 98–142. In: E.O. Salo and T.W. Cundy, editors. Streamside Management: Forestry and Fishery Interactions. Univ. of Washington Inst. of Forest Resour. Contrib. 57, Seattle, WA.

Gibbons, D.R., and Salo, E.O. 1973. An annotated bibliography of the effects of logging on fish of the western United States and Canada. General Technical Report PNW-10. USDA Forest Service, Portland, OR. 145pp.

Harris, D.D. 1973. Hydrologic changes after clearcut logging in a small Oregon coastal watershed. U.S. Geol. Surv. J. Res. 1:487–491.

Harris, D.D. 1977. Hydrologic changes after logging in two small Oregon coastal watersheds. Water-Supply Paper 2037. U.S. Geological Survey Washington, DC. 31pp.

Hawks, A., Beschta, R.L., and Jackson, W.L. 1987. A lightweight, inexpensive foot-bridge for stream sampling. British Geomorphological Research Group, Technical Bulletin 36:8–12.

Hayward, J.A. 1980. Hydrology and stream sediments in a mountain catchment. Univ. of Canterbury, Tussock Grasslands and Mountain Lands Institute Special Pub. 17, Christchurch, NZ. 236pp.

Hayward, J.A., and Sutherland, A.J. 1974. The Torlesse stream vortex-tube sediment trap. J. Hydrol. (New Zealand) 13:41–53.

Helley, E.J., and Smith, W. 1971. Development and calibration of a pressure-differential bedload sampler. U.S. Geological Survey Open File Report, Menlo Park, CA. 18pp.

Hubbell, D.W. 1987. Bed load sampling and analysis, pp. 89–118. In: C.R. Thorne, P.C. Bathurst, and R.D. Hey, editors. Sediment Transport in Gravel-bed Rivers. John Wiley & Sons, Ltd., Chichester, UK.

Iwamoto, R.N., Salo, E.O., Madej, M.A., and McComas, R.L. 1978. Sediment and water quality: a review of the literature including a suggested approach for water quality criteria. EPA 910/9-78-048. U.S. Environmental Protection Agency, Region 10, Seattle, WA. 151pp.

Jackson, W.L. 1981. Bed Material Routing and Streambed Composition in Alluvial Channels. Ph.D. Thesis. Oregon State Univ., Corvallis, OR. 174pp.

Jackson, W.L., and Beschta, R.L. 1982. A model of two-phase bedload transport in an Oregon Coast Range stream. Earth Surf. Process. Landf. 7:517–527.

Jackson, W.L., and Beschta, R.L. 1984. Influences of increased sand delivery on the morphology of sand and gravel channels. Water Resour. Bull. 20:527–533.

Klingeman, P.C. 1971. Evaluation of bedload and total sediment yield processes on small mountain streams, pp. 58–169. In: J.T. Krygier, editor. Studies on Effects of Watershed Practices on Streams. U.S. Environmental Protection Agency Water Pollution Control Research Series 13010 EGA 02/71, Washington, DC.

Milhous, R.T., and Klingeman, P.C. 1973. Sediment transport system in a gravel-bottomed stream, pp. 293–303. In: Hydraulic Engineering and the Environment, Proc. 21st Annual Hydraulics Division Specialty Conference. Amer. Soc. of Civil Eng.

Moring, J.R., and Lantz, R.L. 1975. The Alsea Watershed Study: effects of logging on the aquatic resources of three headwater streams of the Alsea River, Oregon. Part I. Biological studies. Fish. Res. Rep. 9. Oregon Dept. of Fish and Wildlife, Corvallis, OR. 66pp.

O'Leary, S.J. 1980. Bedload Transport in an Oregon Coast Range Stream. M.S. Thesis. Oregon State Univ., Corvallis, OR. 107pp.

O'Leary, S.J., and Beschta, R.L. 1981. Bedload transport in an Oregon Coast Range stream. Water Resour. Bull. 17:886–894.

Paustian, S.J. 1977. The Suspended Sediment Regimes of Two Small Streams in Oregon's Coast Range. M.S. Thesis. Oregon State Univ., Corvallis, OR. 122pp.

Paustian, S.J., and Beschta, R.L. 1979. The suspended sediment regime of an Oregon Coast Range stream. Water Resour. Bull. 15:144–154.

Platts, W.S., Shirazi, M.A., and Lewis, D.H. 1979. Sediment Particle Sizes Used by Salmon for Spawning with Methods for Evaluation. EPA-600/3-79-043. U.S. Environmental Protection Agency, Corvallis, OR. 39pp.

Shirazi, M.A., and Seim, W.K. 1981. Stream system evaluation with emphasis on spawning habitat for salmonids. Water Resour. Res. 17:592–594.

Sorenson, D.L., McCarth, M.M., Middlebrooks, E.K., and Porcella, D.B. 1977. Suspended and dissolved solids effects on biota: a review. EPA-600/3-77-042. U.S. Environmental Protection Agency Research Lab, Corvallis, OR. 64pp.

Stack, W.R. 1989. Factors Influencing Pool Morphology in Oregon Coastal Streams. M.S. Thesis. Oregon State Univ., Corvallis, OR. 115pp.

Stack, W.R., and Beschta, R.L. 1989. Factors influencing pool morphology in Oregon coastal streams, pp. 401–411. In: Proceedings of symposium on Headwaters Hydrology. Amer. Water Resour. Assoc., Bethesda, MD.

Stelczer, K. 1981. Bed-load Transport: Theory and Practice. Water Resources Pubs., Littleton, CO. 295pp.

Sullivan, K., Lisle, T.E., Dolloff, C.A., Grant, G.E., and coauthors. 1987. Stream channels: the link between forests and fishes, pp. 39–97. In: E.O. Salo and T.W. Cundy, editors. Streamside Management: Forestry and Fishery Interactions. College of Forest Resources, Univ. of Washington Inst. of Forest Resour. Contrib. 57, Seattle, WA.

Van Sickle, J. 1981. Long-term distributions of annual sediment yields from small watersheds. Water Resour. Res. 17:659–663.

Van Sickle, J., and Beschta, R.L. 1983. Supply-based models of suspended sediment transport in streams. Water Resour. Res. 19:768–778.

Vanoni, V.A. 1975. Sedimentation Engineering, Manuals and Reports on Engineering Practice 54. Amer. Soc. Civil Eng., New York, NY. 424pp.

Walkotten, W.J. 1973. A freezing technique for sampling stream bed gravel. Research Note PNW-205. USDA Forest Service, Pacific Northwest Forest and Range Experiment Station, Portland, OR. 7pp.

Chapter 13
Woody Debris from the Streamside Forest and its Influence on Fish Habitat

Charles W. Andrus

Streams are often shaped by trees growing near the channel. Woody debris originating from the streamside forest and root masses of trees growing at water's edge can alter the geometry and surface substrate of a channel (Keller and Swanson 1979; Hogan 1986; Sullivan 1986). Tree boles and rootwads of fallen trees shape the channel by deflecting the flow of water. This can result in features such as deep pools, zones of low water velocity, multiple channels, and sorted gravels (Bisson et al. 1987; Nakamura and Swanson 1993). The potential of a stream to support fish of different species and age classes, in turn, is often influenced by the abundance of these channel features (Bisson and Sedell 1984; Hicks 1989; Fausch and Northcote 1992).

During the last century timber harvesting and conversion of forest land to other uses have influenced the amount of woody debris in streams and the composition of streamside vegetation in the Oregon Coast Range. Today, many streams are bordered by trees that are younger and of different species than those which existed prior to European settlement. The result is a change in the volume and type of woody debris in streams and, consequently, a change in the quantity and type of fish habitat.

There is increasing interest in managing streamside forests in ways that improve debris-related fish habitat features. For example, timber harvesting regulations now require that a portion of the streamside forest along certain streams be retained when adjacent slopes are harvested. A primary reason for retaining these buffer trees is to provide a source of woody debris within streams that is both large and persistent. Nevertheless, attempts to create stands of older streamside trees that yield this persistent woody debris have been frustrated at times by premature mortality due to windthrow and sparse conifer regeneration within streamside areas. Attempts to increase woody debris levels in streams by placing logs in the channel have also been limited at times by inappropriate designs and high costs.

Charles W. Andrus
Adolfson Associates, Portland, OR 97204

J. D. Stednick (ed.), *Hydrological and Biological Responses to Forest Practices.*

The interest to manage streamside forests to increase woody debris levels in streams is not without conflict. While conifer trees retained along streams, especially those of large size, produce the most persistent woody debris, these also represent valuable timber. Placing logs in the channel that were obtained outside the streamside area can also be costly for either forest landowners or the public.

This chapter examines the interactions among streamside forests, woody debris, and fish habitat. Changes in streamside forests and woody debris over the last century will be examined, as well as, the likely consequences of forest management practices on the composition of future streamside stands and woody debris in the channel. Included is an evaluation of common strategies to improve woody debris levels by retaining trees along streams within timber harvest units, promoting conifer establishment near streams, and adding woody debris directly to streams.

Connections Between Woody Debris and Fish Habitat

Anadromous fish in the Pacific Northwest such as coho salmon (*Oncorhynchus kisutch*) or steelhead (*O. mykiss*) spend a year or more in streams before migrating to the ocean. In contrast, some species such as sockeye salmon (*O. nerka*) or Chinook salmon (*O. tshawytscha*) occupy streams only a short time before moving downstream to lakes, large rivers, or estuaries. For resident fish and anadromous fish that live in streams for extended periods certain stream features are important to survival and growth.

The abundance of pools can influence feeding opportunities and survival during the summer, a time when streams are shallow and temperature stress is high. Pools provide fish large areas from which to stage feeding forays and avoid competition with more aggressive fish (Nielsen 1992; Harvey and Nakamoto 1996). In addition, thermally stratified water sometimes occurs in deep pools providing refuge zones of cool water during warm summer days (Matthews et al. 1994; Nielsen et al. 1994). During the summer, juvenile coho salmon densities can be two to six times higher in pools than in nearby glides or riffles (Fig. 13.1). Other salmonids are less dependent on pools during the summer. Young trout and one-year-old steelhead often prefer glides and riffles as much as pools during the summer (Hicks 1989; Nickelson et al. 1992).

Unique habitat features can influence survival and feeding opportunities in the winter even more so than during other times of the year (Murphy et al. 1986; Nickelson et al. 1992). Winter storms in the Pacific Northwest periodically create high flows that can displace fish downstream if refuge areas do not exist (Tschaplinksi and Hartman 1983; Quinn and Peterson 1996). The slack water found in alcoves, pools behind dams built by beaver (*Castor canadensis*), and other dammed pools provide refuge areas of low-velocity water which are heavily used by fish in the winter (Bustard and Narver 1975; Peterson 1982; Everest et al. 1986; Swales and Levings 1989). The water is swift in plunge pools

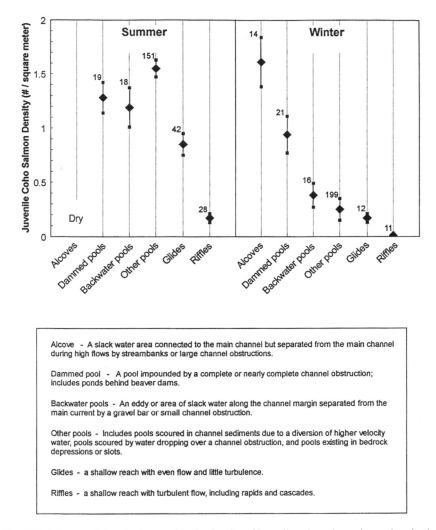

Fig. 13.1 Mean and standard error for the density of juvenile coho salmon in pool and other habitat types for the Oregon Coast Range during summer and winter (after Nickelson et al. 1992). Sample size is shown next to the upper standard error bar. Reproduced from Nickelson 1992, with permission from the Canadian Journal of Fisheries and Aquatic Sciences

and scour pools during the winter, making these features less attractive as refuge areas. For the same reason, glides and riffles are seldom used by fish during winter high flows (Nickelson et al. 1992). During the winter, juvenile coho salmon densities can be two to eight times higher in alcoves and dammed pools than in other habitat types (Fig. 13.1).

Pieces of woody debris are the primary creators of deep pools in most forest streams not dominated by boulder or bedrock substrate (Andrus et al. 1988; Adams and Stack 1989; Carlson et al. 1990; Richmond and Fausch 1995).

Woody debris, acting singly or in jams, can dam the flow of water and scour pools in the streambed. Large pieces of stable woody debris can also provide the structural framework to help keep beaver dams from washing out during high flows (Leidholdt-Bruner et al. 1992). Woody debris also plays an important role in creating areas with low velocity water adjacent to the main channel such as side channels and alcoves. During high flows water diverted by woody debris in the main channel can fill abandoned channel segments or carve out new channels in the flood plain (Sedell et al. 1983; Bryant 1985; House and Boehne 1986).

Another role for woody debris is providing cover for fish occupying pools and off-channel features (Bustard and Narver 1975; Tschaplinksi and Hartman 1983; McMahon and Hartman 1989). Cover can strongly influence the positions selected by young fish of various species (Fausch 1993). Substantial increases in juvenile coho salmon densities can occur when woody debris is added to dammed pools and alcoves that lack cover (Nickelson et al. 1992). Nevertheless, cover may not necessarily influence fish survival and growth in the summer if food abundance is low (Wilzbach et al. 1986), fish are very young (Spalding et al. 1995), or if pool depth is sufficient to minimize predation by birds (Lonzarich 1994).

Woody debris can trap twigs and leaves, especially when logs create jams that span the channel (Bilby and Likens 1980; Lamberti et al. 1991). Nutrients incorporated in twigs and leaves are more thoroughly processed by microorganisms and aquatic insects when they are detained by debris jams (Naiman and Sedell 1980). The source of nitrogen for many small forest streams may be found mostly in coarse wood and fine organic particulate matter (Triska et al. 1984). Carcasses of fish that spawn and die are also a potential source of nutrients and food for the local aquatic community, including young salmonids and aquatic invertebrates (Walter 1984; Bilby et al. 1996). The persistence of fish carcasses in a stream and the degree to which they are processed by stream organisms is conceivably influenced by whether or not the carcasses are retained in the stream for very long. Jams of woody debris may help keep the carcasses in place during freshets.

Streamside Forests and Woody Debris: Past and Present

Woody debris levels are low for many streams flowing through managed forest land when compared to unmanaged forest land (Bilby and Ward 1991; Ralph et al. 1994). Inventories conducted in a large portion of the central Coast Range of Oregon underlain by sedimentary rock indicated that woody debris volume in streams bordered by second-growth forest was only about one-fifth of that in streams flowing through unmanaged forest land bordered by trees 80 to 250 years old (Fig. 13.2). Many streams flowing through managed forest land in the Drift Creek watershed, which includes the Alsea Watershed Study streams, also have low levels of woody debris. An inventory of the Drift Creek

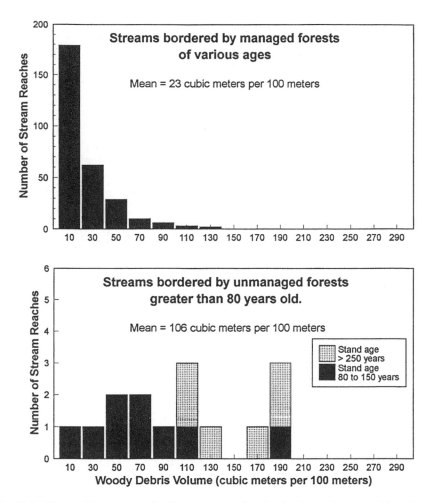

Fig. 13.2 Woody debris volume for 291 stream reaches flowing through managed forest land and bordered by second-growth trees (Oregon Department of Fish and Wildlife, Corvallis, unpublished data) compared to 15 stream reaches flowing through unmanaged forest land and bordered by trees 80 to 250 years old (Ursitti 1990). Streams are in the "sedimentary ecoregion" (Thiele et al. 1992), an area underlain by sedimentary rock that includes a large portion of the central Oregon Coast Range

watershed indicated that the volume of large woody debris within streams where timber had been harvested to the water's edge averaged about half that of the streams flowing through unharvested forests (Veldhuisen 1990).

The removal of streamside trees during harvest of the previous stand, the young age and dominance of red alder trees (*Alnus rubra*) in second-growth stands growing along streams, and intentional removal of debris from streams are likely causes for declines in woody debris for streams in logged drainages.

Conversion to Younger Trees

The age of forests in the Oregon Coast Range has been greatly altered during the last century. Trees under the age of one hundred years now occupy about 96% of nonfederal forest lands in the north half of the coast range.[1] It is estimated that in 1850 only about 30% of forest lands in this three-county area supported timber less than 100 years old (Teensma et al. 1991).

The amount of woody debris that accumulates within a stream generally increases with age of the streamside stand but the rate differs between hard-wood and conifer debris. During the first century, the developing streamside stand yields predominantly hardwood debris with only small amounts of conifer debris (Fig. 13.3, Grette 1985; Heimann 1988). Conifer debris increases rapidly during the second century, while the amount of hardwood woody debris in the stream declines rapidly. Trees are typically clearcut harvested every 50 years in the Oregon Coast Range so little woody debris accumulates from these stands unless buffer trees are retained along the stream during logging and allowed to become old.

Conversion to Alder Trees

Until about 40 years ago few timber harvest areas were replanted. In the absence of fire, deep logging slash and competition from brush and hardwoods often resulted in sparse conifer establishment next to coastal streams. Even later, when planting conifers after logging became a common practice, successful regeneration near streams was difficult to achieve unless preceded by burning of logging slash and brush and application of herbicides.

For a sample of stream reaches in the central Oregon Coast Range, conifer basal area was only about 22% of the total basal area for second-growth streamside stands, but 90% of total basal area for older stands that originated following wildfire (Andrus, unpublished data). A more comprehensive inventory of streams flowing through managed forest land showed that only a small proportion of stream reaches were bordered by stands that consisted mostly of conifer trees (Fig. 13.4).

Conifer trees were more plentiful in streamside stands a century ago probably because many of these stands originated after a wildfire. Fires in the Oregon Coast Range were usually hot, extended over large areas, and often burned even trees at the edge of streams (Teensma et al. 1991). After wildfires burned to the edge of streams, the exposed mineral soil provided a favorable seed bed for regeneration of conifer species such as Douglas-fir (*Pseudotsuga*

---

[1] Unpublished data from the Oregon Department of Forestry, Salem, Oregon. Data were for 1994 and included Clatsop, Tillamook, and Lincoln counties. Most of the northern half of the Coast Range is included within these three counties. The data were for only non-federal land which makes up about 80% of the forest land in the northern half of the Coast Range.

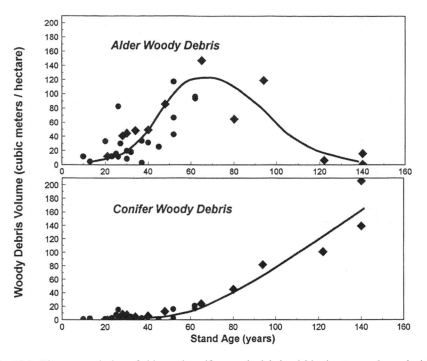

Fig. 13.3 The accumulation of alder and conifer woody debris within the stream channel with increasing age of the streamside trees. Stream reaches measured by Heimann (1988) are shown as diamonds and as circles for Grette (1985). Curves were fitted by eye. Reproduced from both Grette 1985, with permission from Grette Associates and Heimann 1988, with permission from the U.S. Geological Survey

menziesii). Without such a disturbance, conifer regeneration is impaired and competing brush species and hardwoods (mostly red alder trees) can dominate streamside sites for long periods. Some conifers such as western red-cedar (*Thuja plicata*) and western hemlock (*Tsuga heterophylla*) are able to compete with brush and hardwoods better than others, but the time needed to gain dominance over the competing vegetation is many decades. Brush species such as salmonberry (*Rubus spectabilis*) can exclude most other vegetation on wet sites in the Coast Range for hundreds of years at a time (Minore 1990). Some landforms, such as low floodplain terraces and unstable streamside slopes, probably never did support many conifers, even following hot wildfires. Alder and brush are better able to occupy sites subject to periodic flooding and mass soil movement (Rot 1995).

The woody debris from young conifer or alder streamside areas is less persistent in the channel than woody debris from older conifer stands. Pieces are smaller and therefore more easily transported downstream during high flows (Grette 1985; Heimann 1988; Veldhuisen 1990). More important, alder debris decays rapidly. The boles of red alder trees decay within a decade or less

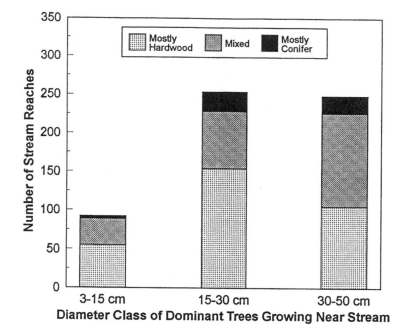

Fig. 13.4 The species composition of second-growth streamside trees for 599 stream reaches throughout coastal Oregon. Stream reaches varied considerably in length but averaged 1.6 km. Data are unpublished and were compiled by the Oregon Department of Fish and Wildlife, Corvallis

(Harmon et al. 1986; Andrus unpublished data) while conifer boles persist for many decades or even centuries for some species such as western red-cedar (Harmon et al. 1986; Murphy and Koski 1989).

Intentional Debris Removal

For an unknown number of coastal streams intentional removal of woody debris has contributed greatly to declines in woody debris abundance. Beginning early in the century, woody debris was removed from some streams during splash damming[2] (Sedell et al. 1991). Woody debris was also removed to promote navigation (Sedell and Luchessa 1982), protect bridges from being damaged by log jams, and to keep water from being diverted at adjacent roads or railroad grades. In addition, there was a period from about 1960 to 1980 when woody debris was considered a serious obstacle to adult fish passage and

---

[2] Splash dams were temporary structures built in the channel to pond water. Logs were placed in the pond and then the dam was released. The resultant flood of water and logs were carried downstream to the sawmill.

landowners were encouraged or required by agencies to remove both logging slash and natural woody debris from channels. Many kilometers of streams in the Pacific Northwest were cleared of large woody debris before the harmful consequences of such actions were recognized (Bilby 1984).

Status of Woody Debris in Managed Forest Streams

Because of the shift to younger streamside stands and to alder trees the volume of woody debris contributed by the second-growth forest is relatively small. In fact, most of the woody debris found today in streams bordered by second-growth stands is actually debris left over from the previous old-growth forest. Various studies show that the proportion of woody debris originating from the previous stand averages 76 to 93% of the total debris volume (Table 13.1). For many streams, the woody debris originating from the previous stand is now entering advanced stages of decay (Grette 1985; Long 1987; Heimann 1988), putting greater emphasis on supplies of woody debris provided by current stands.

The small size of debris originating from second-growth stands and the scarcity of large logs that span the channel can cause much of the woody debris

Table 13.1 Abundance and origin of woody debris within stream reaches bordered by second-growth stands 25 to 100 years old

| Source and location | Origin of Woody Debris | Woody Debris Volume (m^3 ha^{-1})[1] | | | |
| | | Stand age: 25–50 years | | Stand age: 51–100 years | |
		Mean & Standard deviation[2]	% of Total	Mean & Standard deviation	% of Total
Heimann (1988); central Coastal Oregon	Current stand	62 (19)	13%	–	–
	Previous stand	500 (258)	87%	–	–
Andrus (1988); central Coastal Oregon	Current stand	43 (16)	14%	–	–
	Previous stand	274 (140)	86%	–	–
Grette (1985); northern Coastal Washington	Current stand	29 (23)	7%	94 (33)	22%
	Previous stand	367 (107)	93%	339 (161)	78%
Andrus, unpublished data; southern Coastal Washington	Current stand	64 (70)	17%	118 (118)	24%
	Previous stand	302 (336)	83%	366 (619)	76%

[1] Volume of wood residing within the bankfull channel divided by bankfull channel area.
[2] Standard deviation in parenthesis.

in managed streams to be rafted to the channel margins during high flows. Here, it is less likely to interact with the channel much of the year. A study of small streams in western Washington revealed that only 50% of woody debris pieces in logged basins interacted with the low-flow channel, compared to 83% for unharvested basins (Ralph et al. 1994).

Trees Retained Along Streams Following Logging

During the last few decades, the primary strategy for providing a future source of woody debris to streams has been to retain trees along streams when timber harvesting occurred on adjacent slopes. In early years, these buffers consisted mostly of hardwood trees or the low-value conifer trees nearest the stream. In recent years streamside buffers have tended to include more conifers and a wider band of trees each side of the stream.

Considerable debate has accompanied proposals to include more valuable conifers within buffers and expand the width of buffers. As more forest land is included within streamside buffers, and therefore is not available for harvest, questions arise about how much is needed to maintain woody debris levels in streams at desired levels. The processes that cause a streamside tree to become woody debris in the stream are often long-term and keyed to infrequent events such as severe windstorms or landslides. Because buffers have been retained along streams for only a relatively short time, questions about the sufficiency of proposed buffer designs have been difficult to evaluate.

Source Areas of Woody Debris

Recent studies have lead to a better understanding of which trees growing along streams have greater potential for contributing woody debris to streams. The likelihood that the bole or rootwad will enter the stream once a tree is toppled by wind or other forces is primarily a function of how far the tree grows from the stream. A study of windthrown trees retained along streams in the central and northern coast range of Oregon indicated about a 70% chance that the bole of a tree will land in the stream if the tree is growing next to the stream bank (Fig. 13.5). However, the probability is halved when the ratio of distance from stream to tree height is 0.8. The theoretical probability of a tree intersecting a straight stream channel assuming random direction of fall (McDade et al. 1990) is less than observed values (Fig. 13.5). This difference is likely a result of the tendency of trees growing closest to the stream to lean toward the stream. In addition, stream channels are usually sinuous rather than straight and this increases the chance that portions of a fallen tree will intersect the channel at one or more locations.

The likelihood that the rootwad of a fallen tree enters the stream is also dependent on how far the tree grows from the stream. Conifer trees growing

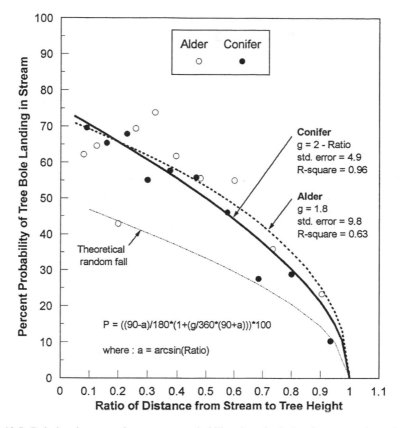

Fig. 13.5 Relation between the percent probability that the bole of a tree ends up in the channel when it falls and the ratio of the distance it grew from the stream to the tree's height. The curves were derived from data on 573 conifer and 599 hardwood windthrown buffer trees in the Oregon Coast Range (Andrus, unpublished data). Data points indicate the observed percentage of trees within discrete ratio classes that fell into the stream. Ten ratio classes were constructed with approximately equal numbers of trees in each class. Trees grew in stands that were 55 to 140 years old. The percent probability associated with theoretical random fall (McDade et al. 1990) is also displayed. Reproduced from McDade et al. (1990) with permission of the Canadian Journal of Forest Research

1.5 m or less from the stream bank have about a 60% chance of landing in the stream when they fall over while the rootwads of trees growing more than 3 m from the stream almost never enter the channel (Fig. 13.6). Rootwads can be an important part of the total woody debris load in a stream. A survey of 63 Pacific Northwest stream reaches showed that the wood within rootwads was nearly 40% of the total woody debris volume being added to channels by stands 20 to 100 years old (Andrus, unpublished data).

The proportion of woody debris contributed by trees growing within a given distance of the stream can be estimated by integrating under the curves illustrated in Fig. 13.5. For example, in a conifer stand of uniform density and 40 m

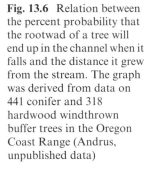

Fig. 13.6 Relation between the percent probability that the rootwad of a tree will end up in the channel when it falls and the distance it grew from the stream. The graph was derived from data on 441 conifer and 318 hardwood windthrown buffer trees in the Oregon Coast Range (Andrus, unpublished data)

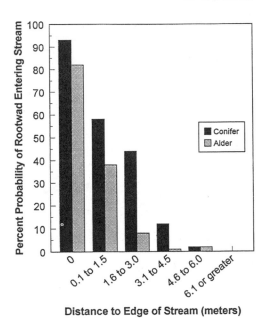

in height, an estimated 90% of tree boles that would someday enter the stream will originate from trees growing ≤30 m from the stream. Similarly, for a stand consisting of hardwoods with a height of 25 m an estimated 90% of the woody debris pieces in the stream are likely to originate from trees growing ≤20 m from the stream. The above example does not include the volume of rootwads within the stream.

These calculated values are similar to observations on fallen streamside trees in unlogged mature (80 to 200 years old) and old-growth conifer stands for western Washington and Oregon (McDade et al. 1990) where 90% of conifer woody debris pieces in the stream originated from trees ≤26 m from the stream. These conifers had an estimated average height of 40 m. For streams bordered by intact stands of old-growth trees in southeast Alaska 90% of conifer woody debris pieces originated from trees that had grown within 20 m of the stream (Murphy and Koski 1989). This narrower woody debris source area can be explained by the shorter height of southeast Alaska old-growth conifers compared to mature or old-growth trees in the Pacific Northwest.

Woody debris can also be delivered to streams from a long distance through debris flows within steep tributary channels (Ketcheson 1978). Logging can increase the occurrence of debris flows, with much of this increase attributable to road failures (O'Loughlin 1972; Amaranthus et al. 1985). However, contributions of large woody debris by debris flows are likely to decrease on managed forest land in the future as woody debris originating from the previous old-growth forest decays within steep tributary channels. Young managed forests will not provide much woody debris to take its place unless trees are retained next to these tributaries during timber harvesting as a future source of large woody debris. Debris flows are limited to only a portion of the Oregon Coast Range where tributary channels are steep.

Longevity of Buffer Trees

Trees within streamside corridors or buffers behave differently than trees growing in an intact streamside stand. Because upslope areas have been clearcut harvested, buffers usually experience increased mortality due to high winds. The rate at which trees within streamside buffers blow down is important because this influences the amount of woody debris within streams, both in the short-term and far in the future. Nevertheless, little is known about the longevity of buffer trees or the frequency of wind storms capable of toppling buffer trees.

Considerable windthrow has been known to occur along exposed clearcut harvest boundaries in the Oregon Coast Range when high winds occur (Curtis 1943; Ruth and Yoder 1953; Gratkowski 1956; Steinbrenner and Gessel 1956; Moore 1977). Extreme wind storms in the Oregon Coast Range have occurred in 1880, 1951, and 1962, each toppling millions of cubic meters of trees (Ruth and Yoder 1953; Lynott and Cramer 1966). Buffers of standing trees were usually not retained along streams at the time when these extreme wind storms occurred. Hence, little is known about the susceptibility of buffers during unusually high winds. Information on buffer tree mortality is limited to recently established buffers and for periods when unusually high winds did not occur.

An evaluation of 30 streamside buffers along the coastal fringe of the central and northern Oregon Coast Range one to six years following harvest indicated that wind mortality was highly variable, ranging from none to nearly three-quarters of the initial basal area[3] (Andrus and Froehlich 1991). For the sites on average, 20% of the basal area of buffer trees had succumbed to wind during the first few years after logging. This is similar to observations made on old-growth streamside buffers in the Oregon Cascade Mountains where on average 25% of the basal area had blown down during the first 20 years after logging (Sherwood 1993).

Among these 30 coastal Oregon streamside buffers, windthrow was greater where more of the streamside area had water-saturated soils (Andrus and Froehlich 1991). Also, windthrow was greater in streamside stands that consisted mostly of conifer trees. Hardwood stands were relatively windfirm, probably because they lacked foliage during the winter when high winds usually occurred. Streams flowing perpendicularly to the prevailing winds had more windthrow than streams flowing parallel to the wind. Finally, sites with

---

[3] Basal area is the cross-sectional area of a tree, commonly measured at breast height.

topographic relief upwind of the stream had less windthrow than sites that were more exposed (Fig. 13.7).

A multiple linear regression equation incorporating these four factors was developed for predicting the percentage of the initial basal area in a buffer that is likely to succumb to high winds during the first few years following harvesting of adjacent slopes. The equation is:

$$WIND = 0.892^*SATUR + 0.421^*CONIFER + 10.8^*ORIENT$$
$$+ 12.8^*TERRAIN - 26.5 \qquad (13.1)$$

$R^2 = 0.57$; $n = 29$; standard error of estimate $= 12.5$, where:

WIND   = Basal area of those buffer trees uprooted or snapped off by the wind expressed as a percentage of the initial basal area.

SATUR   = Percentage of trees in the initial buffer that grow on water-saturated soils.

CONIFER   = Basal area of buffer trees that are conifers expressed as a percentage of total basal area.

ORIENT   = General orientation of the stream reach. ORIENT = 0 if the stream flows northeast or southwest. ORIENT = 1 if the stream flows northwest or southeast.

TERRAIN   = Topography southwest of the stream (Fig. 13.6). TERRAIN = 0 if buffer receives greater protection from the surrounding terrain. TERRAIN = 1 if buffer receives lesser protection from the surrounding terrain.

WIND is assumed to be zero for combinations of independent variables that result in a negative calculated value of WIND. Variables included in the equation were significant at p<0.05.

This study was limited to streams 25 km or less from the ocean and during a period in the 1980's when unusually high winds did not occur. Undoubtedly, the above equation underestimates windthrow during periods when extreme wind storms occur.

Effect of Buffer Trees when they Fall

Buffer trees that fall over can have no effect, a delayed effect, or an immediate effect on a stream. In a study of about 1500 uprooted or snapped-off buffer trees in the Oregon Coast Range, 42% of the uprooted trees landed on streamside slopes outside a vertical projection of the stream's high-water mark (Fig. 13.8). Furthermore, nearly one-third of the windthrown trees ended up suspended 0.6 m or more above the water, having no immediate effect on the stream. Only 11% of the fallen trees landed such that the bole was in the channel or suspended less than 0.6 m above the water (Fig. 13.8). For most trees in this

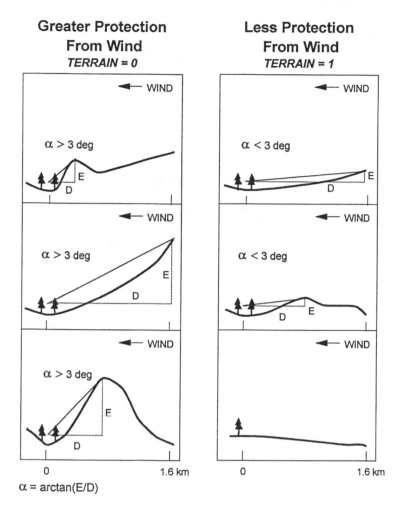

Fig. 13.7 Determination of wind protection offered by terrain southwest of a stream buffer. The degree of wind protection is evaluated by using a topographic map to plot elevation vs. distance for transects drawn from the stream segment midpoint. One transect is drawn for a bearing of S30W and another for S60W and compared to the diagrams above to determine if the value to the variable *TERRAIN* should be 0 (angle less than 3 degrees) or 1 (angle greater than 3 degrees). If one transect rates 0 and the other rates 1, then 1 would be the value entered

study, the distance to the stream was less than the tree's height. Such a large percentage of these 55 to 140 year old trees ended up suspended above the channel because the boles rarely broke upon hitting the ground. The long boles had ample opportunity to become suspended over the channel, especially in narrow valleys. Suspended conifer logs likely remain suspended for decades until decay causes them to collapse into the channel. Old-growth trees break into two or more pieces upon hitting the ground more often than trees from

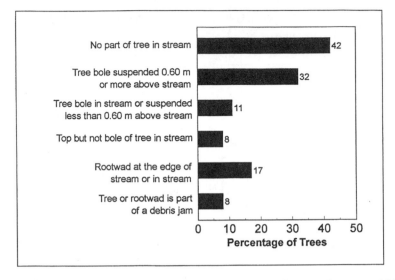

No part of tree in stream — 42

Tree bole suspended 0.60 m or more above stream — 32

Tree bole in stream or suspended less than 0.60 m above stream — 11

Top but not bole of tree in stream — 8

Rootwad at the edge of stream or in stream — 17

Tree or rootwad is part of a debris jam — 8

Percentage of Trees

Fig. 13.8 Percentage of uprooted or snapped-off streamside buffer trees that were within one or more position categories. A total of 1554 trees were evaluated. Reproduced from Andrus and Froehlich 1991, with permission from Oregon State University

younger stands and therefore are more likely to interact immediately with the channel (McDade 1988).

Because of their length and attached rootwads, windthrown trees from second-growth buffers are usually stable in all but the largest streams. Therefore, they provide critical key material for holding back small pieces of woody debris, assuming the bole is not suspended high above the channel (Andrus and Froehlich 1991).

Concern has been raised frequently (Moore 1977; Steinblums et al. 1984) about possible increases in stream sedimentation when wind topples large numbers of trees within a buffer, but rarely has this been quantified. As part of a study on the incidence of windthrow in streamside buffers, observations were made on sediment increases associated with windthrown trees (Andrus and Froehlich 1991). Visual estimates were made of the volume of soil washed from rootwads in or near the stream, soil scoured by water from rootwad holes, streambank sediments mobilized by boles or rootwads in the channel, and soil raveling from rootwad holes on steep slopes. Only 12% of the 1554 windthrown trees were sources of accelerated sedimentation. Increased sedimentation was nearly always associated with the upturning of rootwads in or at the edge of the stream. Uprooted trees on steep slopes did not cause landslides nor were the disturbed portions eroded by rain. Sediment increases were small, averaging only 0.9 m³ for each of the trees identified as a sediment source. Although not quantified, those rootwads and logs that ended up in the stream appeared to trap as much additional sediment as was released by upturned rootwads in the channel.

While at first glance extensive windthrow of buffer trees may appear disruptive, the net effect on the channel may be positive. Accelerated sedimentation is usually minor and, because buffer trees from second-growth stands rarely break when they fall, tight log jams of small pieces of wood that could limit fish movement are not likely to develop. Many streams in the Oregon Coast Range are currently deficient in woody debris, so the immediate flux of windthrown trees can be helpful. Nevertheless, the long-term supply of woody debris can be compromised when widespread windthrow occurs. Many decades must pass before a new stand becomes established in the voids and is old enough to contribute much woody debris to the stream. The windthrown trees may decay long before the new stand is old enough to replenish the stream.

Managing Alder-dominated Streamside Areas

Conifers rarely regenerate in streamside areas dominated by alder trees in the Oregon Coast Range, whether they are intact stands or buffers (Hibbs et al. 1991). The few conifers found growing beneath the canopy of alder stands are usually severely stunted due to lack of sunlight. Often, clearcut harvest areas immediately uphill of streamside buffers in the coast range also lack conifer regeneration, even when efforts are made to plant conifers close to the buffer (Andrus, unpublished data). Aggressive competition from brush and damage from animals are likely causes of sparse conifer regeneration on clearcut land immediately adjacent to buffers.

Because alder trees currently dominate so many streamside areas on managed forest land in the Oregon Coast Range (Fig. 13.4), several strategies are being tried to increase conifer establishment near streams. The goal is to increase conifer establishment now so that it will translate into large and persistent conifer woody debris decades in the future, even if that means sacrificing some of the benefits provided by existing alder trees. Some coastal streams do have riparian areas stocked with young stands of conifers, a result of clearcut harvesting to the edge of the stream, treating logging slash and brush, planting appropriate conifer seedlings, and controlling animal damage. Efforts are now being made to emulate these successful regeneration treatments in streamside areas that currently support only alder trees, while retaining at least a portion of the alder tree canopy to provide shade, streambank stability, and inputs of litter.

Evaluations of narrow alder buffers and aggressive conifer regeneration in streamside areas show mixed results. Preliminary results indicate that maximum summer water temperatures commonly increased 1.5° to 2.5°C for streams flowing through recent clearcuts for which only narrow buffers of alders were retained along the channel (Oregon Department of Forestry, unpublished data). Conifers planted in clearcut areas near streams display variable survival and growth depending on site conditions and the degree of brush control. Third-year results of a study that included six sites in the Oregon Coast Range indicated significantly higher survival, basal diameter growth, and height growth for conifers at sites where brush was sprayed or pulled during the first year (Maas and Emmingham 1995). Others have noted extensive mortality of planted conifers by beaver irrespective of brush control efforts (Newton, M. unpublished data, Oregon State Univ.). The beaver population in the Coast Range of Oregon has risen dramatically during the last several decades and may present a serious obstacle to regenerating conifers in streamside areas.

Planting conifers beneath heavily thinned streamside alder stands has shown some promise. Survival and early growth of conifers planted in heavily thinned areas have even surpassed that of conifers planted in areas where all alder trees were cut (Maas and Emmingham 1995). Nevertheless, it is expected that the crowns of residual alder trees quickly expand to fill the gaps in the canopy, requiring further thinning of the alder overstory.

Direct Placement of Woody Debris

Several strategies have been devised to supplement woody debris in streams where the streamside stand is not expected to yield much woody debris in the near-term. These strategies include rearranging woody debris already in the stream so that it is more effective at creating habitat features, enhancing the effectiveness of existing habitat features, and bringing in woody debris from other areas.

Rearranging Existing Woody Debris

One of the more direct ways to improve fish habitat in streams is to rearrange woody debris in or near the channel so that more of the pieces interact with the channel. Among these techniques is to simply cut suspended tree boles into one or more pieces so that the logs drop and come in contact with the active channel. Another technique involves rearranging logs and rootwads already in or near the channel so that they are better positioned to create pools or other desired habitat features. A promising although untested activity is to position logs so that they provide structural support at beaver dam sites.

Cutting suspended logs to bring them in contact with the active channel is probably most applicable within stream reaches where numerous buffer trees have fallen. Because about one-half of the buffer trees that fall into a stream end up with their boles suspended above the stream (Fig. 13.8), there is ample opportunity to increase woody debris loading in the channel. Cutting tree boles so that they drop into the channel is inexpensive, involving only a chain saw, and can be used even in terrain that is difficult to access with heavy machinery.

Rearranging existing woody debris in or near the channel usually requires the use of machinery to lift or drag the pieces. Consequently, this method is often restricted to areas where machinery such as crawler tractors or hydraulic excavators can be maneuvered next to the stream. However, if cable logging equipment or helicopters are used to move the pieces, then this method can be used in areas inaccessible to ground-based machines.

A variation on this method of manipulating existing woody debris in streams is that of increasing the stability of natural log jams by cabling together the larger pieces in the jam. Compared to engineered structures (discussed below) these reinforced natural structures are more likely to function effectively during high flows (Frissell and Nawa 1992).

Enhancing Existing Habitat Features

One effective strategy for improving fish habitat is to add woody debris to existing habitat features. Substantial increases in fish densities have been achieved by adding both large and fine woody material to alcoves, beaver ponds, and other dammed ponds (House et al. 1989; McMahon and Hartman 1989; Nickelson et al. 1992). This method is commonly used for increasing cover within large pools and alcoves. Material added to these features can be young alders, tops of larger downed trees, small trees cut during stand thinning operations, and even discarded Christmas trees. Because most of this woody debris is relatively small, the work can be accomplished without machinery and is an activity suitable for volunteers who are interested in improving streams. The benefit achieved from adding wood to alcoves and ponds probably approaches a threshold as some level of cover is reached. Beyond this point additional wood may have no effect on fish growth and survival or even have a detrimental influence on feeding (Wilzbach et al. 1986).

Bringing in Woody Debris from Elsewhere

A more costly method for increasing the supply of woody in streams is to bring in large pieces of woody debris from surrounding slopes or elsewhere. Woody debris used for this purpose commonly includes sound old-growth logs that were left behind during harvest of the previous stand, adjacent windthrown buffer trees, stumps excavated during construction of a nearby road, or conifer trees harvested from nearby slopes.

Many early efforts to add large pieces of wood to streams involved constructing highly-engineered structures. Logs were placed in specific configurations deemed necessary to create some type of habitat feature and then cabled or otherwise anchored to bedrock, large boulders, or streambank trees. The utility of this approach has been questioned recently, not only because of the high cost but because such structures seem less capable of functioning as intended following flood flows. The success rate five to ten years following construction of these wood structures can be very low (Frissell and Nawa 1992). Engineered approaches to adding woody debris to streams seem least successful in larger streams, particularly when the length of the introduced logs is considerably less than the bankfull width of the stream (Everest et al. 1986; Andrus and Froehlich 1991; Frissell and Nawa 1992). Another criticism of the highly engineered structures is that they seldom create habitat as usable to fish as that created by natural jams of large woody debris (Nickelson et al. 1992). Complex winter habitat that allows fish to find refuge from high flows seems most deficient in many of the structure designs (Andrus and Froehlich 1991).

Recent efforts to reduce the expense and increase the effectiveness of large woody debris placed in the stream have focused on adding pieces that are naturally stable in the channel and foregoing expensive engineering or cabling. The length of a piece of wood with respect to the bankfull width of the stream strongly influences whether or not it is able to withstand downstream movement during floods. A study of the movement of old-growth logs in a third-order stream in the Oregon Cascade Mountains indicated that only pieces shorter than the stream's bankfull width moved during an unusually high flow (Lienkaemper and Swanson 1987). Similarly, log movement in a fourth-order stream in the Coast Range of Washington during unusually high flows was restricted to logs that were shorter than the average bankfull width (Bilby 1984). Guidelines developed by the Oregon Department of Forestry and Department of Fish and Wildlife for forest landowners who choose to place woody debris in streams require that logs be at least two times the bankfull width or one and one-half times the width if a rootwad is attached (Oregon Dept. of Forestry 1995).

Conclusion: Dealing with the Past and Planning for the Future

The majority of streams in the coastal portions of Oregon and Washington flow through managed forest land. Due to a variety of past practices many of these streams now have low levels of woody debris. Furthermore, most of the wood found in coastal streams originated from the previous forest and is now entering advanced stages of decay. Habitat features such as deep pools, cover, and structurally complex channels are commonly formed by woody debris and because these features are important for fish survival and growth in coastal streams, fish populations have probably suffered as a result of this decline in woody debris.

Stands that currently grow along streams are generally younger and include a higher proportion of alder trees than the unmanaged streamside stands that they succeeded. Second-growth stands yield less woody debris and the pieces are smaller and decay faster than woody debris from older conifer stands. Streams can experience a pulse of woody debris following clearcut harvesting of second-growth stands as high winds blow over some of the trees that were retained along streams. However, a majority of these windthrown buffer trees never hit the stream or remain suspended above the stream for long periods. Accelerated sedimentation caused by a large number of buffer trees being blown down at once is usually not serious, but it may result in a period of low woody debris in the stream several decades in the future when existing debris has decayed and the new trees growing in their place are not yet old enough to yield much woody debris.

Streamside buffers composed mostly of alder trees provide shade to the stream, guard against disturbance of streamside slopes, provide some woody debris, and yield detritus and insects for aquatic organisms to feed upon. Yet, they also frustrate attempts to boost levels of conifer debris far in the future. Conifer regeneration rarely occurs beneath alder buffers, and the presence of a streamside buffer can even impede conifer regeneration within clearcut areas adjacent to the buffer. Natural establishment of conifers in streamside areas that are dominated by alders may take many decades or perhaps centuries. Clearcut harvesting, heavy thinning, or patch cutting of some portions of alder-dominated streamside stands may be necessary to establish conifers near coastal streams.

Placing large woody debris in streams, rearranging existing woody debris, or adding small woody debris to enhance existing pools and alcoves can help increase preferred fish habitat features in streams currently lacking woody debris. The number of coastal Oregon and Washington streams deficient in wood is so great that most methods currently being used to accomplish this may be too expensive to improve many streams. More streams could be treated if the cost were lower. Lesser expensive techniques include simply cutting existing suspended logs to get them in contact with the stream, adding logs of sufficient length so they are stable in the stream without expensive engineering or cabling, and taking advantage of opportunities to add large woody debris when logging equipment is already set up near the stream. An indirect method for increasing the fish-rearing capacity of a stream is to place logs in locations that promote stable dam sites for beaver. Stable dams are more likely to hold young fish in an upstream pond during high flows.

Decisions about managing specific streams for woody debris need to include a careful look at both the present and the distant future. Actions that achieve short-term gains in woody debris abundance may not also create long-term gains. Similarly, manipulations of streamside areas that create conditions favorable for woody debris recruitment a century from now may come with a cost of declining habitat quality in the near term. A complicating factor is that we have no way to know when extreme events such as wind storms or floods will truncate processes that we are counting on to move streams toward a certain goal. Ideally, sound management of streamside stands should incorporate the consequences of extreme events if and when they occur.

Literature Cited

Adams, P.W., and Stack, W.R. 1989. Streamwater quality after logging in Southwest Oregon, project completion report. Supplement No. PNW-87-400. USDA Forest Service, Pacific Northwest Forest and Range Experiment Station. 19pp.

Amaranthus, M.P., Rice, R.M., Barr, N.R., and Ziemer, R.R. 1985. Logging and forest roads related to increased debris slides in southwestern Oregon. J. For. 83:229–233.

Andrus, C.W., and Froehlich, H.A. 1991. Wind damage in streamside buffers and its effects on accelerated sedimentation in coastal Oregon streams. COPE Report 5(1): 7–9.

Andrus, C.W., Long, B.A., and Froehlich, H.A. 1988. Woody debris and its contribution to pool formation in a coastal stream 50 years after logging. Can. J. Fish. Aquat. Sci. 45:2080–2086.

Bilby, R.E. 1984. Removal of woody debris may affect stream channel stability. J. For. 82:609–613.

Bilby, R.E., Fransen, B.R., and Bisson, P.A. 1996. Incorporation of nitrogen and carbon from spawning coho salmon into the trophic system of small streams: evidence from stable isotopes. Can. J. Fish. Aquat. Sci. 53:164–173.

Bilby, R.E., and Likens, G.E. 1980. Importance of organic debris dams in the structure and function of stream ecosystems. Ecology 61:1107–1113.

Bilby, R.E., and Ward, J.W. 1991. Characteristics and function of large woody debris in streams draining old-growth clear-cut, and second-growth forests in southwestern Washington. Can. J. Fish. Aquat. Sci. 48:2499–2508.

Bisson, P.A., Bilby, R.E., Bryant, M.D., Dolloff, C.A., and coauthors. 1987. Large woody debris in forest streams in the Pacific Northwest: past, present, and future, pp.143–190. In: E.O. Salo and T.W. Cundy, editors. Streamside Management: Forestry and Fishery Interactions. Univ. of Washington Inst. of Forest Resour. Contrib. 57, Seattle, WA.

Bisson, P.A., and Sedell, J.R. 1984. Salmonid populations in streams in clearcut vs. old-growth forests of western Washington, pp. 121–129. In: W.R. Meehan, T.R. Merrell, Jr., and T.A. Hanley, editors. Fish and Wildlife Relationships in Old-Growth Forests. American Institute of Fishery Research Biologists, Juneau, AK.

Bryant, M.D. 1985. Changes 30 years after logging in large woody debris, and its use by salmonids, pp. 329–334. In: Riparian Ecosystems and their Management: Reconciling Conflicting Uses. General Technical Report RM-120. USDA Forest Service.

Bustard, D.R., and Narver, D.W. 1975. Aspects of the winter ecology of juvenile coho salmon (*Oncorhynchus kisutch*) and steelhead trout (*Salmo gairdneri*). J. Fish. Res. Board Can. 32:667–680.

Carlson, J.Y., Andrus, C.W., and Froehlich, H.A. 1990. Woody debris, channel features, and macroinvertebrates of streams with logged and undisturbed riparian timber in northeastern Oregon, U.S.A. Can. J. Fish. Aquat. Sci. 47:1103–1111.

Curtis, J.D. 1943. Some observations on wind damage. J. For. 41:877–887.

Everest, F.H., Reeves, G.H., Sedell, J.R., Wolfe, J., and coauthors. 1986. Abundance, behavior, and habitat utilization by coho salmon and steelhead trout in Fish Creek, Oregon, as influenced by habitat enhancement. 1985 Annual Report U.S. Dept. Energy, Bonneville Power Administration, Division of Fish and Wildlife, Portland, OR. 100pp.

Fausch, K.D. 1993. Experimental analysis of microhabitat selection by juvenile steelhead (*Oncorhynchus mykiss*) and coho salmon (*O. kisutch*) in a British Columbia stream. Can. J. Fish. Aquat. Sci. 50:1198–1207.

Fausch, K.D., and Northcote, T.G. 1992. Large woody debris and salmonid habitat in a small coastal British Columbia stream. Can. J. Fish. Aquat. Sci. 49:682–693.

Frissell, C.A., and Nawa, R.K. 1992. Incidence and causes of physical failure of artificial habitat structures in streams of western Oregon and Washington. N. Amer. J. Fish. Manage. 12:182–197.

Gratkowski, H.J. 1956. Windthrow around staggered settings in old-growth Douglas-fir. Forest Sci. 2:60–74.

Grette, G.B. 1985. The Role of Large Organic Debris in Juvenile Salmonid Rearing Habitat in Small Streams. M.S. Thesis. Univ. of Washington, Seattle, WA. 105pp.

Harmon, M.E., Franklin, J.F., Swanson, F.J., Sollins, P., and coauthors. 1986. Ecology of coarse woody debris in temperate ecosytems. Adv. Ecol Res 15:133–302.

Harvey, B.C., and Nakamoto, R.J. 1996. Effects of steelhead density on growth of coho salmon in a small coastal California stream. Trans. Amer. Fish. Soc. 125:237–243.

Heimann, D.C. 1988. Recruitment Trends and Physical Characteristics of Coarse Woody Debris in Oregon Coast Range Streams. M.S. Thesis Oregon State Univ., Corvallis, OR. 121pp.

Hibbs, D.E., Giordano, P., and Chan, S. 1991. Vegetation dynamics in managed coastal riparian areas. COPE Report 4:3–5. Coastal Oregon Productivity Enhancement Program, Newport, OR.

Hicks, B.J. 1989. The Influence of Geology and Timber Harvest on Channel Morphology and Salmonid Populations in Oregon Coast Range Streams. Ph.D. Thesis. Oregon State Univ., Corvallis, OR. 212pp.

Hogan, D.L. 1986. Channel morphology of unlogged, logged, and debris torrented streams in the Queen Charlotte Islands. Land Management Report 49. British Columbia Ministry of Forests and Lands, Victoria, BC. 85pp.

House, R., and Boehne, P.L. 1986. Effects of instream structure on salmonid habitat and populations in Tobe Creek, Oregon. N. Amer. J. Fish. Manage. 6:38–46.

House, R., Crispin, V., and Monthey, R. 1989. Evaluation of stream rehabilitation projects – Salem District (1981–1988). Technical Note OR-6. USDI Bureau of Land Management, Portland, OR. 50pp.

Keller, E.A., and Swanson, F.J. 1979. Effects of large organic material on channel form and fluvial processes. Earth Surf. Process. Landf. 4:361–380.

Ketcheson, J.L. 1978. Hydrologic Factors and Environmental Impacts of Mass Soil Movements in the Oregon Coast Range. M.S. Thesis. Oregon State Univ., Corvallis, OR. 55pp.

Lamberti, G.A., Gregory, S.V., Ashkenas, L.R., Wildman, R.C., and coauthors. 1991. Stream ecosystem recovery following a catastrophic debris flow. Can. J. Fish. Aquat. Sci. 48:196–208.

Leidholdt-Bruner, K., Hibbs, D.E., and McComb, W.C. 1992. Beaver dam locations and their effects on distribution and abundance of coho salmon fry in two coastal Oregon streams. Northwest Science 66:218–223.

Lienkaemper, G.W., and Swanson, F.J. 1987. Dynamics of large woody debris in streams in old-growth Douglas-fir forests. Can. J. For. Res. 17:150–156.

Long, B.A. 1987. Recruitment and Abundance of Large Woody Debris in an Oregon Coastal Stream System. M.S. Thesis. Oregon State Univ., Corvallis, OR. 68pp.

Lonzarich, D.G. 1994. Stream Fish Communities in Washington: Patterns and Processes. Ph.D. Thesis Univ. of Washington, Seattle, WA. 121pp.

Lynott, R.E., and Cramer, O.P. 1966. Detailed analysis of the 1962 Columbus Day windstorm in Oregon and Washington. Mon. Weather Rev. 94:105–117.

Maas, K., and Emmingham, B. 1995. Third-year survival and growth of conifers planted in red alder-dominated riparian areas. COPE Report 8:5–7. Coastal Oregon Productivity Enhancement Program, Newport, OR.

Matthews, K.R., Berg, N.H., and Azuma, D.L. 1994. Cool water formation and trout habitat use in a deep pool in the Sierra Nevada, California. Trans. Amer. Fish. Soc. 123:549–564.

McDade, M.H. 1988. The Source Area of Coarse Woody Debris in Small Streams in Western Oregon and Washington. M.S. Thesis Oregon State Univ., Corvallis, OR. 69pp.

McDade, M.H., Swanson, F.J., McKee, W.A., Franklin, J.F., and coauthors. 1990. Source distances for coarse woody debris entering small streams in western Oregon and Washington. Can. J. For. Res. 20:326–330.

McMahon, T.E., and Hartman, G.F. 1989. Influence of cover complexity and current velocity on winter habitat use by juvenile coho salmon (*Oncorhynchus kisutch*). Can. J. Fish. Aquat. Sci. 46:1551–1557.

Minore, D. 1990. Riparian tree seedlings, shrubs, and environment in the Oregon Coast Range. Northwest Sci. 64:96.

Moore, M.K. 1977. Factors contributing to blowdown in streamside leave strips on Vancouver Island. Land Management Report 3. Province of British Columbia, Ministry of Forests, Victoria, BC.

Murphy, M.L., Heifetz, J., Johnson, S.W., Koski, K.V., and coauthors. 1986. Effects of clear-cut logging with and without buffer strips on juvenile salmonids in Alaskan streams. Can. J. Fish. Aquat. Sci. 43:1521–1533.

Murphy, M.L., and Koski, K.V. 1989. Input and depletion of woody debris in Alaska streams and implications for streamside management. N. Amer. J. Fish. Manage. 9:427–436.

Naiman, R.J., and Sedell, J.R. 1980. Relationships between metabolic parameters and stream order in Oregon. Can. J. Fish. Aquat. Sci. 37:834–847.

Nakamura, F., and Swanson, F.J. 1993. Effects of coarse woody debris on morphology and sediment storage of a mountain stream system in western Oregon. Earth Surf. Process. Landf. 18:43–61.

Nickelson, T.E., Rodgers, J.D., Johnson, S.L., and Solazzi, M.F. 1992. Seasonal changes in habitat use by juvenile coho salmon (*Oncorhynchus kisutch*) in Oregon coastal streams. Can. J. Fish. Aquat. Sci. 49:783–789.

Nielsen, J.L. 1992. Microhabitat-specific foraging behavior, diet, and growth of juvenile coho salmon. Trans. Amer. Fish. Soc. 121:617–634.

Nielsen, J.L., Lisle, T.E., and Ozaki, V. 1994. Thermally stratified pools and their use by steelhead in northern California streams. Trans. Amer. Fish. Soc. 123:613–626.

O'Loughlin, C.L. 1972. An Investigation of the Stability of the Steepland Forest Soils in the Coast Mountains, British Columbia. Ph.D. Thesis. Univ. of British Columbia, Vancouver, BC. 147pp.

Oregon Dept. of Forestry. 1995. A guide to placing large wood in streams. Oregon Dept. of Forestry, Salem, OR. 15pp.

Peterson, N.P. 1982. Immigration of juvenile coho salmon (*Oncorhynchus kisutch*) into riverine ponds. Can. J. Fish. Aquat. Sci. 39:1308–1310.

Quinn, T.P., and Peterson, N.P. 1996. The influence of habitat complexity and fish size on over-winter survival and growth of individually marked juvenile coho salmon (*Oncorhynchus kisutch*) in Big Beef Creek, Washington. Can. J. Fish. Aquat. Sci. 53:1555–1564.

Ralph, S.C., Poole, G.C., Conquest, L.L., and Naiman, R.J. 1994. Stream channel morphology and woody debris in logged and unlogged basins of western Washington. Can. J. Fish. Aquat. Sci. 51:37–51.

Richmond, A.D., and Fausch, K.D. 1995. Characteristics and function of large woody debris in subalpine Rocky Mountain streams in northern Colorado. Can. J. Fish. Aquat. Sci. 52:1789–1802.

Rot, B.W. 1995. The Interaction of Valley Constraint, Riparian Landform, and Riparian Plant Community Size and Age Upon Channel Configuration of Small Streams of the Western Cascade Mountains, Washington. M.S. Thesis. Univ. of Washington, Seattle, WA. 67pp.

Ruth, R.H., and Yoder, R.A. 1953. Reducing wind damage in the forests of the Oregon Coast Range. Research Paper 7. USDA Forest Service, Pacific Northwest Forest and Range Experiment Station, Portland, OR. 30pp.

Sedell, J.R., Leone, F.N., and Duval, W.S. 1991. Water transportation and storage of logs, pp. 325–368. In: W.R. Meehan, editor. Influence of Forest and Rangeland Management on Salmonid Fishes and Their Habitats. American Fisheries Society Special Publ. 19.

Sedell, J.R., and Luchessa, K.J. 1982. Using the historical record as an aid to salmonid habitat enhancement, pp. 210–223. In: N.B. Armantrout, editor. Acquisition and Utilization of Aquatic Habitat Inventory Information. Western Division, American Fisheries Society, Bethesda, MD.

Sedell, J.R., Yuska, J.E., and Speaker, R.W. 1983. Study of westside fisheries in Olympic National Park, Washington. Final Report CX-9000-0-E-081. U.S. National Park Service, Port Angeles, WA.

Sherwood, K. 1993. Buffer Strip Dynamics in the Western Oregon Cascades. M.S. Thesis. Oregon State Univ., Corvallis, OR. 108pp.

Spalding, S., Peterson, N.P., and Quinn, T.P. 1995. Summer distribution, survival, and growth of juvenile coho salmon under varying experimental conditions of brushy instream cover. Trans. Amer. Fish. Soc. 124:124–130.

Steinblums, I.J., Froehlich, H.A., and Lyons, J.K. 1984. Designing stable buffer strips for stream protection. J. For. 82:49–52.

Steinbrenner, E.C., and Gessel, S.P. 1956. Windthrow along cutlines in relation to physiography on the McDonald tree farm. Weyerhaeuser Timber Co. Forestry Research Note 15, Centralia, WA. 19pp.

Sullivan, K. 1986. Hydraulics and Fish Habitat in Relation to Channel Morphology. Ph.D. Thesis. Johns Hopkins Univ., Baltimore, MD. 431pp.

Swales, S.F., and Levings, C.D. 1989. Role of off-channel ponds in the life cycle of coho salmon (*Oncorhynchus kisutch*) and other juvenile salmonids in the Coldwater River, British Columbia. Can. J. Fish. Aquat. Sci. 46:232–242.

Teensma, P., Rienstra, J., and Yieter, M. 1991. Prehistoric old growth timber in Northwest Oregon, Unpublished document and maps. Bureau of Land Management, Salem, OR.

Thiele, S.A., Kiilsgaard, C.W., and Omernik, J.M. 1992. The subdivision of the Coast Range ecoregion of Oregon and Washington. U.S. Environmental Protection Agency Research Lab, Corvallis, OR. 17pp.

Triska, F.J., Sedell, J.R., Cromak Jr., K., Gregory, S.V., and coauthors. 1984. Nitrogen budget for a small coniferous forest stream. Ecol. Monogr. 54:119–140.

Tschaplinksi, P.J., and Hartman, G.F. 1983. Winter distribution of juvenile coho salmon (*Oncorhynchus kisutch*) before and after logging in Carnation Creek, British Columbia, and some implications for overwinter survival. Can. J. Fish. Aquat. Sci. 40:452–461.

Ursitti, V.L. 1990. Riparian Vegetation and Abundance of Woody Debris in Streams of Southwestern Oregon. M.S. Thesis Oregon State Univ., Corvallis, OR. 115pp.

Veldhuisen, C.N. 1990. Coarse Woody Debris in Streams of the Drift Creek Basin, Oregon. M.S. Thesis. Oregon State Univ., Corvallis, OR. 109pp.

Walter, R.A. 1984. A stream ecosystem in an old-growth forest in southeast Alaska: Part 2: structure and dynamics of the periphyton community, pp. 57–69. In: W.R. Meehan, T.R. Merrell, Jr., and T.A. Hanley, editors. Fish and Wildlife Relationships in Old-Growth Forests. American Institute of Fishery Research Biologists, Juneau, AK.

Wilzbach, M.A., Cummins, K.W., and Hall, J.D. 1986. Influence of habitat manipulations on interactions between cutthroat trout and invertebrate drift. Ecology 67:898–911.

Chapter 14
Long-term Trends in Habitat and Fish Populations in the Alsea Basin

Stanley V. Gregory, John S. Schwartz, James D. Hall, Randall C. Wildman, and Peter A. Bisson

The Alsea Watershed Study (AWS) was the earliest long-term basin study to document effects of timber harvest practices on stream habitat quality and salmonid populations (Hall and Lantz 1969; Moring and Lantz 1975). The 16-year project was initiated in 1958. Three coastal basins with mature forests of Douglas-fir (*Pseudotsuga menziesii*), western redcedar (*Thuja plicata*), and western hemlock (*Tsuga heterophylla*) were selected in the upper Drift Creek drainage of the Alsea River (See Chapter 1). Flynn Creek, a 202-ha basin, was the reference stream. Deer Creek, a 303-ha basin, was clearcut in three patches, each approximately 25 ha in area, with mixed deciduous/conifer buffers along the stream. Needle Branch, a 71-ha basin, was completely logged by clearcutting with no buffers. Logging continued in the Deer Creek basin after the original AWS was completed in 1973. An additional 45 ha were harvested in three clearcut units in 1978, 1987, and 1988.

Salmonid communities in these study streams are dominated by two species, coho salmon (*Oncorhynchus kisutch*) and coastal cutthroat trout (*O. clarkii*). During the AWS, small numbers of steelhead (*O. mykiss*) were observed in Deer Creek (Moring and Lantz 1975). A few straying Chinook salmon (*O. tshawytscha*) passed through the upstream trap in Deer Creek but all returned downstream without spawning. Nonsalmonid fishes found in these streams are the reticulate sculpin (*Cottus perplexus*), Pacific lamprey (*Lampetra tridentata*), and western brook lamprey (*Lampetra richardsoni*).

The original AWS included a 7-year prelogging phase (1959–1965) to document the natural annual variation in environmental factors and salmonid populations, a year of logging (1966), and a 7-year post-logging phase (1967–1973) to measure effects on fish populations and habitat. Limited sampling of salmonid populations was continued in the summer of 1974. Responses of fish assemblages in the early pre- and post-logging periods

Stanley V. Gregory
Department of Fisheries and Wildlife, Oregon State University, Corvallis, OR 97331
stanley.gregory@oregonstate.edu

J. D. Stednick (ed.), *Hydrological and Biological Responses to Forest Practices.* 237
© Springer 2008

have been reported in numerous publications (e.g., Chapman 1961, 1965; Lowry 1964, 1965; Hall and Lantz 1969; Au 1972; Lindsay 1975; Moring 1975; Moring and Lantz 1975; Knight 1980; Hall et al. 1987) and are summarized in Chapter 5.

Pretreatment and posttreatment studies such as the AWS and the Carnation Creek watershed study (Hartman and Scrivener 1990) have documented changes in habitat and fish populations in the first decade after forest harvest. In general, these studies have observed short-term effects (1–4 years) on habitat quality, including increased maximum stream temperature, decreased dissolved oxygen, and increased suspended sediments. Densities of juvenile coho salmon generally increased after logging, and cutthroat trout populations decreased.

Basin comparisons such as the AWS provide an important context for evaluating responses of salmonid populations to land-use practices across ranges of natural variation. We extended the time frame for this type of land-use experiment by reexamining fish populations in the AWS streams 22–30 years after harvest. This chapter builds on an analysis of data collected in 1988 and 1989 as part of a comparison to the earlier work on the AWS (Schwartz 1991). From 1988 to 1996, we measured habitat conditions and fish populations during summer in Flynn Creek, Deer Creek, and Needle Branch. Long-term fluctuations in juvenile coho salmon and cutthroat trout populations, even in the reference stream, demonstrate the value of long-term investigations such as this one.

Methods

Measurements of stream habitat and fish populations that were used from 1959 to 1974 are described briefly in Chapter 5 and in greater detail in publications from the original study. We did not attempt to follow the original measurement protocols, but we used the spatial framework of the AWS to extrapolate our reach measurements to the full stream length for each basin.

Stream Habitat Measurements

During 1988–1996, physical habitat was measured by a visual estimation method (Hankin and Reeves 1988) during the second or third week of August each year. Major bedforms or channel units were classified as pool, riffle, glide, rapid, cascade, step, or side channel. Physical habitat measurements at each channel unit included wetted channel length, wetted channel width, active channel width, mean depth, and maximum depth. The active channel was visually estimated by observing high-flow markings. We estimated mean width of the wetted channel from three measurements in each unit and mean depth from nine measured depths (three at each width transect). We calculated means for each reach by weighting channel unit dimensions according to the proportion of reach length represented by channel unit length.

Salmonid Population and Biomass Estimates

During 1988–1996, we estimated salmonid populations by the two-pass removal method, using backpack electroshockers as described by Armour et al. (1983). Peterson and Cederholm (1984) had previously shown that estimates of coho salmon populations made by the removal method were similar to estimates made by the mark-recapture method when at least 1 hour was allowed between electrofishing passes. We allowed approximately 1 hour or more to elapse between electrofishing passes at most sites while we were measuring and weighing fish from the preceeding pass.

Abundances of cutthroat trout were estimated by stratifying the population into fry (age 0) and older trout (age 1+). Age groups for each stream were determined by length-frequency analysis. Trout were grouped into 2-mm size intervals and assigned to either age 0 or age 1+ by evaluation of breaks between size classes. Demarcations between age-0 and 1+ trout differed between years, ranging from approximately 70 to 90 mm. Size classes for these age groups are similar to ranges reported by Sumner (1962) and Lowry (1964), which were based on scale analysis. Estimates of numbers for the two age groups were computed separately. Based on freshwater age of returning adult coho salmon (Moring and Lantz 1975), we estimated that about 95% of the juveniles left the streams as smolts after 1 year of rearing. Because such a small percentage remained in the streams for an additional year, we combined all juvenile coho salmon in one group for estimating population size.

We estimated salmonid populations and biomass for the entire fish-bearing stream length using stratified sampling by reach type. Within each stream, four to five electrofishing sites were selected randomly in morphologically different sections to stratify the sampling by reach-scale characteristics. Major reach types included constrained (valley floor width <2 active channel widths) and unconstrained (valley floor width >2 active channel widths) (Gregory et al. 1991). Lengths of reach types were measured in each stream. We electrofished approximately 20% of the total length in each stream.

Areal and lineal estimates of fish density and biomass for each stream reach were derived by dividing the estimate of population number at an electrofishing site by the proportion of the area or length of the major reach type represented by the sampled site. Reach estimates were summed to obtain estimates for the entire stream. Trout biomass (g m^{-2}) was estimated for each site by multiplying the population estimates for age-0 and 1+ fish by the average fish weight for each age group and dividing by the sampled site area. We used coefficient of variation [(standard deviation/mean) · 100] as a measure of interannual variation in population abundance. This statistic, expressed as a percentage and abbreviated as CV, provides a measure of relative variation, independent of absolute abundance.

Interannual comparisons of fish abundance required accurate estimates of reach lengths and areas to convert population estimates from this study and the original AWS to lineal and areal densities. We reviewed all original field notes to confirm estimates of reach lengths and areas from 1959 to 1974. We also recalculated all population estimates for salmon and trout for 1959 to 1974 from original data, resulting in some corrections to values reported in Moring and Lantz (1975) and Hall et al. (1987).

During the original AWS, researchers designated standard study lengths for each stream that encompassed nearly all the stream length accessible to anadromous fish. Estimates of density and biomass were based on average areas during the study period. There were minor year-to-year variations in average width of the three streams, but in contrast to the 1988–1996 period, these variations were not sufficient to require modification of average area for a given length of stream, even during the post-logging period in Needle Branch. Fish populations were estimated in lengths of stream that differed somewhat by year and species. In Flynn Creek, estimates were made over a length of 1430 m and area of 2660 m^2 for coho salmon smolts in all years, juvenile salmon for 1959–1968, and juvenile cutthroat trout for 1962–1963. All other estimates in Flynn Creek covered a length of 1310 m and area of 2540 m^2, except for trout in 1964, when the respective values were 1460 m and 2700 m^2. In Deer Creek, estimates were made over a length of 2320 m and area of 4720 m^2 for all years and both species. In Needle Branch, estimates were made for a length of 970 m and area of 1060 m^2 for salmon smolts for all years, juvenile salmon for 1959–1968, and juvenile trout for 1962–1963. All other estimates through 1974 were made over a length of 870 m and area of 930 m^2.

Stream width was measured each year from 1988 to 1996, but we standardized the total stream length to which reach estimates were applied. Total stream lengths were close to those for the original AWS: Flynn Creek – 1310 m, Deer Creek – 2070 m (East Fork of Deer Creek was not sampled after 1974), and Needle Branch – 870 m.

Needle Branch was not sampled in 1988 and 1992 because of drought conditions in the Oregon Coast Range. In late summer of 1988, approximately 65% of the channel length was dry from the fish trap to the first falls. Electrofishing in the remaining shallow wetted pools would have excessively stressed the fish remaining in the isolated sections. Consequently, we did not estimate fish populations for 1988, but their abundance must have been quite low. In 1992, only one residual pool contained water in the same distance of stream. No fish were observed in this pool, thus the population in Needle Branch was assumed to be zero in 1992.

Results

Stream Habitat

Stream habitat composition and dimensions during the 1988–1996 period differed substantially from prelogging habitat conditions for Flynn Creek and Deer Creek, and minor changes occurred in Needle Branch (Table 14.1). Stream depths were greater in this period than depths recorded in 1959–1962 for all streams, but differences in measurement protocols between these periods limit comparisons. Beaver activity in Flynn Creek approximately doubled the stream area in the sampled stream reaches from 1991–1994, compared to 1988–1990. Average width more than doubled and average depth increased by about 80%. Stream area returned to earlier values when a flood during 1994–1995 removed the beaver dams. Beaver activity in Deer Creek, which varied greatly from year to year, caused considerable fluctuation in stream area. The average for mean widths during 1988–1996 was somewhat greater than the width during 1959–1962, and widths varied between years, ranging from 2.2 m to 3.0 m. The area of Needle Branch remained relatively more constant because of the absence of beaver activity. However, because of drought from 1994 to 1996, wetted channel area was 19% (range of 9% to 28%) less than the wetted area from 1989 to 1991. Composition of habitat types was similar in all streams, with average pool-riffle ratios close to 1:1. Though the percentage of the stream length in pool habitat varied considerably from year to year in each stream during 1988–1996, the averages were quite similar to those for 1959–1962.

Salmonid Population and Biomass

Cutthroat trout populations decreased markedly in Needle Branch, the clearcut stream, in the period immediately following forest harvest in the AWS, whereas trout abundance increased in both Deer Creek and Flynn Creek (See Chapter 5). During 1989–1996, trout numbers (all ages combined) per meter of stream in Needle Branch recovered to levels similar to prelogging populations (Table 14.2). These lineal densities are closely related to estimates of populations for the entire study streams that were made during 1962–1974 and provide a basis for comparison with the earlier period. Both mean numbers (Table 14.2) and median numbers (Fig. 14.1) approached prelogging populations. Numbers of trout per stream length increased immediately after harvest in Deer Creek and Flynn Creek, but lineal densities for 1988–1996 returned to near prelogging levels.

Abundance estimates of trout for 1988–1996 adjusted for stream area (Table 14.2) differ from measures of abundance based on stream length, because stream area changed greatly over the 9-year period. Numbers of trout per m^2

Table 14.1 Stream habitat composition and dimensions for the AWS streams from 1988 to 1996. Estimates of width and depth are expressed in meters and area in square meters. Estimates of percent pools are based on stream length. Field crews estimated that habitat conditions had not changed in Deer Creek during 1991 and 1992 and in Needle Branch in 1991, and habitat was not remeasured. All reaches were dry in Needle Branch in 1992. Areas marked with an asterisk indicate significant beaver influence. Also shown are means for years 1959–1962 (from Moring 1975) and 1988–1996

	Flynn Creek				Deer Creek				Needle Branch			
Year	% pools	Mean width	Mean depth	Wetted area	% pools	Mean width	Mean depth	Wetted area	% pools	Mean width	Mean depth	Wetted area
1988	54.6	1.85	0.10	2420	58.1	3.01	0.17	*6240	68.3	1.30	0.10	Dry[1]
1989	38.8	1.85	0.09	2420	40.3	2.45	0.12	*5080	67.7	1.33	0.14	1160
1990	42.7	1.85	0.15	2420	44.2	2.18	0.15	4510	54.9	1.33	0.10	1160
1991	60.2	4.65	0.24	*6090	44.2	2.18	0.15	4510	54.9	1.33	0.10	1160
1992	48.1	3.29	0.21	*4310	44.2	2.18	0.15	4510	–	–	–	Dry
1993	62.7	4.09	0.20	*5360	50.9	2.92	0.17	*6050	60.0	1.37	0.12	1190
1994	53.6	3.50	0.16	*4590	44.9	2.39	0.15	*4950	68.7	0.95	0.11	830
1995	53.4	2.06	0.13	2700	51.8	2.62	0.15	*5420	66.8	1.22	0.09	1060
1996	45.8	1.83	0.11	2400	62.3	2.65	0.17	*5480	58.7	1.07	0.13	930
59–62	54.1	1.86	0.11	2660	56.2	2.03	0.11	4720[2]	59.7	1.09	0.07	1060
88–96	51.1	2.77	0.15	3630	49.0	2.51	0.15	5190	62.5	1.24	0.11	1070

[1] Channel was dry along 65% of its length
[2] Area of East Fork Deer Creek was included during 1959–1962, but not during 1988–1996.

Table 14.2 Estimates of cutthroat trout populations in AWS streams from 1962 to 1996, expressed as lineal density (# m^{-1}), areal density (# m^{-2}), and biomass (g m^{-2}). Dash indicates that stream was not sampled. (1962–1965: prelogging, 1966–1974: early post-logging, and 1988–1996: later post-logging)

Year	Number per meter			Number per square meter			Grams per square meter		
	Flynn	Deer	Needle	Flynn	Deer	Needle	Flynn	Deer	Needle
1962	0.55	0.46	0.20	0.30	0.23	0.18	6.23	4.26	3.58
1963	0.43	0.36	0.39	0.23	0.18	0.35	2.72	1.68	2.80
1964	0.45	0.37	0.35	0.24	0.18	0.33	3.20	2.71	4.01
1965	0.36	0.32	0.23	0.18	0.16	0.21	2.45	2.29	2.98
1966	0.47	0.36	0.07	0.24	0.18	0.06	2.41	1.98	1.40
1967	0.55	0.27	0.11	0.28	0.18	0.10	3.48	2.36	0.72
1968	0.82	0.39	0.10	0.42	0.19	0.09	3.39	2.23	1.61
1969	0.71	0.59	0.20	0.37	0.29	0.19	4.29	3.79	2.95
1970	0.64	0.58	0.06	0.33	0.29	0.06	3.28	2.97	1.11
1971	0.66	0.52	0.12	0.34	0.26	0.11	3.64	3.15	1.22
1972	0.58	0.42	0.08	0.30	0.20	0.07	3.60	3.11	1.97
1973	0.44	–	0.06	0.23	–	0.06	3.17	–	1.51
1974	0.55	–	0.24	0.28	–	0.22	3.63	–	2.93
1988	0.58	0.14	–	0.31	0.05	–	3.39	0.96	–
1989	0.51	0.34	0.15	0.28	0.14	0.11	3.11	2.42	0.76
1990	0.34	0.32	0.15	0.19	0.15	0.11	2.15	2.02	2.06
1991	0.69	0.30	0.32	0.15	0.14	0.24	2.40	1.93	2.65
1992	0.53	0.42	0.00	0.16	0.19	0.00	1.88	2.80	0.00
1993	0.52	0.39	0.37	0.13	0.14	0.27	1.10	1.83	0.94
1994	0.56	0.47	1.09	0.16	0.20	1.14	1.22	2.14	3.72
1995	0.35	0.66	0.21	0.17	0.25	0.17	2.34	2.45	2.42
1996	0.35	0.36	0.24	0.19	0.14	0.22	1.88	2.31	1.77

(continued)

Table 14.2 (continued)

Year	Number per meter			Number per square meter			Grams per square meter		
	Flynn	Deer	Needle	Flynn	Deer	Needle	Flynn	Deer	Needle
Mean and standard deviation									
1962–65 mean	0.446	0.377	0.290	0.239	0.186	0.267	3.650	2.735	3.343
S.D.	0.079	0.062	0.093	0.046	0.030	0.086	1.748	1.101	0.556
1966–74 mean	0.601	0.445	0.114	0.311	0.226	0.107	3.432	2.799	1.713
S.D.	0.120	0.120	0.062	0.062	0.049	0.058	0.498	0.635	0.776
1988–96 mean	0.492	0.378	0.316	0.193	0.155	0.283	2.163	2.096	1.790
S.D.	0.121	0.140	0.333	0.061	0.055	0.357	0.764	0.521	1.191
Ratio of means (%)									
1966–74/1962–65	134.8	118.0	39.5	130.3	121.8	40.0	94.0	102.3	51.3
1988–96/1962–65	110.3	100.2	109.1	80.9	83.6	105.8	59.3	76.6	53.5
Coefficient of variation									
1962–65	17.7	16.4	32.2	19.1	16.3	32.2	47.9	40.3	16.6
1966–74	19.9	27.1	54.1	19.9	21.7	53.9	14.5	22.7	45.3
1988–96	24.6	37.2	105.1	31.6	35.4	126.4	35.3	24.8	66.5

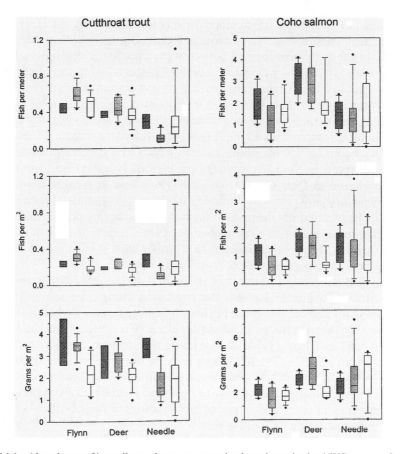

Fig. 14.1 Abundance of juvenile cutthroat trout and coho salmon in the AWS streams for the periods 1959–1965, 1966–1974, and 1988–1996. Bars within boxes are median values, boxes represent 25th to 75th percentiles, whiskers show 10th and 90th percentiles, and dots show extreme outliers. Whiskers are omitted for small sample sizes. Dark shading indicates 1962–1965 for trout and 1959–1965 for salmon (prelogging), light shading indicates 1966–1974 (early post-logging), and no shading indicates 1988–1996 (later post-logging)

in Needle Branch approached prelogging levels (Table 14.2, Fig. 14.1). In contrast to the density measure, estimates of biomass of trout per m^2 in Needle Branch remained close to the low levels observed immediately after logging. Trout biomass (g m^{-2}) in Flynn Creek and Deer Creek decreased to 60–75% of the biomass observed during the prelogging years.

The fact that trout numbers in Needle Branch during 1989–1996 recovered to levels approaching prelogging populations but biomass did not increase indicates that the recovery was based on responses of age-0 trout. This increase in numbers of young fish masked a continued decrease in older fish and potential spawners in the population. Separation of age classes of trout reveals that there were major increases in production of cutthroat trout fry in Needle Branch but no significant change in fry numbers in Flynn Creek (Fig. 14.2). In sharp contrast, mean numbers of age-1+ trout in Needle Branch, which had decreased by 60% in the early post-logging period, were even lower during 1989–1996, averaging about 20% of prelogging abundance. Mean and median numbers of age-1+ trout in Flynn Creek were just slightly lower than prelogging values.

Abundances of juvenile coho salmon did not change appreciably between prelogging and early post-logging periods in the two logged streams, but there was a notable decrease in Flynn Creek, the reference stream (Table 14.3, Fig. 14.1). From 1988–1996, abundances in Flynn Creek were similar to populations observed in the early postlogging period. Numbers and biomass of juvenile salmon in Deer Creek from 1988–1996 were approximately 60% of the early postlogging values. In contrast, numbers and biomass of coho salmon in Needle Branch did not decrease in the early post-logging period, and during 1988–1996 remained equal to or higher than abundances of salmon observed prior to logging. These results indicate that timber harvest had no clear effect on summer populations of juvenile coho salmon.

Though numbers and biomass of individual species changed in the AWS streams following timber harvest, the combined biomass of salmonids (i.e., cutthroat trout and coho salmon) per length of stream did not change substantially in either the treatment watersheds or reference watershed over the 30 years between 1966 and 1996 (Table 14.4). Grams per meter is considered the best index of overall biomass in each stream because of the substantial variation in stream widths during 1988–1996. Biomass of salmonids per meter of stream was greatest in the two larger basins, Flynn Creek and Deer Creek. Averages for the lineal estimates of combined salmonid biomass for the periods 1966–1974 and 1988–1996 were within 10% of the preharvest averages for 1962–1965, with the exception of the 17% decrease in Needle Branch for 1966–1974, which was due to the substantial decrease in older cutthroat trout.

This extended study provided an opportunity to document long-term variation of salmonid populations in both undisturbed and disturbed habitats. Interannual variation in abundance of juvenile coho salmon and cutthroat trout was high in these small streams (Tables 14.2 and 14.3). In the undisturbed Flynn Creek, the overall CV from 1959–1996 was 49% for juvenile salmon and 24% for trout, based on lineal density. Over this 38-year period, densities of juvenile salmon in late summer differed by an order of magnitude (Table 14.3). Substantial variation in salmon density was also observed in prelogging data from Deer Creek and Needle Branch, which had CVs of 26% and 49%, respectively. Trout densities from prelogging years in Deer Creek and Needle Branch showed CVs of 16% and 32%, respectively. In general, juvenile salmon densities were more variable than trout densities. Estimates of biomass of juvenile salmon almost always had lower CVs than estimates of density, for all streams and periods of record (Table 14.3). For cutthroat trout, this relationship was consistent only for Needle Branch (Table 14.2).

The clearcut logging in Needle Branch appeared to cause substantial increases in variation in density and biomass of cutthroat trout. In the early

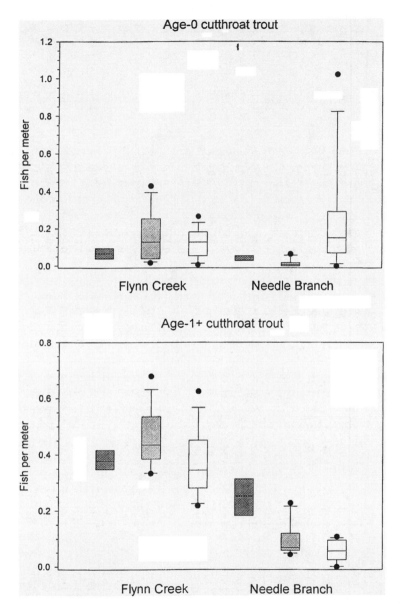

Fig. 14.2 Lineal density of age-0 and age-1+ cutthroat trout in Flynn Creek and Needle Branch for the periods 1962–1965, 1966–1974, and 1988–1996. Bars within boxes are median values, boxes represent 25th to 75th percentiles, whiskers show 10th and 90th percentiles, and dots show extreme outliers. Whiskers are omitted for small sample sizes. Dark shading indicates 1962–1965 (prelogging), light shading indicates 1966–1974 (early post-logging), and no shading indicates 1988–1996 (later post-logging)

Table 14.3 Estimates of coho salmon populations in AWS streams from 1959 to 1996, expressed as lineal density (# m^{-1}), areal density (# m^{-2}), and biomass (g m^{-2}). Dash indicates that stream was not sampled. (1959–1965: prelogging, 1966–1974: early post-logging, and 1988–1996: later post-logging)

Year	Number per meter			Number per square meter			Grams per square meter		
	Flynn	Deer	Needle	Flynn	Deer	Needle	Flynn	Deer	Needle
1959	2.51	3.40	0.93	1.46	1.66	0.97	3.05	3.06	1.88
1960	2.72	2.28	1.55	1.41	1.12	1.35	2.66	2.25	1.97
1961	2.30	2.75	0.78	1.24	1.36	0.71	2.23	3.19	1.84
1962	3.21	4.09	2.38	1.73	2.01	2.17	2.39	3.34	3.08
1963	1.01	3.27	1.66	0.55	1.61	1.51	2.15	3.27	2.84
1964	1.22	1.94	0.54	0.66	0.95	0.50	1.64	2.33	1.33
1965	1.40	3.96	2.17	0.75	1.95	1.98	1.56	3.61	3.47
1966	1.99	2.84	1.27	1.07	1.40	1.16	2.74	3.72	3.31
1967	2.51	4.60	4.22	1.35	2.27	3.84	2.44	6.03	7.30
1968	1.62	3.27	1.92	0.87	1.61	1.75	2.03	4.03	4.05
1969	0.23	1.73	0.62	0.12	0.63	0.54	0.39	2.22	2.18
1970	–	–	0.24	–	–	0.22	–	–	0.64
1971	–	–	0.18	–	–	0.17	–	–	0.98
1972	0.64	2.08	1.27	0.33	1.02	1.18	0.73	2.67	2.48
1973	0.65	–	0.08	0.34	–	0.08	1.24	–	0.74
1974	1.19	–	0.85	0.62	–	0.79	1.51	–	1.87
1988	1.62	2.36	–	0.88	0.78	–	2.48	2.65	–
1989	1.67	1.94	3.38	0.90	0.79	2.52	2.00	2.31	4.70
1990	1.72	1.69	2.46	0.93	0.78	1.83	2.22	1.91	4.20
1991	2.98	1.48	0.89	0.64	0.68	0.66	2.10	1.61	2.22
1992	0.73	0.85	0.00	0.23	0.39	0.00	0.86	1.54	0.00
1993	2.61	5.23	3.31	0.64	1.79	2.41	1.35	4.29	4.94
1994	1.57	1.45	0.93	0.45	0.61	0.97	1.40	1.90	4.60
1995	1.13	1.40	0.44	0.55	0.54	0.36	1.72	1.60	1.47
1996	1.08	1.66	1.35	0.59	0.63	1.26	1.59	2.00	3.81

Table 14.3 (continued)

Year	Number per meter			Number per square meter			Grams per square meter		
	Flynn	Deer	Needle	Flynn	Deer	Needle	Flynn	Deer	Needle
Mean and standard deviation									
1959–65 mean	2.05	3.10	1.43	1.11	1.52	1.31	2.24	3.01	2.34
S.D.	0.84	0.81	0.71	0.46	0.40	0.63	0.53	0.52	0.78
1966–74 mean	1.26	2.90	1.46[1]	0.67	1.39	1.33[1]	1.58	3.73	3.13[1]
S.D.	0.82	1.13	1.35	0.45	0.62	1.23	0.87	1.48	2.12
1988–96 mean	1.68	2.01	1.60	0.65	0.78	1.25	1.75	2.20	3.24
S.D.	0.72	1.28	1.30	0.23	0.40	0.93	0.50	0.86	1.80
Ratio of Means (%)									
1966–74/1959–65	61.4	93.7	102.1	60.3	91.0	101.7	70.7	124.2	133.6
1988–96/1959–65	82.0	64.8	111.9	58.6	51.3	95.5	78.1	73.1	138.6
Coefficient of variation									
1959–65	41.1	26.3	49.3	41.1	26.3	47.9	23.7	17.2	33.5
1966–74	65.1	38.8	92.1	66.4	44.7	92.0	55.0	39.7	67.6
1988–96	42.9	63.7	81.3	35.4	51.3	74.5	28.6	39.1	55.7

[1] Means do not include 1970 and 1971 because there was no estimate for the reference stream.

Table 14.4 Estimates of combined biomass of juvenile coho salmon and cutthroat trout in the AWS streams from 1962 to 1996, expressed as grams per meter of stream length. Dashes indicate that stream was not sampled for both species

	Year	Flynn Creek	Deer Creek	Needle Branch
Prelogging Period	1962	16.03	15.46	7.28
	1963	9.13	10.07	6.16
	1964	8.97	10.25	5.74
	1965	7.65	12.00	6.98
Early Post-logging Period	1966	9.77	11.60	5.11
	1967	11.29	17.07	8.75
	1968	10.35	12.74	6.15
	1969	9.07	12.23	5.48
	1970	–	–	–
	1971	–	–	–
	1972	8.40	11.76	4.76
	1973	8.55	–	2.41
	1974	9.97	–	5.13
Later Post-logging Period	1988	10.84	10.88	–
	1989	9.44	11.61	7.28
	1990	8.07	8.56	8.35
	1991	20.92	7.71	6.49
	1992	9.01	9.46	0.00
	1993	10.02	17.89	8.04
	1994	9.18	9.66	7.94
	1995	8.37	10.60	4.73
	1996	6.36	11.41	5.96
Statistics for Study Periods	Mean			
	1962–65	10.45	11.95	6.54
	1966–74	9.63	13.08	5.40
	1988–96	10.25	10.86	6.10
	1962–96	10.07	11.72	5.93
	CV			
	1962–65	36.2	20.9	10.9
	1966–74	10.7	17.4	34.9
	1988–96	41.0	27.0	45.1
	1962–96	31.6	23.1	35.4

post-logging period, CVs increased by about 70% to 170% over the prelogging values (Table 14.2). During 1989–1996 CVs increased even further, reaching 3–4 times the prelogging values. Variation in the other two streams either decreased in the post-logging period, or increased by a substantially smaller amount. For coho salmon there was no clear relationship between disturbance and the coefficient of variation. The CVs tended to increase in both post-logging periods in all three streams. However, changes were variable, often greater in Deer Creek, the patchcut watershed that experienced almost no disturbance, than in Needle Branch (Table 14.3). Variance in combined salmonid biomass per length of stream was similar between study periods for Flynn Creek and Deer Creek, but variance (CV) in Needle Branch was 3–4 times greater after timber harvest than before.

Discussion

There was no clear evidence of a change in stream channel dimensions due to logging. The average pool-riffle ratio has not changed markedly in the last 30 years. Changes in habitat structure were not detectable at a channel-unit scale. Microhabitat structure, such as cover from undercut banks, boulders, and large wood, was not measured during the AWS, and thus cannot be compared to recent data. However, preexisting large wood was nearly totally removed by stream cleaning after logging in Needle Branch, and wood volumes in Needle Branch remain lower than in the other streams in the AWS (Veldhuisen 1990). Blowdown of alder in Needle Branch in recent years has begun to provide undercut banks and rootwads, creating more complex habitat for salmonids.

Beaver activity caused major changes in stream habitat area and complexity in Flynn Creek and Deer Creek during 1988–1996. Beaver also moved into Deer Creek during the early post-logging period, but a desire to focus solely on logging effects caused researchers to remove beaver from Deer Creek during the original study. Both width and depth increased in Flynn Creek from 1991 to 1994 after beaver dammed the lower study reaches. Changes in Deer Creek were generally less extensive and more variable from year to year. In general, lineal abundances of both juvenile coho salmon and cutthroat trout (numbers per meter and grams per meter) were greater in both streams during years of beaver activity. At the same time, abundances and biomass per square meter often were lower during these years. Results were extremely variable. With the exception of the increase in area in both streams and in grams per meter of both coho salmon and the two combined species in Deer Creek, none of the changes were statistically significant. Other studies in the Pacific Northwest and Alaska have suggested the importance of beaver as an agent of habitat formation for salmonids (Sanner 1987; Leidholdt-Bruner et al. 1992; Nickelson et al. 1992).

Beaver potentially increase food production and area and volume of stream habitat (Naiman et al. 1984; Naiman et al. 1986), which potentially influence abundance of fish. If fish populations are habitat limited but not food limited, any increase in habitat should increase total numbers of fish (i.e., numbers m^{-1}). Under such conditions, density (number m^{-2}) would increase only if depth increased. Fish abundance could also increase if beaver create a different type of habitat that was not present previously, such as accumulations of large wood. If fish populations are food limited and beaver activity causes an increase in channel dimensions but does not change the rate of food production per unit area, total abundance (numbers m^{-1}) would increase but density (number m^{-2}) would not change.

The cutthroat trout population in Needle Branch has not recovered over the long term, though in recent years (1989–1996) there have been intermittent increases in abundance of age-0 trout. The failure of age-1 and older trout to recover 30 years after logging is remarkable. The occasional high numbers of fry indicate that recruitment is possible, but survival of age-0 trout after the summer season is unusually low.

There are several possible explanations for the failure of older cutthroat trout to rebound. The initial decrease may have been due in part to the substantial increase in stream temperature during the first few years following logging. In Carnation Creek, British Columbia, temperature changes as a result of riparian cover removal influenced trout abundance and shifted life-history patterns of salmonids (Hartman and Scrivener 1990). However, temperatures in Needle Branch returned to prelogging levels within a few years and cannot be a factor in the long-term decline. Cover associated with large wood was removed by the stream-cleaning operations during the AWS, and undercut banks were destroyed. Thus the long-term detrimental effects on trout in Needle Branch may reflect degraded habitat.

Interaction with juvenile coho salmon, which remained abundant after logging, may also have influenced trout abundance in Needle Branch. In general, similarity of preharvest and post-harvest estimates of combined salmonid biomass (Table 14.4) reflects the interaction of cutthroat trout and coho salmon and potential compensatory effects of competition for food and habitat. The stream's small size and altered morphology may exacerbate such interaction. During late summer the stream often flows subsurface through the low-gradient gravel riffles, leaving only isolated small pools occupied by both trout and salmon. In such habitat, the body morphology of juvenile coho salmon may give them an advantage over age-0 cutthroat trout (Bisson et al. 1988). In addition, coho salmon have generally been thought to be more aggressive than cutthroat trout (Glova 1986, 1987). However, more recent work has shown that size-matched cutthroat trout may be equally competitive with coho salmon (Sabo and Pauley 1997). In our streams coho salmon emerge from the gravel earlier and at a larger size than trout and thus may gain an early advantage. In four of the seven recent years, juvenile coho salmon were on average larger than age-0 cutthroat trout at the time of the summer population estimates (Table 14.5). But even in years when the average size of trout fry was larger than the salmon, the trout were so outnumbered that substantial numbers of juvenile salmon were larger than any age-0 trout. However, this circumstance also prevailed in the prelogging period, when total biomass of cutthroat trout in Needle Branch exceeded that of the juvenile salmon. If competition has played a role in the continued low population of larger cutthroat trout, factors other than size-related dominance alone must have been responsible for competitive effects.

Table 14.5 Estimated number and size of age-0 cutthroat trout and juvenile coho salmon during August in Needle Branch, 1989–1996 (stream was dry in 1992). The population estimate is for the entire 870-m length of stream

Population properties	Year						
	1989	1990	1991	1993	1994	1995	1996
Cutthroat trout							
Estimated pop.	87	35	193	313	889	97	165
Mean length (mm)	66.8	48.2	56.2	60.9	57.3	66.9	68.2
Std. deviation (mm)	4.97	6.70	10.75	12.62	8.83	5.56	9.39
Size range	56–77	39–58	38–69	34–78	35–79	54–74	38–79
No. measured	33	13	30	72	186	18	28
Coho salmon							
Estimated pop.	2,930	2,130	770	2,870	806	384	1,170
Mean length (mm)	54.3	55.1	64.5	53.2	68.1	70.1	60.8
Std. deviation (mm)	9.95	10.28	9.90	9.49	13.53	8.66	8.03
Size range	38–101	39–94	49–107	36–91	48–108	53–86	44–88
No. measured	520	440	180	560	135	69	222

Reeves et al. (1997) put forth one hypothesis that may help to explain the result. They suggested that a complex pool environment, which existed in Needle Branch prior to clearcut logging, would create physical heterogeneity, allowing trout and salmon to be segregated within pools. Removal of large wood and pool simplification after logging would increase potential for interactions, to the possible disadvantage of the smaller cutthroat trout. Data for the AWS streams are not adequate to support or refute these competitive mechanisms for change in trout population size. However, there is little doubt that the initial reduction in trout abundance was in part related to the logging, burning, and stream channel disturbance in the Needle Branch watershed. The persistent reduction in numbers of larger cutthroat trout may be related to stream size, but further work will be required to clarify the mechanisms.

Juvenile coho salmon showed no long-term shifts in density or biomass beyond the range of natural variation as a result of timber harvest. In the original AWS, summer populations of juvenile salmon did not decline significantly in the two logged streams, though smolt outmigration decreased, particularly in the reference stream (See Chapter 5). In all AWS streams, density and biomass of juvenile salmon exhibited greater variability in the post-logging periods than in the prelogging period. From 1989–1996, salmon density and biomass in Needle Branch were high compared to the average of AWS years. The high variability of juvenile salmon abundance makes interpretation of land-use impacts difficult, but increased variation of populations in streams in harvested basins was consistent both for the early post-harvest period and for the later period 22–30 years after harvest.

Trends in coho salmon populations in the AWS are affected by many factors in addition to freshwater habitat conditions, including ocean conditions, commercial and sport harvest, hatchery operations, and predation by marine mammals and birds. Natural trends in climate and ocean conditions have raised many management questions about coho salmon populations over the last two decades (Francis and Sibley 1991; Pearcy 1992). Changes in numbers of spawning adults could strongly influence patterns of juvenile salmon abundance the following summer. In the AWS streams, average numbers of female spawners did not change substantially from prelogging to post-logging (1958–1964: Flynn Creek – 19.5, Deer Creek – 26.2, Needle Branch – 10.8; 1965–1971: Flynn Creek – 17.6, Deer Creek – 28.0, Needle Branch – 14.3). For these same periods, spawner estimates for a standard survey reach in an adjacent stream, Horse Creek, were 13.3 spawners for 1958–1964 and 12.9 spawners for 1965–1971 (Oregon Department of Fish & Wildlife, unpublished data). Estimates of spawner numbers are not available for the AWS streams during the 1987–1995 period, but numbers of spawners estimated in the Horse Creek reach decreased by 30% to an average of 9.2. Decreases in coastwide escapement were even greater. From 1987–1995, two spawners or fewer were observed in the Horse Creek survey reach for four of the nine years. Such low numbers of spawners had been observed in only two years of the 32-year record for Horse Creek prior to 1987. It is likely that the lower density and biomass of juvenile coho salmon that occurred in some years (e.g., 1992, 1995) in all AWS streams from 1988–1996 were related in part to regional changes in climate and ocean conditions that reduced marine survival of adult coho salmon.

In addition to disturbance, tributary size may have influenced the degree of annual variation in salmonid populations. During both the prelogging and post-logging periods, Needle Branch, the smallest stream, had the greatest annual variation in population density and the lowest lineal abundance of salmonids compared to the two larger AWS streams. Correspondingly, CVs for juvenile coho salmon and cutthroat trout prior to harvest were lowest in Deer Creek, the largest stream. Large year-to-year fluctuations in salmonid populations have been observed in other studies, thus long-term data are essential for adequate evaluation of land-use effects (Hall and Knight 1981; Platts and Nelson 1988; House 1995, See Chapter 15).

Responses of salmonids to timber harvest practices are complex and reflect the array of environmental factors (e.g., elevation, temperature, precipitation, water chemistry), geological factors (e.g., parent geology, topography, floodplain development, groundwater supply), and biotic factors (e.g., riparian vegetation, litter inputs, algal production, invertebrate assemblages, fish assemblages, other vertebrates) that influence salmonids in streams (Gregory et al. 1987; Gregory et al. 1991; Bisson et al. 1992; Naiman et al. 1992). Many studies discuss potential impacts of forest practices on salmonid populations in streams, but few have directly measured responses of fish populations to land-use practices (Hicks et al. 1991). Overall, timber harvest has the potential to change fish populations over several decades. In the two longest-running studies of effects of forest harvesting, the AWS and Carnation Creek, British Columbia, fish populations in harvested basins were more variable than populations in undisturbed forests. During periods of increased risks to depressed salmonid stocks (e.g., poor ocean conditions, drought, frequent floods, high commercial harvest), the potential for detrimental effects of timber harvest increases the need for efforts to protect streams and basins from habitat degradation.

Literature Cited

Armour, C.L., Burnham, K.P., and Platts, W.S. 1983. Field Methods and Statistical Analyses for Monitoring Small Salmonid Streams. FWS/OBS-83/33. USDI Fish and Wildlife Service. 200pp.

Au, D.W.K. 1972. Population Dynamics of the Coho Salmon and its Response to Logging in Three Coastal Streams. Ph.D. Thesis. Oregon State Univ., Corvallis, OR. 245pp.

Bisson, P.A., Quinn, T.P., Reeves, G.H., and Gregory, S.V. 1992. Best management practices, cumulative effects, and long-term trends in fish abundance in Pacific Northwest river systems, pp. 189–232. In: R.J. Naiman, editor. Watershed Management: Balancing Sustainability and Environmental Change. Springer-Verlag, New York.

Bisson, P.A., Sullivan, K., and Nielsen, J.L. 1988. Channel hydraulics, habitat use, and body form of juvenile coho salmon, steelhead, and cutthroat trout in streams. Trans. Amer. Fish. Soc. 117:262–273.

Chapman, D.W. 1961. Factors Determining Production of Coho Salmon (*Oncorhynchus kisutch*) in Three Oregon Streams. Ph.D. Thesis. Oregon State Univ., Corvallis, OR. 227pp.

Chapman, D.W. 1965. Net production of juvenile coho salmon in three Oregon streams. Trans. Amer. Fish. Soc. 94:40–52.

Francis, R.C., and Sibley, T.H. 1991. Climate change and fisheries: what are the real issues? Northwest Environ. J. 7:295–307.

Glova, G.J. 1986. Interaction for food and space between experimental populations of juvenile coho salmon (*Oncorhynchus kisutch*) and coastal cutthroat trout (*Salmo clarki*) in a laboratory stream. Hydrobiologia 131:155–168.

Glova, G.J. 1987. Comparison of allopatric cutthroat trout stocks with those sympatric with coho salmon and sculpins in small streams. Environ. Biol. Fish. 20:275–284.

Gregory, S.V., Lamberti, G.A., Erman, D.C., Koski, K.V., and coauthors. 1987. Influences of forest practices on aquatic production, pp. 233–255. In: E.O. Salo and T.W. Cundy, editors. Streamside Management: Forestry and Fishery Interactions. Univ. of Washington Inst. of Forest Resour. Contrib. 57, Seattle, WA.

Gregory, S.V., Swanson, F.J., McKee, W.A., and Cummins, K.W. 1991. An ecosystem perspective of riparian zones. BioScience 41:540–551.

Hall, J.D., Brown, G.W., and Lantz, R.L. 1987. The Alsea Watershed Study: a retrospective, pp. 399–416. In: E.O. Salo and T.W. Cundy, editors. Streamside Management: Forestry and Fishery Interactions. Univ. of Washington Inst. of Forest Resour., Seattle, WA.

Hall, J.D., and Knight, N.J. 1981. Natural Variation in Abundance of Salmonid Populations in Streams and its Implications for Design of Impact Studies. EPA-600/S3-81-021. U.S. Environmental Protection Agency, Corvallis, OR. 85pp.

Hall, J.D., and Lantz, R.L. 1969. Effects of logging on the habitat of coho salmon and cutthroat trout in coastal streams, pp. 355–375. In: T.G. Northcote, editor. Symposium on Salmon and Trout in Streams. Univ. of British Columbia, H.R. MacMillan Lectures in Fisheries, Vancouver, BC.

Hankin, D.G., and Reeves, G.H. 1988. Estimating total fish abundance and total habitat area in small streams based on visual estimation methods. Can. J. Fish. Aquat. Sci. 45:834–844.

Hartman, G.F., and Scrivener, J.C. 1990. Impacts of forest practices on a coastal stream ecosystem, Carnation Creek, British Columbia. Can. Bull. Fish. Aquat. Sci. 223. 148pp.

Hicks, B.J., Hall, J.D., Bisson, P.A., and Sedell, J.R. 1991. Responses of salmonids to habitat changes, pp. 483–518. In: W.R. Meehan, editor. Influences of Forest and Rangeland Management on Salmonid Fishes and their Habitats. Amer. Fish. Soc. Spec. Publ. 19.

House, R. 1995. Temporal variation in abundance of an isolated population of cutthroat trout in western Oregon, 1981–1991. N. Amer. J. Fish. Manage. 15:33–41.

Knight, N.J. 1980. Factors Affecting the Smolt Yield of Coho Salmon (*Oncorhynchus kisutch*) in Three Oregon Streams. M.S. Thesis. Oregon State Univ., Corvallis, OR. 105pp.

Leidholdt-Bruner, K., Hibbs, D.E., and McComb, W.C. 1992. Beaver dam locations and their effects on distribution and abundance of coho salmon fry in two coastal Oregon streams. Northwest Sci. 66:218–223.

Lindsay, R.B. 1975. Distribution and Survival of Coho Salmon Fry after Emigration from Natal Streams. M.S. Thesis. Oregon State Univ., Corvallis, OR. 41pp.

Lowry, G.R. 1964. Net Production, Movement, and Food of Cutthroat Trout (*Salmo clarki clarki* Richardson) in Three Oregon Coastal Streams. M.S. Thesis. Oregon State Univ., Corvallis, OR. 72pp.

Lowry, G.R. 1965. Movement of cutthroat trout, *Salmo clarki clarki* (Richardson) in three Oregon coastal streams. Trans. Amer. Fish. Soc. 94:334–338.

Moring, J.R. 1975. The Alsea Watershed Study: effects of logging on the aquatic resources of three headwater streams of the Alsea River, Oregon. Part II. Changes in environmental conditions. Fish. Res. Rep. 9. Oregon Dept. of Fish and Wildlife, Corvallis, OR. 39pp.

Moring, J.R., and Lantz, R.L. 1975. The Alsea Watershed Study: effects of logging on the aquatic resources of three headwater streams of the Alsea River, Oregon. Part I. Biological studies. Fish. Res. Rep. 9. Oregon Dept. of Fish and Wildlife, Corvallis, OR. 66pp.

Naiman, R.J., Beechie, T.J., Benda, L.E., Berg, D.R., and coauthors. 1992. Fundamental elements of ecologically healthy watersheds in the Pacific Northwest Coastal Ecoregion, pp. 127–188. In: R.J. Naiman, editor. Watershed Management: Balancing Sustainability and Environmental Change. Springer-Verlag, New York, NY.

Naiman, R.J., McDowell, D.M., and Farr, B.S. 1984. The influence of beaver (*Castor canadensis*) on the production dynamics of aquatic insects. Internationale Vereinigung für Theoretische und Angewandte Limnologie, Verhandlungen 22:1801–1810.

Naiman, R.J., Melillo, J.M., and Hobbie, J.E. 1986. Ecosystem alteration of boreal forest streams by beaver. Ecology 67:1254–1269.

Nickelson, T.E., Rodgers, J.D., Johnson, S.L., and Solazzi, M.F. 1992. Seasonal changes in habitat use by juvenile coho salmon (*Oncorhynchus kisutch*) in Oregon coastal streams. Can. J. Fish. Aquat. Sci 49:783–789.

Pearcy, W.G. 1992. Ocean Ecology of North Pacific Salmonids. Univ. of Washington Press, Seattle, WA. 179pp.

Peterson, N.P., and Cederholm, C.J. 1984. A comparison of the removal and mark-recapture methods of population estimation for juvenile coho salmon in a small stream. N. Amer. J. Fish. Manage. 4:99–102.

Platts, W.S., and Nelson, R.L. 1988. Fluctuations in trout populations and their implications for land-use evaluation. N. Amer. J. Fish. Manage. 8:333–345.

Reeves, G.H., Hall, J.D., and Gregory, S.V. 1997. The impact of land-management activities on coastal cutthroat trout and their freshwater habitats, pp. 138–144. In: J.D. Hall, P.A. Bisson, and R.E. Gresswell, editors. Sea-Run Cutthroat Trout: Biology, Management, and Future Conservation. Oregon Chapter, American Fisheries Society, Corvallis, OR.

Sabo, J.L., and Pauley, G.B. 1997. Competition between stream-dwelling cutthroat trout (*Oncorhynchus clarki*) and coho salmon (*Oncorhynchus kisutch*): effects of relative size and population origin. Can. J. Fish. Aquat. Sci. 54:2609–2617.

Sanner, C.J. 1987. Effects of Beaver on Stream Channels and Coho Salmon Habitat, Kenai Peninsula, Alaska. M.S. Thesis. Oregon State Univ., Corvallis, OR. 81pp.

Schwartz, J.S. 1991. Influence of Geomorphology and Land Use on Distribution and Abundance of Salmonids in a Coastal Oregon Basin. M.S. Thesis. Oregon State Univ., Corvallis, OR. 207pp.

Sumner, F.H. 1962. Migration and growth of coastal cutthroat trout in Tillamook County, Oregon. Trans. Amer. Fish. Soc. 91:77–83.

Veldhuisen, C.N. 1990. Coarse Woody Debris in Streams of the Drift Creek Basin, Oregon. M.S. Thesis. Oregon State Univ., Corvallis, OR. 109pp.

Chapter 15
The Alsea Watershed Study: A Comparison with Other Multi-year Investigations in the Pacific Northwest

Peter A. Bisson, Stanley V. Gregory, Thomas E. Nickelson, and James D. Hall

The Alsea Watershed Study (AWS) was the first long-term fisheries research project to address the effects of forestry operations on salmonid populations in the Pacific Northwest using a watershed approach. To this day it remains one of a very limited number of investigations that have provided long-term information on salmon and trout responses to forestry operations. These studies have had a significant impact on the development of state and provincial forest practices regulations. The initial results of the AWS, which demonstrated negative impacts of logging on salmonid spawning and rearing (Hall and Lantz 1969), contributed to some of the first laws regulating forestry operations adjacent to small streams.

Over the past several decades, different study designs have been used to evaluate the effects of forest management activities in Pacific Northwest streams. The first type of study has involved intensive, long-term evaluations of experimentally applied forestry operations within a single area. This has been the approach used in both the AWS (Hall et al. 1987) and the Carnation Creek watershed study (Hartman and Scrivener 1990) in British Columbia. The AWS examined experimentally controlled timber harvest in three small adjacent watersheds, with multi-year pre and posttreatment evaluation periods. In the Carnation Creek watershed study, harvest treatments were applied along different reaches of one stream rather than in adjacent watersheds. However, the pre and posttreatment monitoring approach was generally similar to the AWS (Hartman and Scrivener 1990). In each of these studies, treatments were not replicated in other watersheds.

Using a second type of study design, other investigators followed stream recovery after forest management had already occurred, with nearby unmanaged watersheds or stream reaches acting as control sites. In most studies there was little or no pretreatment sampling and the treatments themselves were not

Peter A. Bisson
USDA Forest Service, Pacific Northwest Research Station, Olympia, WA 98512-9193
pbisson@fs.fed.us

J. D. Stednick (ed.), *Hydrological and Biological Responses to Forest Practices.*   259
© Springer 2008

applied to several watersheds. Such an approach was used in the Clearwater River basin of Washington's Olympic Peninsula (Cederholm and Reid 1987). A third design has involved substituting spatial replication (multiple sites) for temporal replication (multiple years) by employing geographically extensive surveys of streams whose watersheds have undergone forestry operations or forestry-related disturbances over similar time periods (Murphy and Hall 1981; Murphy et al. 1981; Hawkins et al. 1983; Bisson and Sedell 1984).

Each of the three approaches has usually involved some type of control or reference condition such as an unmanaged watershed, upstream undisturbed stream reach, or pretreatment sampling period. Few studies, however, have attempted to replicate experimental treatments among different streams in a preplanned manner, i.e., where replication of treatments at different sites has been systematically included in hypothesis testing. Given the considerable interest in experimental design of ecological studies (e.g., Hurlburt 1984; Walters et al. 1988, 1989) lack of replication has limited our ability to extrapolate results from these studies to other watersheds in the region or to separate treatment effects from concurrent extraneous factors such as climate cycles. Effects attributable to forest management apply only to those reaches or watersheds measured and only during the period in which they were studied.

Although both the Alsea Watershed and Carnation Creek studies lasted more than 15 years, the majority of postdisturbance monitoring studies and extensive surveys lasted less than ten years, and many have taken place over intervals of only a year or two. Most examinations of streams and their fish populations have occurred from the 1970s to 1990s (Hicks et al. 1991b), but there has been little temporal overlap among studies within Pacific Northwest ecoregions, and this time interval has witnessed significant climate and oceanic change (Francis and Sibley 1991; Pearcy 1992; Beamish and Bouillon 1993). Investigations of the impacts of forestry practices on salmonid populations in coastal and western Cascade watersheds of the Pacific Northwest have thus differed in approach, in geographical location, and over time periods when climate was changing. Perhaps it is not surprising, therefore, that consistent and unequivocal conclusions concerning the impacts of forest management on fish populations in Pacific Northwest streams have failed to emerge (Hicks et al. 1991b; Bisson et al. 1992).

Other multi-year fish population studies in the region have attempted to evaluate salmonid restoration and enhancement projects (Everest et al. 1984; Johnston et al. 1990; Ward and Slaney 1993) or supplementation of wild populations with hatchery-produced fry (Nickelson et al. 1986). These studies have often been carried out in watersheds where some logging has occurred, but the primary intent of the investigations has not been to examine forestry-related impacts. Nevertheless, such studies constitute another source of long-term salmonid population data for the region. Syntheses of the results of multi-year investigations of salmonids (Lichatowich and Cramer 1979; Elliott 1985, 1994; Hall et al. 1987; Hartman and Scrivener 1990; Hilborn and Winton 1993) have shown that considerable interannual variation in population abundance is commonplace and that high levels of natural variability pose an obstacle to understanding how anthropogenic disturbances or other management activities affect fish.

The objective of this paper is to compare results of the AWS with other multi-year watershed investigations in the Pacific Northwest. In particular, we address the question of how much interannual variation exists in the abundance of stream-dwelling salmonids and the extent to which this variation limits our ability to detect population responses to forestry-related disturbances and other management activities. Comparisons are limited to studies of 5 or more consecutive years duration, conducted in coastal or Cascade Mountain streams during the period of summer low flow, when populations tend to reflect summer carrying capacities (Fransen et al. 1993). Information on the variability of adult brood-year escapement and smolt production is also used where available. We further limited comparisons to small and mid-sized watersheds with either non-migratory trout or a combination of resident and anadromous salmonids, as these circumstances most closely resemble those of the AWS. We conclude with a discussion of the importance of long-term research projects like the AWS, and the importance of identifying fish population measures that are especially sensitive to management activities.

Alsea Watershed and Forest Practices

Within the states of Oregon and Washington the pattern of forest land ownership is mixed, but federally owned lands occupy the overall majority of forest area in the region. State-owned and private commercial forest lands make up a lower overall percentage of coastal and west slope Cascade Mountain forests but tend to dominate coastal river basins from central Oregon to the west coast of Washington's Olympic Peninsula (Pease 1993). Federal forestry practices have been regulated by regional operating guidelines of the USDA Forest Service and the USDI Bureau of Land Management, while state and private forestry practices have been regulated by an evolving set of state forest practice regulations that were initiated in the early 1970s.

Environmental protection standards applied to state and private forest lands have differed somewhat from those applied to federal lands over the previous decades. In general, watersheds within state and private ownerships have greater percentages of recent timber harvest, higher road densities, and narrower buffer strips along fish-bearing streams than watersheds within predominantly federal ownership, due in part to designated wilderness or research natural areas in federal forests and to stricter federal environmental guidelines (Thomas 1993). The AWS, which involved three small watersheds with mixed federal and private industrial ownership, included one site (Needle Branch) that was logged and burned without streamside protection and another site (Deer Creek) that was patchcut and included streamside buffers. A third watershed (Flynn Creek) served as the control. Although it is tempting to extrapolate the Needle Branch results to logging impacts in state and private forests and the Deer Creek results to federal forests, such extrapolation would probably not be valid. Needle Branch was treated in a manner more severe than would currently be permitted under any set of regulations, although similar treatments did occur in other small watersheds on private forest land at the time it was logged. The pattern of patchcutting and buffer strips in the Deer Creek watershed may now be found in most types of forest ownership. No attempts have been made to distinguish ownership type in our comparisons. Different locations have had different natural disturbance histories, varying logging practices, and sometimes other land uses such as agriculture and grazing. Thus it is virtually impossible to relate habitat conditions and fish populations to specific disturbance events, whether natural or anthropogenic. Unless the watershed has been designated as a wilderness area or a late-successional forest reserve, we assumed that every site has experienced some changes from forestry operations.

Study Locations, Objectives, and Methods

Locations and dates of multi-year salmonid population studies selected for comparison of interannual variation are in Table 15.1. Study sites were scattered along the Pacific Northwest coast from the northern end of Vancouver Island to the Coos River system in southwestern Oregon (Fig. 15.1). Most studies selected for comparison with the Alsea watershed were located in coastal drainages and contained anadromous salmonids. A few sites (McKenzie River tributaries, Fish Creek, Huckleberry Creek) were located in the Cascade Mountains. These sites tended to be of higher elevation than the coastal sites, but still had primarily rainfall-dominated hydrologic regimes. The two McKenzie River tributaries (Mack Creek and Johnson Creek) did not contain anadromous salmonids, although juvenile Chinook salmon (*Oncorhynchus tshawytscha*) were occasionally found near the mouth of Johnson Creek.

Most study sites contained populations of coho salmon (*O. kisutch*), steelhead (*O. mykiss*), and coastal cutthroat trout (*O. clarkii clarkii*). Investigators usually made no attempt to distinguish anadromous from resident cutthroat trout, and juvenile rainbow trout were normally assumed to be anadromous steelhead. Objectives of individual studies sometimes limited evaluations to certain species (e.g., Nickelson et al. 1986). We found the majority of long-term investigations, i.e., those with 5 or more consecutive years of population data, were concerned with salmonid enhancement, including fry supplementation (Nickelson et al. 1986) and habitat restoration (Bilby unpublished; Reeves et al. 1990; Ward and Slaney 1993). Most sites also contained a variety of non-salmonid fishes, especially cottids and cyprinids, but their abundances were rarely reported. Several of the studies were ongoing and have not been published as technical reports or papers in scientific journals

Table 15.1 Long-term salmonid population study sites in Pacific Northwest watersheds used in this comparison. Those without superscript indicate coho, steelhead, and cutthroat trout.

State/ ProvinceRiver basin/Watershed Stream	Species	Years of study	Adults	Summer juveniles	Smolts	Source(s)
Oregon						
Alsea Watershed	Coho,					
Study	cutthroat	1959–1974;				Chapter 5;
Flynn Creek		1988–1996	x[1]	x	x[1]	Chapter 14
Deer Creek			x[1]	x	x[1]	
Needle Branch			x[1]	x	x[1]	
Oregon coastal						
streams	Coho	1980–1985				Nickelson et al. (1986)
Siuslaw River						
Panther Creek			x	x		
Rogers Creek			x	x		
Misery Creek			x	x		
Doe Creek			x	x		
Dogwood Creek			x	x		
Billie Creek			x	x		
Beaver Creek						
No. Fk. Beaver						
Creek			x	x		
Alsea River						
Horse Creek			x	x		
Drift Creek			x	x		
Yaquina River						
Hayes Creek			x	x		
Salmon Creek			x	x		
Deer Creek			x	x		
Neskowin Creek			x	x		
Little Nestucca						
River						
Louie-Baxter						
Creek			x	x		
Bear Creek			x	x		
Oregon coastal	Coho	1988–1993				T.E. Nickelson
streams						(unpublished)
Alsea River						
East Fork						
Lobster Creek			x	x		
Upper Lobster						
Creek			x	x		
Nestucca River						
East Fork Creek			x	x		
Moon Creek			x	x		

(continued)

Table 15.1 (continued)

State/Province River basin/Watershed Stream	Species	Years of study	Adults	Summer juveniles	Smolts	Source(s)
Clackamas River (Cascades)	Steelhead, coho	1983–1990				Reeves et al. (1990)
Fish Creek			x	x	x	
McKenzie River (Cascades)						
Mack Creek	Cutthroat	1973–1996		x		S.V. Gregory (unpublished)
Johnson Creek	Cutthroat	1987–1992		x		P.A. Bisson (unpublished)
Coos River (Coast Range)						
Lost Creek	Steelhead	1986–1993		x		R.E. Bilby (unpublished)
Washington						
Olympic						
Peninsula streams	Steelhead, coho,					Johnson and Cooper (1986)
Siebert Creek	sea-run	1981–1986	x^2		x	
McDonald Creek	cutthroat		x^2		x	
Deschutes River						
Huckleberry Creek	Coho	1986–1992	x	x		Washington Dept. Fish & Wildlife and P. A. Bisson (unpublished)
British Columbia						
Vancouver Island						
Carnation Creek	Coho, steelhead	1970–1987	x	x	x	Hartman and Scrivener (1990)
Keogh River	Steelhead	1976–1982	x	x	x	Ward and Slaney (1993)

^{1}coho salmon
^{2}steelhead

(Table 15.1). Although their methods and results have not been peer-reviewed, the number of unpublished studies constituted a significant fraction of the long-term studies in the region, so they have been included in our comparison.

Objectives of the studies, as noted earlier, generally related either to the effects of logging on fish populations or to productivity enhancement through fry supplementation or creation of habitat. In the latter case we excluded study sites in which hatchery fish had been planted, but included unstocked control streams (Nickelson

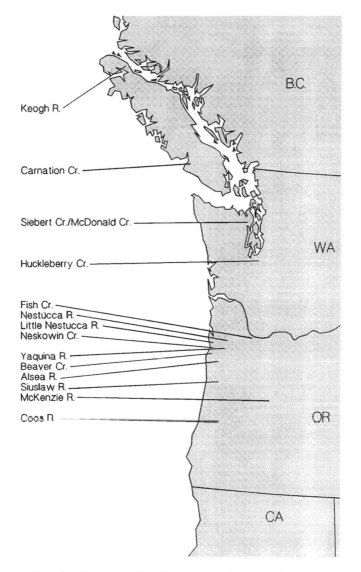

Fig. 15.1 Location of multi-year studies of salmon populations in the Pacific Northwest that were used in the comparison

et al. 1986). To be considered for comparison we required that studies (1) included quantitative estimates of salmonid abundance during the period of summer low flow, (2) had been sampled at least 5 consecutive years, (3) used similar sampling methods (but not snorkeling) from year to year, (4) possessed information on at least one of the following species: coho salmon, steelhead, or cutthroat trout, and (5) had taken place in small or mid-size watersheds (2nd–5th order) over the same time period as the AWS, i.e., 1959–1996.

In most studies, including the AWS, fish populations were censused by mark-recapture, multiple-pass electrofishing, or seining. Actual measures of abundance, however, often differed. Some investigators reported density as numbers per unit of stream area (no. m^{-2}); others chose numbers per unit of channel length (no. $100m^{-1}$) while still others presented estimates of the total population inhabiting the streams based on stratified sampling of representative habitat types. Differences in the method of reporting fish abundance made density comparisons among study sites difficult, because it was usually not possible to convert each estimate to a common density measure. Thus we were not able to determine whether the absolute abundance of salmonids in Alsea Watershed streams departed significantly from results of some of the other long-term investigations. However, it was possible to compare the interannual coefficients of variation [(standard deviation/mean)·100] in salmonid density among all sites. Interannual coefficient of variation (CV) is a statistic describing the relative variation in population abundance from year to year, independent of the absolute magnitude of the density estimates. Likewise, we compared the CV of adult spawners and smolts among sites where these data were available. Because the relationship between numbers of brood-year adults and juvenile progeny was relevant to the question of whether variation in numbers of rearing juveniles was influenced primarily by rearing conditions or adult escapement, we plotted trends in these parameters over time.

A total of 11 multi-year fisheries studies from the Pacific Northwest were compared (Table 15.1), about half of which were still in progress at the time this paper was written or unpublished and are therefore not reviewed here in detail. In only four studies (Alsea Watershed, Fish Creek, Carnation Creek, Keogh River) were migrating adults, juveniles, and smolts concurrently censused. Other investigations monitored either adults and juveniles or juveniles and smolts.

Comparison of Study Results

Studies of Logging Effects

Detailed findings of the AWS are given in Chapters 5 and 14, and are also discussed by Hall and Lantz (1969), Moring and Lantz (1975), and Hall et al. (1987). Populations of both juvenile coho salmon and cutthroat trout were monitored from 1959 until the early 1970s (cutthroat trout were not censused until 1962) and again from 1988 through 1996. Densities have been highly variable in all three watersheds over these intervals (Fig. 15.2), and inferences about the effects of logging on salmonid abundance have been somewhat difficult to draw. The cutthroat trout population in Needle Branch, the clearcut watershed, was depressed after logging (Hall et al. 1987), but densities of cutthroat trout (all age groups combined) in the late 1980s and early 1990s were comparable to prelogging levels during most years. However, densities of trout age 1 and older were even further depressed than they had been immediately after logging (See Chapter 14). Flow in lower Needle Branch had always been low in late summer during the original AWS, with the stream flowing subsurface between pools. However, in 1988 and 1992 the flow was exceptionally low; in late summer the stream was reduced to one or only very few isolated pools. During these 2 years few or no salmon or trout were present (Fig. 15.2). The cause of the recent increase in intermittency of Needle Branch in summer is not known, but may be related in some way to the history of forest operations. The summer low flow of a stream that had been clearcut in the Oregon Cascades dropped below prelogging values during the period 10–25 years after logging (Hicks et al. 1991a). Vigorous regrowth of riparian vegetation was suggested as the cause for reduced summer flow in that watershed.

Production of coho salmon smolts declined in each stream, including the unlogged control, during the early post-logging period (1967–73) relative to the prelogging period (1959–66) (See Chapter 5). The numbers of adult salmon returning to the streams (Fig. 15.3) were quite variable over both the pre- and post-logging periods, but the averages after logging were similar to the prelogging values in all three streams. Although the correlation between the number of brood-year females and summer fry densities was not strong, the density of juvenile coho salmon in the AWS sites did seem to be influenced by low spawner escapement (Fig. 15.3); low summer fry densities followed from years of unusually poor adult escapement. Hall et al. (1987) computed the total production of migrant fry per spawning female to provide an index of survival from egg to emergence. They found that abundance of migrant fry in Needle Branch declined after logging, but did not decline significantly in either the patchcut or control watersheds (Fig. 15.4). This observation suggested that survival of coho salmon during emergence and in early post-emergent environments was reduced after logging in the Needle Branch watershed.

The Carnation Creek study documented a number of physical and biological changes to the watershed after timber harvest and silvicultural treatments that included slash burning and herbicide application (Table 15.2). Increases in solar radiation striking the stream surface, summer and winter water temperatures, nutrient levels, water yield from small tributaries (but not from the mainstem of Carnation Creek), channel erosion, and fine sediment concentrations in spawning areas were documented after logging, burning, and herbicide treatments. Overall reductions were found in both large and small woody debris, as well as litter inputs, and the biota of Carnation Creek was generally reduced and more variable after timber harvest (Hartman and Scrivener 1990). Macroinvertebrates declined by almost half. Coho salmon declined slightly (although there were both positive and negative effects of logging on survival and growth of juveniles), chum salmon declined significantly (although much of the decline was caused by

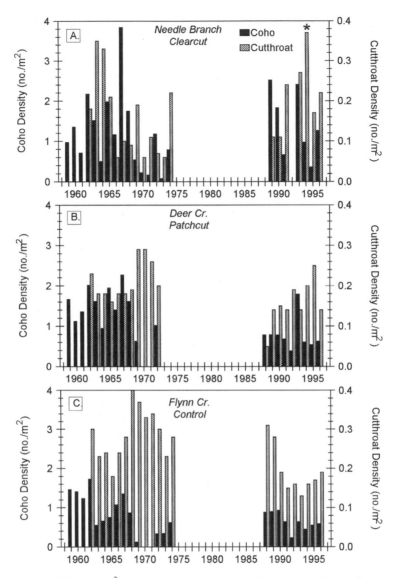

Fig. 15.2 Density (no. m^{-2}) of coho salmon and cutthroat trout during late summer (August–September) in the three Alsea Watershed streams. Data from Chapter 14. Asterisk denotes that trout density for Needle Branch in 1994 was 1.14 fish per m^2 (94% age 0)

poor ocean conditions), the steelhead population decreased, and cutthroat trout were largely unchanged (combined trout are shown in Fig. 15.5a).

Escapement of adult coho salmon to Carnation Creek dropped sharply in the late 1980s (Fig. 15.5b), raising the possibility that reduced numbers of juveniles may have been caused primarily by oceanic factors unrelated to

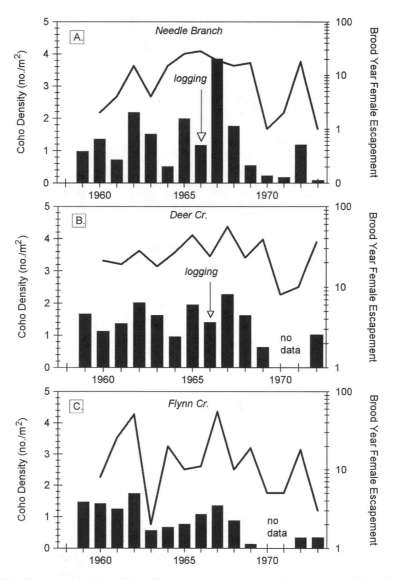

Fig. 15.3 Summer densities of juvenile coho salmon (bars) and escapement of brood-year females (line) to each of the AWS streams from 1959 to 1972. Data from Chapter 5

logging effects. Using data from the Carnation Creek study, we calculated the number of downstream migrant fry produced per spawning female before, during, and after logging in order provide an index of survival to emergence to compare with the results from the AWS. Declines in numbers of migrant fry per adult female in Carnation Creek during and after logging (Fig. 15.4d) were remarkably similar to the reduction in migrant fry observed after

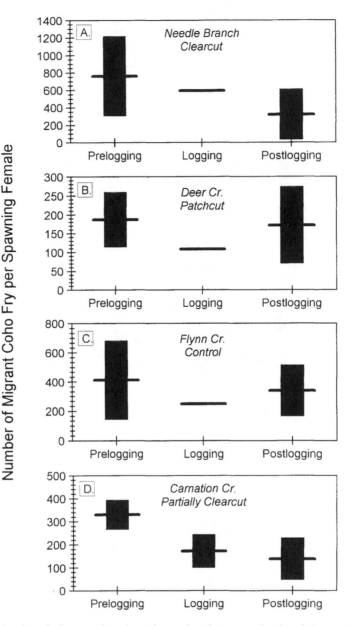

Fig. 15.4 Number of migrant coho salmon fry produced per spawning female (mean ± 1 SD) in the AWS streams and Carnation Creek before, during, and after logging. Note differences in the scale of the vertical axis for each graph. Data from Chapter 5 and Hartman and Scrivener (1990)

logging in Needle Branch (Fig. 15.4a); both streams exhibited a reduction of about 60%. However, the actual number of migrant fry per female was about twofold greater in Needle Branch than in Carnation Creek. It is possible that

Table 15.2 Summary of physical and biological changes in the Carnation Creek watershed after timber harvest and silvicultural treatments. Biological changes (right column) were likely related to a combination of the physical changes in the left column. From Hartman and Scrivener (1990, pages 124–125)

Physical changes	Biological changes
Light intensity on the stream surface doubled or more than doubled following logging.	Macroinvertebrate densities were reduced 40–50% following streamside logging and silvicultural treatments.
Diurnal and seasonal variability of stream temperature increased. Mean temperatures were 3°C higher during summer and 0.5°C higher during winter for the first decade following logging.	Coho salmon egg-to-fry survival, numbers of age-0 fish in autumn, and numbers of 2-yr-old smolts declined after logging; fry emerged earlier producing a longer period for growth and a larger size for parr; numbers of age-1 smolts and female adults increased then decreased at double the prelogging interannual variability.
Nutrient levels increased 40–80%, at least during high flows, for 2–4 yr following logging and for 1–2 yr following herbicide application.	
Water yield increases were detected in tributary watersheds >95% clearcut, but they were not significant for the total basin (41% clearcut). Groundwater levels increased in the floodplain. Duration of the period of higher groundwater levels was at least a decade.	Chum salmon egg-to-fry survival, fry size, and adult returns were reduced and more variable following logging and the fry emigrated earlier to the ocean. The 90% reduction of adult chum salmon was caused by poor ocean conditions (64%) and by logging (26%).
Fine woody debris increased in the stream, but it was lost within 2 yr.	The steelhead population decreased following logging.
Large woody debris became more clumped, and it was reduced to ∼30% of prelogging volume within 2 yr in areas that were logged to the stream bank. Stability and piece size of large woody debris decreased within 2 yr. Instability of large woody debris continued for at least a decade.	The cutthroat trout population was unchanged following logging.
Channel erosion and change in channel location began within 4 yr of the onset of logging and continued throughout the study.	
Litter input to the stream was reduced to 25–50% of prelogging levels after logging and silvicultural treatments. About half of this loss had recovered within a decade.	
Pea gravel and sand content of the streambed doubled during the decade since logging and continued to change throughout the study.	

the number of fry per female was influenced by watershed area in some way. Fewer fry were produced per female from the largest watersheds—Carnation Creek and Deer Creek—than from the two smaller watersheds—Flynn Creek and Needle Branch. The mechanisms causing this apparently significant difference in fry per female between watersheds of different sizes are not

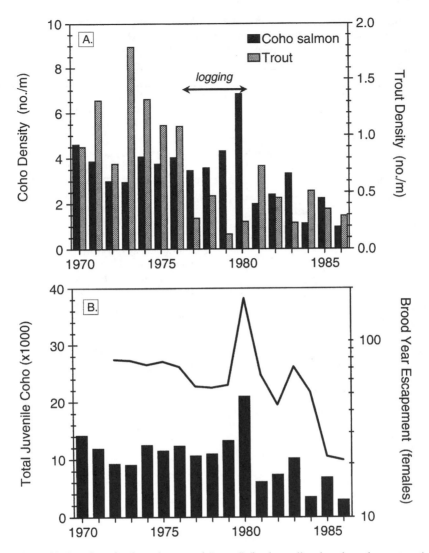

Fig. 15.5 (a) Density of coho salmon and "trout" (both steelhead and cutthroat trout) in Carnation Creek from 1970 to 1986. (b) Estimated number of juvenile coho salmon (bars) in Carnation Creek and the number of brood-year females (line). Data from Hartman and Scrivener (1990)

well understood, but may have included the effect of distance from spawning sites to downstream traps.

Mack Creek, a small (3.2-m average width) headwater tributary of the McKenzie River in western Oregon, contains a population of resident cut-throat trout that has been studied since the early 1970s as part of long-term ecological investigations in the H.J. Andrews Experimental Forest. A portion of the Mack Creek watershed was logged in 1965, and reference sites have been established in the clearcut and old-growth forested reaches of the stream. The trout population in the old-growth reach of Mack Creek has been less variable than trout populations in the AWS sites, but in all but one year in which both old-growth and clearcut reaches have been sampled, densities were greater in the reach in which the riparian zone had been clearcut (Fig. 15.6). This trend was consistent with other comparisons of salmonids in clearcut and old-growth streams of the Cascade Mountains (Murphy and Hall 1981; Hawkins et al. 1983; Bisson and Sedell 1984; Bilby and Bisson 1987) and differs from the results of the AWS, in which a significant decrease in trout populations was found after logging. The high-gradient Cascade streams in Oregon showed minimal increases in sediment and temperature after clearcutting. This contrast, along with increased food production and higher foraging efficiency of trout in the clearcut Cascade streams, compared with those in old-growth (Wilzbach et al. 1986), may have contributed to the different response.

Studies of Supplementation

Nickelson et al. (1986 and unpublished) examined the effects of supplementing naturally spawned coho salmon populations in Oregon coastal streams with hatchery-produced fry. They measured summer populations in 15 stocked and 15 unstocked control sites from 1980 to 1985, and in four unstocked sites from 1988 to 1993. In the 1980–1985 period both juveniles and brood-year adults were censused, while in the 1988–1993 period juveniles and smolts were censused. Only trends from the 15 naturally spawned (unstocked) populations are reported here, as these are most relevant to the AWS. From 1980 to 1984, summer densities of coho salmon tended to decline in the unstocked control sites, followed by an increase at all sites in 1985 (Fig. 15.7). During this period, fish populations in the AWS sites were not monitored; thus, we were not able to determine if a trend similar to that observed by Nickelson et al. (1986) occurred at the Alsea small watersheds over the same interval. Two larger streams in the Alsea River basin (Horse Creek and Drift Creek) were included in the Nickelson et al. (1986) study, and presumably reflected trends in the Alsea small watersheds. Summer coho salmon densities in Carnation Creek also exhibited erratic declines from 1980 to 1986 (Fig. 15.5). The early 1980s, therefore, appeared to be a period of overall declining abundance of coho salmon in the region.

In the Oregon coastal streams there was a general correspondence between numbers of brood-year adults observed in spawning surveys and corresponding juvenile densities the following summer (Fig. 15.7b), suggesting, as in the AWS, that numbers of juvenile coho salmon rearing in the streams studied by Nickelson et al. (1986) were somewhat influenced by adult escapement. Returns of naturally spawning adult coho salmon to the Oregon coast over the past two decades have been reported to be below levels needed to adequately colonize

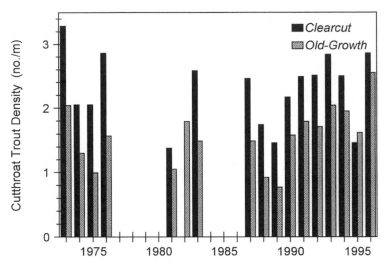

Fig. 15.6 Cutthroat trout densities in old-growth, forested, and clearcut sections of Mack Creek, Oregon, a stream with an average summer wetted width of 3.2 m (Moore and Gregory 1988). Data from several studies summarized by S. Gregory, R. Wildman, and L. Ashkenas (unpublished)

available habitats in coastal river systems (Nickelson et al. 1992). The positive association between number of spawners and juvenile densities supports the hypothesis that during the early 1980s low numbers of returning adults produced too few offspring to fully use available rearing space.

Studies of Habitat Enhancement

The Fish Creek watershed in the Mt. Hood National Forest of Oregon has been the site of a large-scale attempt to rehabilitate stream habitat throughout a 5th-order drainage system with approximately 16.7 km of habitat used by anadromous salmonids (Reeves et al. 1990). Fish Creek is a tributary of the Clackamas River, a subbasin of the Willamette River, draining the northwestern foothills of Mt. Hood in the Cascade Range of Oregon. The biophysical features of the stream differ in some important respects from those of the AWS sites. The predominant rock type in Fish Creek is volcanic, stream channels tend to be relatively steep with abundant coarse and little fine sediment, and much of the watershed is within the transient winter snow zone. Winter steelhead and, occasionally, fall Chinook salmon spawn in Fish Creek. Both species are essentially absent from the AWS sites, although they do occur in larger streams within the Alsea basin. Fish Creek also supports coho salmon and cutthroat trout, but both species tend to be far outnumbered by juvenile steelhead.

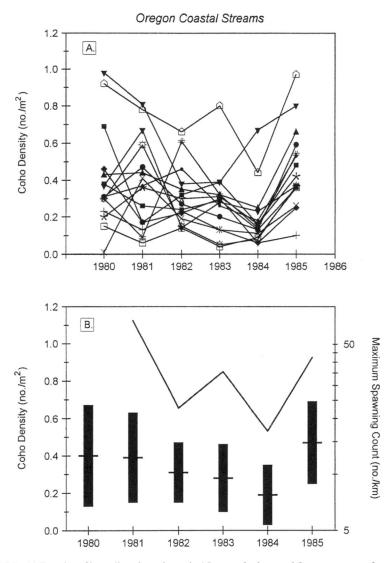

Fig. 15.7 (a) Density of juvenile coho salmon in 15 unstocked coastal Oregon streams from 1980 to 1985. (b) Densities of juvenile coho salmon (mean ± 1 SD) and mean peak number of adults (line) for the same streams. Data from Nickelson et al. (1986) and T. Nickelson (unpublished)

Beginning in 1983 and continuing until 1988, over 1,400 structures were placed in the mainstem and selected tributaries of Fish Creek. Most of these structures were combinations of logs and boulders that were placed to create pool habitat, store coarse sediment, improve cover, and increase hydraulic complexity. Off-channel ponds were created in lower Fish Creek in the early

1980s to increase winter rearing habitat, and road crossings known to block fish migrations were repaired to allow both adult and juvenile salmonid passage.

Estimated total numbers of juvenile anadromous salmonids in Fish Creek during summer were highest in the middle 1980s but dropped steadily during the latter part of the decade despite the extensive effort to improve habitat conditions (Fig. 15.8). Total declines were made up in large part by reduced numbers of age-0 steelhead. Age-1 steelhead and age-0 coho salmon were about as abundant in the late 1980s as in the early part of the decade. Pool habitat in Fish Creek doubled as a result of enhancement activities (Reeves et al. 1990). It is possible that the reduction in age-0 steelhead, which tend to prefer riffles (Bisson et al. 1988), may have been related in part to conversion of riffle habitat to pools. There was no apparent association between summer juvenile density and the strength of brood-year adult escapement to Fish Creek for either steelhead or coho salmon (Fig. 15.9) during the 1980s. There did appear to be a weak inverse relationship between the numbers of salmon and steelhead smolts from 1985 to 1990 (Fig. 15.10), hinting at the possibility of interspecific competition. Such a relationship did not occur in the Alsea Watershed streams because steelhead did not inhabit these small drainages.

Habitat enhancement was done in the Keogh River of northern Vancouver Island, British Columbia, in 1977 (Ward and Slaney 1979). The Keogh River is much larger than any of the AWS sites, with a watershed area of 129 km^2 and a mean annual discharge of 9.24 m^3 s^{-1}. Like Fish Creek, the salmonid community is dominated by juvenile steelhead. Most of the structures placed in the

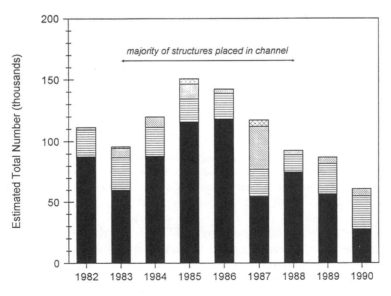

Fig. 15.8 Estimated numbers of juvenile steelhead, coho salmon, and chinook salmon rearing in Fish Creek, Oregon, from 1982 to 1990. Data from Reeves et al. (1990)

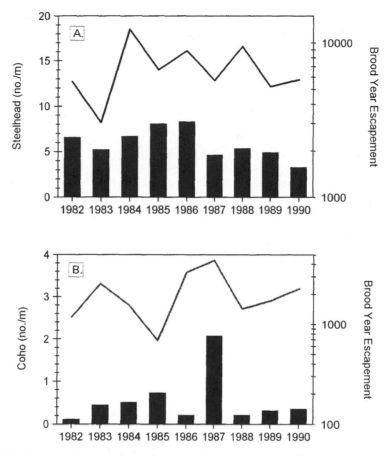

Fig. 15.9 Density of (a) juvenile steelhead and (b) coho salmon in Fish Creek (bars) and the number of brood-year adults (lines) passing a dam on the North Fork of the Clackamas River downstream from the mouth of Fish Creek. Data from Reeves et al. (1990)

channel of the Keogh River were boulder clusters or combinations of boulders and cabled logs, some of which broke loose quickly after initial placement. In addition, nutrients were continuously released in the upper Keogh River during summer in 1981 (Perrin et al. 1987). Annual surveys of the abundance of juvenile salmonids in the stream have shown that most boulder structures remained intact after 12–15 years. There was a net increase in carrying capacity for steelhead of about one fish per boulder (approximately 0.8 age-1 steelhead m^{-2}), and an increase in carrying capacity for juvenile coho salmon to about 0.3 fry m^{-2} in those sites where scour around the boulder clusters improved pool habitat (Ward and Slaney 1993).

Although the summer rearing density of juvenile salmonids in the Keogh River was much lower than at the Alsea study sites (Fig. 15.2), the positive response of both steelhead and coho salmon to habitat enhancement is

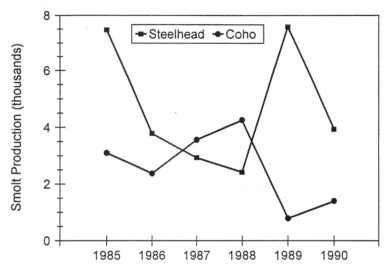

Fig. 15.10 Total production of steelhead and coho salmon smolts in Fish Creek from 1985 to 1990. Data from Reeves et al. (1990)

consistent with the hypothesis that habitat quality can influence the capacity of streams to support juvenile salmonids even when rearing densities are low. In spite of the relatively low salmonid densities reported for the Keogh River, the stream may have been at or near its carrying capacity. There was no obvious association between steelhead fry abundance and the abundance of brood-year spawners (Fig. 15.11a). It appeared that in the 1970s and 1980s sufficient numbers of adults were returning to the stream for their progeny to fully populate available rearing habitat, and that other factors (e.g., winter storms or other severe environmental disturbances) were controlling juvenile abundance. From 1977 to 1982, the period immediately after physical enhancement, increases were recorded in both steelhead fry and smolts produced per female relative to 1976, the year prior to structure placement (Fig. 15.11b). Ward and Slaney (1993) reported that these increases continued into the early 1990s, with the greatest increases accompanying stream fertilization (40–50% increase in numbers of salmon smolts, 62% increase in numbers of steelhead smolts, 40% increase in end-of-summer salmon fry weights). Thus, the Keogh River study provided early evidence that habitat enhancement and nutrient enrichment can successfully improve salmonid carrying capacity in an oligotrophic stream.

During the middle and late 1990s, however, anadromous salmonid populations in the Keogh River declined sharply to very low levels (Ward and McCubbing 1998). By 1998, adult steelhead returns numbered only 30 naturally spawned fish (33 additional adults were hatchery strays) and wild steelhead smolt production was less than 1,000 fish. The production of coho salmon smolts in 1998 was one-third of the 20-year average, and anadromous Dolly Varden (*Salvelinus malma*) populations were likewise in dramatic decline. Ward and

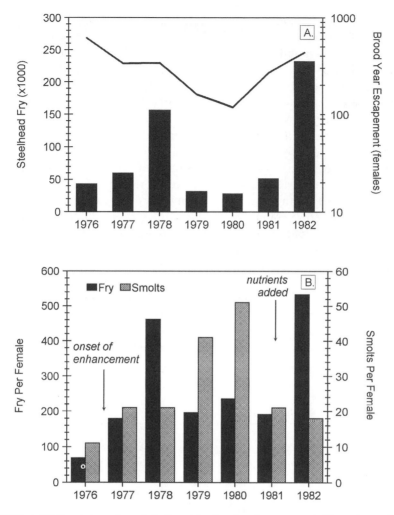

Fig. 15.11 (a) Estimated number of steelhead fry (bars) and escapement of brood-year female steelhead (line) to the Keogh River, British Columbia, from 1976 to 1982. (b) Number of fry and smolts produced per female steelhead. Data from Ward and Slaney (1993)

McCubbing (1998) believed that declines were related to low flows during spring, low numbers of spawning adults, and reductions in recruits per spawner. These trends suggest that the long-term benefits of habitat restoration can be over-ridden by other factors during the life cycle.

Other Multi-year Studies

In at least one other study the abundance of juvenile coho salmon appeared to be influenced by adult escapement, independent of habitat conditions.

Densities of coho salmon in lower Huckleberry Creek, a small tributary of the Deschutes River in southern Puget Sound, fluctuated sharply from 1986 to 1992 (B.R. Fransen and P. A. Bisson, unpublished) and tended to track the number of brood-year adults returning to the stream (Fig. 15.12). The Huckleberry Creek watershed is forested with second-growth Douglas-fir. The study site was not significantly altered during this period, although a debris torrent did occur above the study reach in January 1990, which increased suspended sediment and may have discouraged adult salmon from entering the stream that year.

Johnson Creek, like Mack Creek, is a small headwater tributary of Oregon's McKenzie River and was affected by a large debris torrent in 1986. The debris torrent originated in a first-order channel and swept through the entire length of Johnson Creek before coming to rest near the stream mouth. Summer trout population surveys began in 1988 and continued until 1992. A few cutthroat trout survived the debris torrent and non-anadromous rainbow trout have recolonized the lower reaches of Johnson Creek up to an impassable falls. In 1988, 2 years after the debris torrent, cutthroat trout density in the stream was approximately twice the average density observed in any of the Alsea study sites (Fig. 15.13), with most of the fish being underyearlings. In subsequent years the density of underyearlings declined sharply, and in 1992 the total cutthroat trout population in Johnson Creek had declined to a density less than half the average in the Alsea streams. The cause of the decline in Johnson Creek is not known,

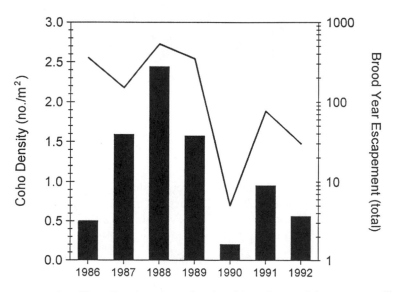

Fig. 15.12 Density of juvenile coho salmon (bars) and brood-year adult escapement (line) in lower Huckleberry Creek, Washington from 1986 to 1992. Data from B. Fransen and P. Bisson (unpublished) and Washington Department of Fish and Wildlife (unpublished)

but may have been related to unstable spawning substrate and to a reduction in food availability related to establishment of dense stands of young red alder (*Alnus rubra*) that heavily shaded the stream, as well as a dramatic proliferation of the snail *Juga plicifera* that could have sequestered much of the stream's primary production.

Interannual Variation

The interannual variability of adults, juveniles, and smolts in most of the multi-year studies (Table 15.3) was relatively high; the coefficient of variation was often 50% or greater. High levels of interannual variation in salmonid abundance have also been noted by other investigators (Lichatowich and Cramer 1979; Hall and Knight 1981; Grossman et al. 1990). Using two-way ANOVA, we detected no significant differences in variability among species or age groups of anadromous salmonids, i.e., no species or age groups appeared to be inherently more variable than others, although there were far fewer long-term studies of steelhead or cutthroat trout than of coho salmon in the Pacific Northwest. The interannual variability of adult coho salmon was significantly greater than the interannual variability of either juveniles or smolts (Table 15.4; $p = 0.02$, one-way ANOVA), but adult steelhead were not significantly more variable from year to year than juveniles or smolts.

It was difficult to determine from available data if anthropogenic disturbance resulted in increased interannual variation in juvenile salmonid populations, a major conclusion of the Carnation Creek study (Hartman and

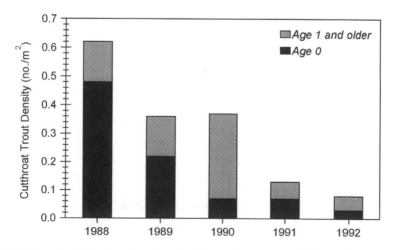

Fig. 15.13 Density of age-0 and age-1 and older cutthroat trout in Johnson Creek, Oregon. Data from P. Bisson and B. Fransen (unpublished)

Table 15.3 Coefficient of variation of the interannual abundance of salmonids in multi-year Pacific Northwest stream studies. Sources of data are given in Table 15.1

Streams and years	Coefficient of variation (%)								
	Coho salmon			Steelhead			Cutthroat trout		
	Adult	Juvenile	Smolt	Adult	Juvenile	Smolt	Adult	Juvenile	Smolt
Alsea Watershed Study (1959–74)									
Flynn Creek	96	55	58					23	
Deer Creek	55	33	35					22	
Needle Branch	79	71	50					64	
(1988–96)									
Flynn Creek		36						32	
Deer Creek		52						36	
Needle Branch		74						126	
Oregon coastal streams (1980–85)									
Panther Creek	39	37							
Rogers Creek	42	47							
Misery Creek	67	34							
Doe Creek	117	86							
Dogwood Creek	85	74							
Billie Creek	80	54							
N. Fk. Beaver Creek	42	34							
Horse Creek	47	25							
Drift Creek	61	52							
Hayes Creek	88	36							
Salmon Creek	85	53							
Deer Creek	69	81							
Neskowin Creek	49	64							
Louie-Baxter Creek	127	32							
Bear Creek	85	54							

Table 15.3 (continued)

Streams and years	Coefficient of variation (%)								
	Coho salmon			Steelhead			Cutthroat trout		
	Adult	Juvenile	Smolt	Adult	Juvenile	Smolt	Adult	Juvenile	Smolt
Oregon coastal streams (1988–93)									
East Fk. Lobster Creek		35	26						
Upper Lobster Creek		45	59						
East Fork Creek		51	48						
Moon Creek		66	32						
Clackamas River									
Fish Creek (1983–90)	54	109	53	40	28	48			
McKenzie River									
Mack Creek (1973–96)									
Clearcut								25	
Old-growth								30	
Johnson Creek (1987–92)								69	
Coos River									
Lost Creek (1986–93)					62				
Olympic Peninsula streams (1981–86)									
Siebert Creek			75		78	26			73
McDonald Creek			28		59	60			64
Deschutes River									
Huckleberry Creek (1986–92)	93	70							
Vancouver Island									
Carnation Creek (1970–87)	54	42	31	88	76	91	92	112	55
Keogh River (1976–82)				52	91	23			

Table 15.4 Average coefficient of variation of the interannual abundance of salmonid adults, juveniles, and smolts based on the data of Table 15.3. An asterisk denotes a significant difference ($p \leq 0.05$, single classification ANOVA) in comparison with other life stages of that species. The number of populations is shown in parentheses

Species	Coefficient of variation (%)		
	Adult	Juvenile	Smolt
Coho salmon	72* (21)	54 (28)	50 (11)
Steelhead	60 (3)	66 (6)	50 (5)
Cutthroat trout	92 (1)	54 (10)	64 (3)

Scrivener 1990). Except for the AWS, most experimental designs and objectives precluded such a determination; however, the interannual variability of both juvenile coho salmon and cutthroat trout in Needle Branch (the clearcut watershed) exceeded juvenile variability in both the patchcut and control watersheds (Table 15.3), a finding consistent with the Carnation Creek results. Moreover, variability of juvenile cutthroat trout in the clearcut stream was substantially greater during both post-logging periods than during prelogging for all measures of abundance (no. m^{-1}, no. m^{-2}, and g m^{-2}). There was no consistent trend in the other two watersheds. Variability of juvenile coho salmon in the clearcut stream also showed the same pattern of increase in both post-logging periods, however variability sometimes also increased in the other two streams (See Chapter 14). Additionally, the interannual variation of resident cutthroat trout in Johnson Creek, the stream altered by a large debris torrent, was over two-fold greater than in the old-growth forested section of nearby Mack Creek, which did not experience a major disturbance over the monitoring period.

The high level of interannual variability among salmonid populations in Pacific Northwest streams poses serious challenges for research seeking to detect significant responses to either environmental degradation or habitat improvement (Lichatowich and Cramer 1979; Hall and Knight 1981). Based on a fairly conservative estimate of interannual CV of 50% for anadromous salmonids (Table 15.4), it is possible to determine how many years of monitoring would be required to be 80% certain of detecting a given difference in population abundance between pre- and posttreatment means at a significance level of $p \leq 0.10$ (Sokal and Rohlf 1981). For treatments resulting in a 50% change in populations, a monitoring period of 26 years (13 years pretreatment and 13 years posttreatment) would be required. For treatments resulting in a 30% change in populations, a monitoring period of 70 years would be required, clearly an unrealistic requirement. Though improvements have since been made in experimental designs (Walters et al. 1988; Underwood 1994), it appears that long-term monitoring can, at best, detect only large changes in salmonid population abundance, and it is therefore not surprising that the two most definitive multi-year fishery investigations in the region, the AWS and the Carnation Creek study, have spanned about two decades. We agree with Lichatowich and Cramer (1979) that 20 years should be considered the minimum time necessary to evaluate the influence of land-use practices or stream habitat restoration on salmonid populations in experimentally treated watersheds and that even longer studies will be necessary to significantly improve our ability to detect true differences.

Conclusions

The AWS remains one of the most comprehensive long-term fishery investigations in the Pacific Northwest. The research has been successful in part because scientists participating in the study realized the value of multi-year investigations in an environment prone to considerable interannual variability. Subsequent investigations in other streams have generally supported key findings of the Alsea fishery studies: (1) salmonid populations change as a result of anthropogenic disturbance, often in unpredictable ways, and (2) population abundance can become more variable after stream habitat is altered. The issue of increased variability with anthropogenic disturbance will become ever more important as we assess population viability for salmonids at risk of extinction. Both trends in abundance and interannual variability play important roles in population viability; relatively slight downward trends (as might be caused by long-term climate cycles), combined with increasing variability, significantly increase the risk of population extirpation (Lawson 1993).

In terms of actual standing stocks, salmonids in the Alsea tributary watersheds were often more abundant on an areal basis than populations in other streams, particularly those in northern Washington and on Vancouver Island, British Columbia. It seems possible that regional differences in biological productivity are mediated by differences in biogeoclimatic regimes, especially temperature and nutrient levels. The dramatic increase in juvenile salmonid productivity after nutrient enrichment of the Keogh River suggests that northern streams are food limited, a condition that may be somewhat less significant in western Oregon where temperatures and nutrient levels are generally higher.

Another factor that influenced anadromous salmonid abundance at some sites was the escapement of adults. Juvenile coho salmon in particular were often affected by the number of returning spawners, suggesting that population levels were recruitment limited. Most of the multi-year investigations took place after 1976, during a period of widespread decline of coho salmon populations in Washington and Oregon (Nickelson et al. 1992; Washington Department of Fisheries et al. 1993). Low escapements of naturally spawning adults and the subsequent inability of progeny to fully populate available rearing space hindered assessment of the effects of habitat alteration on anadromous salmonids in some studies. Until numbers of adults sufficient for adequate habitat colonization are permitted to return to natal streams, our ability to detect the effects of habitat degradation or to evaluate the effectiveness of restoration programs will be severely limited. Indeed, the presence of abundant adult carcasses may itself be an important factor controlling the production of subsequent offspring (Bilby et al. 1996).

Production of juveniles migrating from experimentally controlled watersheds, expressed as numbers of migrants per adult female, proved to be a useful measure of the effects of logging on salmon populations in both the Alsea Watershed and Carnation Creek studies. These two parameters, adult escapement and numbers of downstream migrants, are often neglected in multi-year investigations because of the necessity of two-way fish traps and the time and expense of daily trap cleaning and checking. However, these traps yielded valuable data that were relatively immune to variations in year-to-year abundance, and we recommend that two-way fish traps be incorporated into other multi-year studies where possible. Recently, the importance of movement to resident salmonid populations has been documented (Gowan et al. 1994) as an important means of dispersal for mobile population members. Because knowledge of movements is critical to understanding any long-term study of salmonid ecology (Fausch and Young 1995), two-way traps should be employed in all long-term studies whether of anadromous or resident populations.

Finally, our comparison of multi-year studies revealed the value of continuous monitoring for periods of decades rather than years. Many of the studies lasting from 5 to 10 years did not produce reasonably clear answers to the questions they were designed to address. One of the most daunting problems has proved to be interannual variations in population abundance on the order of 50% or greater for all life history stages of coho salmon, steelhead, and cutthroat trout. This relatively high level of variability will require continuous monitoring for at least two decades, as well as creative experimental designs that partition variation due to yearly climatic and other differences (Walters et al. 1988, 1989), in order to detect even coarse-scale changes in population abundance. Over the course of their multi-decade histories, studies such as those at the Alsea Watershed and Carnation Creek have provided some of the most valuable information about salmonid ecology in the Pacific Northwest. Such studies should be continued, as there is no reason to believe that they have yet yielded all there is to be learned at these sites.

Acknowledgment We thank all those who provided published and unpublished information for our comparison. In particular, we are grateful to J. C. Scrivener, Bruce Ward, Thom Johnson, Brian Fransen, Robert Bilby, Randy Wildman, Linda Ashkenas, Gordon

Reeves, and Bruce Hansen. We also thank the USDA Forest Service PNW Research Station, Oregon State University, the Oregon Department of Fish and Wildlife, and the Weyerhaeuser Company for support. Kurt Fausch offered very helpful comments on the manuscript and John Stednick exercised considerable editorial patience during manuscript preparation.

Literature Cited

Beamish, R.J., and Bouillon, D.R. 1993. Pacific salmon production trends in relation to climate. Can. J. Fish. Aquat. Sci. 50:1002–1016.

Bilby, R.E., and Bisson, P.A. 1987. Emigration and production of hatchery coho salmon (*Oncorhynchus kisutch*) stocked in streams draining an old-growth and a clear-cut watershed. Can. J. Fish. Aquat. Sci. 44:1397–1407.

Bilby, R.E., Fransen, B.R., and Bisson, P.A. 1996. Incorporation of nitrogen and carbon from spawning coho salmon into the trophic system of small streams: evidence from stable isotopes. Can. J. Fish. Aquat. Sci. 53:164–173.

Bisson, P.A., Quinn, T.P., Reeves, G.H., and Gregory, S.V. 1992. Best management practices, cumulative effects, and long-term trends in fish abundance in Pacific Northwest river systems, pp. 189–232. In: R.J. Naiman, editor. Watershed Management: Balancing Sustainability and Environmental Change. Springer-Verlag, New York, NY.

Bisson, P.A., and Sedell, J.R. 1984. Salmonid populations in streams in clearcut vs. old-growth forests of western Washington, pp. 121–129. In: W.R. Meehan, T.R. Merrell, Jr., and T.A. Hanley, editors. Fish and Wildlife Relationships in Old-Growth Forests. American Institute of Fishery Research Biologists, Juneau, AK.

Bisson, P.A., Sullivan, K., and Nielsen, J.L. 1988. Channel hydraulics, habitat use, and body form of juvenile coho salmon, steelhead, and cutthroat trout in streams. Trans. Amer. Fish. Soc. 117:262–273.

Cederholm, C.J., and Reid, L.M. 1987. Impact of forest management on coho salmon (*Oncorhynchus kisutch*) populations of the Clearwater River, Washington: a project summary, pp.373–398. In: E.O. Salo and T.W. Cundy, editors. Streamside Management: Forestry and Fishery Interactions. Univ. of Washington Inst. of Forest Resour. Contrib. 57, Seattle, WA.

Elliott, J.M. 1985. Population dynamics of migratory trout, *Salmo trutta*, in a Lake District stream, 1966–83, and their implications for fisheries management. J. Fish Biol 27 (Suppl. A):35–43.

Elliott, J.M. 1994. Quantitative Ecology and the Brown Trout. Oxford Univ. Press, New York. NY. 304pp.

Everest, F.H., Sedell, J.R., Reeves, G.H., and Wolfe, J. 1984. Fisheries enhancement in the Fish Creek Basin – an evaluation of in-channel and off-channel projects, 1984. Report RWU-1705. USDA Forest Service, Pacific Northwest Forest and Range Experiment Station, Portland, OR.

Fausch, K.D., and Young, M.K. 1995. Evolutionarily significant units and movement of resident stream fishes: a cautionary tale. Amer. Fish. Soc. Symp. 17:360–370.

Francis, R.C., and Sibley, T.H. 1991. Climate change and fisheries: what are the real issues? Northwest Environ. J. 7:295–307.

Fransen, B.R., Bisson, P.A., Bilby, R.E., and Ward, J.W. 1993. Physical and biological constraints on summer rearing of juvenile coho salmon (*Oncorhynchus kisutch*) in small western Washington streams, pp. 271–288. In: L. Berg and P. Delaney, editors. Proceedings of a Workshop on Coho Salmon. Canada Dept. Fish. Oceans, Vancouver, BC.

Gowan, C., Young, M.K., Fausch, K.D., and Riley, S.C. 1994. Restricted movement in resident stream salmonids: a paradigm lost? Can. J. Fish. Aquat. Sci. 51:2626–2637.

Grossman, G.D., Dowd, J.F., and Crawford, M. 1990. Assemblage stability in stream fishes: a review. Environ. Manage. 14:661–671.

Hall, J.D., Brown, G.W., and Lantz, R.L. 1987. The Alsea Watershed Study: a retrospective, pp. 399–416. In: E.O. Salo and T.W. Cundy, editors. Streamside Management: Forestry and Fishery Interactions. Univ. of Washington Inst. of Forest Resour., Seattle, WA.

Hall, J.D., and Knight, N.J. 1981. Natural Variation in Abundance of Salmonid Populations in Streams and its Implications for Design of Impact Studies. EPA-600/S3-81-021. U.S. Environmental Protection Agency, Corvallis, OR. 85pp.

Hall, J.D., and Lantz, R.L. 1969. Effects of logging on the habitat of coho salmon and cutthroat trout in coastal streams, pp. 355–375. In: T.G. Northcote, editor. Symposium on Salmon and Trout in Streams. Univ. of British Columbia, H.R. MacMillan Lectures in Fisheries, Vancouver, BC.

Hartman, G.F., and Scrivener, J.C. 1990. Impacts of forest practices on a coastal stream ecosystem, Carnation Creek, British Columbia. Can. Bull. Fish. Aquat. Sci. 223. 148pp.

Hawkins, C.P., Murphy, M.L., Anderson, N.H., and Wilzbach, M.A. 1983. Density of fish and salamanders in relation to riparian canopy and physical habitat in streams of the northwestern United States. Can. J. Fish. Aquat. Sci. 40:1173–1185.

Hicks, B.J., Beschta, R.L., and Harr, R.D. 1991a. Long-term changes in streamflow following logging in western Oregon and associated fisheries implications. Water Resour. Bull. 27:217–226.

Hicks, B.J., Hall, J.D., Bisson, P.A., and Sedell, J.R. 1991b. Responses of salmonids to habitat changes, pp. 483–518. In: W.R. Meehan, editor. Influences of Forest and Rangeland Management on Salmonid Fishes and their Habitats. Amer. Fish. Soc. Spec. Publ. 19.

Hilborn, R., and Winton, J. 1993. Learning to enhance salmon production: lessons from the Salmonid Enhancement Program. Can. J. Fish. Aquat. Sci. 50:2043–2056.

Hurlburt, S.A. 1984. Pseudoreplication and the design of ecological field experiments. Ecol. Monogr. 54:187–211.

Johnson, T.H., and Cooper, R. 1986. Snow Creek anadromous fish research. Report Number 86–18. Fisheries Management Division, Washington State Game Dept., Olympia, WA.

Johnston, N.T., Perrin, C.J., Slaney, P.A., and Ward, B.R. 1990. Increased juvenile salmonid growth by whole-river fertilization. Can. J. Fish. Aquat. Sci. 47:862–872.

Lawson, P.W. 1993. Cycles in ocean productivity, trends in habitat quality, and the restoration of salmon runs in Oregon. Fisheries 18(8):6–10.

Lichatowich, J., and Cramer, S. 1979. Parameter selection and sample sizes in studies of anadromous salmonids. Information Report Series, Fisheries Number 80–1. Oregon Dept. of Fish and Wildlife, Portland, OR.

Moore, K.M.S., and Gregory, S.V. 1988. Response of young-of-the-year cutthroat trout to manipulation of habitat structure in a small stream. Trans. Amer. Fish. Soc. 117:162–170.

Moring, J.R., and Lantz, R.L. 1975. The Alsea Watershed Study: effects of logging on the aquatic resources of three headwater streams of the Alsea River, Oregon. Part I. Biological studies. Fish. Res. Rep. 9. Oregon Dept. of Fish and Wildlife, Corvallis, OR. 66pp.

Murphy, M.L., and Hall, J.D. 1981. Varied effects of clear-cut logging on predators and their habitat in small streams of the Cascade Mountains, Oregon. Can. J. Fish. Aquat. Sci. 38:137–145.

Murphy, M.L., Hawkins, C.P., and Anderson, N.H. 1981. Effects of canopy modification and accumulated sediment on stream communities. Trans. Amer. Fish. Soc. 110:469–478.

Nickelson, T.E., Nicholas, J.W., McGie, A.M., Lindsay, R.B., and coauthors. 1992. Status of Anadromous Salmonids in Oregon Coastal Basins. Oregon Dept. of Fish and Wildlife, Research and Development Section and Ocean Salmon Management, Corvallis, OR. 83pp.

Nickelson, T.E., Solazzi, M.F., and Johnson, S.L. 1986. Use of hatchery coho salmon (*Oncorhynchus kisutch*) presmolts to rebuild wild populations in Oregon coastal streams. Can. J. Fish. Aquat. Sci. 43:2443–2449.

Pearcy, W.G. 1992. Ocean Ecology of North Pacific Salmonids. Univ. of Washington Press, Seattle, WA. 179pp.

Pease, J.R. 1993. Land use and ownership, pp.31–39. In: P.L. Jackson and A.J. Kimerling, editors. Atlas of the Pacific Northwest. Oregon State Univ. Press, Corvallis, OR.

Perrin, C.J., Bothwell, S.L., and Slaney, P.A. 1987. Experimental enrichment of a coastal stream in British Columbia: effects of organic and inorganic additions on autotrophic periphyton production. Can. J. Fish. Aquat. Sci. 44:1247–1256.

Reeves, G.H., Burnett, K.M., Everest, F.H., Sedell, J.R., and coauthors. 1990. Responses of Anadromous Salmonid Populations and Physical Habitat to Stream Restoration in Fish Creek, Oregon. Project No. 84–11. USDA Forest Service, Pacific Northwest Forest and Range Experiment Station, Portland, OR.

Sokal, R.R., and Rohlf, F.J. 1981. Biometry. W.H. Freeman and Co., San Francisco, CA. 859pp.

Thomas, J.W., editor. 1993. Forest Ecosystem Management: an Ecological, Economic, and Social Assessment. Report of the Forest Ecosystem Management Assessment Team. USDA Forest Service, Portland, OR.

Underwood, A.J. 1994. On beyond BACI: sampling designs that might reliably detect environmental disturbances. Ecol. Appl. 4:3–15.

Walters, C.J., Collie, J.S., and Webb, T. 1988. Experimental designs for estimating transient responses to management disturbances. Can. J. Fish. Aquat. Sci. 45:530–538.

Walters, C.J., Collie, J.S., and Webb, T. 1989. Experimental designs for estimating transient responses to habitat alteration: is it practical to control for environmental interactions? pp. 13–20. In: C.D. Levings, L.B. Holtby, and M.A. Anderson, editors. Proceedings of the National Workshop on Effects of Habitat Alteration on Salmonid Stocks. Can. Spec. Pub. Fish. Aquat Sci. 105.

Ward, B.R., and McCubbing, D.J.F. 1998. Adult steelhead trout and salmonid smolts at the Keogh River during spring 1998 and comparison to the historic record. Fish. Tech. Circ. 102. British Columbia Ministry of Fisheries, Victoria, BC. 20pp.

Ward, B.R., and Slaney, P.A. 1979. Evaluation of in-stream enhancement structures for the production of juvenile steelhead trout and coho salmon in the Keogh River: Progress 1977 and 1978. Fish. Tech. Circ. 45. British Columbia Ministry of Environment, Victoria, BC. 47pp.

Ward, B.R., and Slaney, P.A. 1993. Habitat manipulations for the rearing of fish in British Columbia, pp.142–148. In: G. Shooner et S. Asselin, editors. Le Développement du Saumon Atlantique au Québec: Connaître les Règles du Jeu pour Réussir. Colloque International de la Fédération Québécoise pour le Saumon Atlantique. Québec, decembre 1992. Collection Salmo salar no.1.

Washington Department of Fisheries, Washington Department of Wildlife, and Western Washington Treaty Indian Tribes. 1993. 1992 Washington State Salmon and Steelhead Stock Inventory. Washington Dept. of Fish and Wildlife, Olympia, WA.

Wilzbach, M.A., Cummins, K.W., and Hall, J.D. 1986. Influence of habitat manipulations on interactions between cutthroat trout and invertebrate drift. Ecology 67:898–911.

Chapter 16
Watershed Management

Paul W. Adams

The Watershed Management Approach

Watershed management represents a unique approach to managing natural resources in that it gives particular attention to soil and water resources in a drainage basin context. Historically, this approach has been used primarily to protect and maintain municipal water sources, but more recently it has been used for broader resource objectives. Because several general concepts in resource management can be especially important when managing watersheds, they are briefly reviewed first.

Resource Management Concepts

Sound resource management and decision-making begin with good planning, including both a short- and a long-term perspective (sometimes referred to as tactical and strategic planning, respectively), the latter being especially appropriate for forests and related resources. A simple conceptual model for watershed management planning is shown in Fig. 16.1. The elements shown in the three boxes at the left provide much of the foundation and direction for management.

Despite the general proliferation of natural resources information and increasingly sophisticated means of accessing and displaying it, effective management still often requires both updated resource inventories and careful organization of relevant existing information. In watershed management, some key categories of inventory include: watershed boundaries; property boundaries and uses; terrain, geology and soils; climate; hydrology; water quality;

Paul W. Adams
Department of Forest Engineering, Oregon State University, Corvallis, OR 97331
paul.adams@oregonstate.edu

J. D. Stednick (ed.), *Hydrological and Biological Responses to Forest Practices.*   291
© Springer 2008

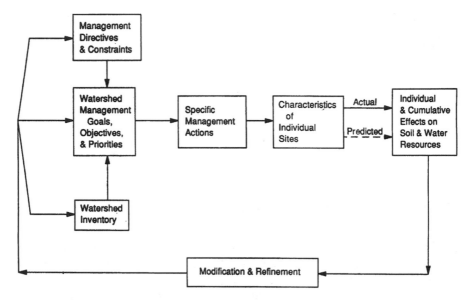

Fig. 16.1 Conceptual model for wildland watershed management planning. Reproduced from Satterlund and Adams 1992, with permission from John Wiley & Sons, Inc

vegetation; construction features; and socio-economic considerations (Satterlund and Adams 1992). The costs of inventory and information organization can be significant, but they should be carefully weighed against benefits of improved decision-making.

Historical knowledge long has been used in the social sciences, and its application to natural resource management has expanded considerably in recent years. In the Pacific Northwest, one of the most enlightening areas of recent study has been the patterns of wildfire. The Douglas-fir forest region of western Oregon and Washington, for example, apparently experienced very large (i.e., tens of thousands of hectares), high-severity fires every 150 years or so (Agee 1990). Such fires are known to have major effects on soil and water resources (e.g., increased erosion and sedimentation), as well as on forest ecology. Understanding these and other important natural influences (floods, droughts, volcanoes, etc.) can help managers make better judgments about the relative effects of human activities on natural resources.

Resource management decisions for both private and public lands are strongly shaped by the basic values of the organization or individuals involved, often expressed through statements of goals or objectives. Laws or policies such as the Oregon Forest Practices Act or the National Forest Management Act provide further direction or limits to management, as does the social or cultural environment. Although the importance of these controlling influences is widely recognized, their detailed and explicit revelation may be a critical early step in effective management planning and decision making.

Decisions about specific management actions usually require information about the specific sites involved, with more detail than typically provided by a general resource or watershed inventory. Effects of actions then can be predicted by formal (models, etc.) or informal (local experience, etc.) means, and a decision made to proceed with or revise plans. Concerns about the cumulative effects of past, present, and future management actions are now receiving much attention, although the means for evaluating such effects remain relatively unrefined. Even where there is considerable confidence that management effects have been accurately predicted, monitoring of results and refinement of actions are additional key steps in management planning (Fig. 16.1).

Finally, as technology increases our ability to both predict and identify environmental effects that result from management practices, we are especially challenged to very carefully interpret the significance of these effects. Both profound and subtle environmental changes occur over short and long time scales even in pristine watersheds. Depending upon the specific resource of interest, management practices can have positive, negative, neutral, or some combination of effects, and a widely variable duration of influence. Because the phrase "management impacts" typically infers only negative effects, use of this terminology when assessing effects may limit creative and scientific (i.e., critical) thinking about natural resources that inherently are much more diverse and complex in their responses to management.

Watersheds as a Basis for Resource Management

The 1990s could be considered the "watershed decade" in that never before had so much attention been given to watersheds as a basis for resource management. In the Pacific Northwest, concerns about such issues as declining wild fish stocks, cumulative effects, and domestic water supplies have often been framed with a watershed perspective. The attention to watersheds paralleled interest in "ecosystem management," which often itself is cited as requiring a basin or landscape perspective. Indeed, many concepts of watershed and ecosystem management seem to go hand in hand, particularly with respect to aquatic ecosystems, whose structure and function may extend from a high mountain spring to the mouth of an estuary.

While attractive conceptually, however, both watershed and ecosystem management can present formidable challenges in implementation (Adams and Atkinson 1993). In many larger watersheds, varying land use and ownership patterns result in diverse management objectives and constraints. For example, the Oregon Forest Practices Rules require significant riparian protection measures on forest lands (See Chapter 6), whereas adjacent agricultural and other lands generally have had more limited requirements. Even where ownership or administration is consistent, physical and biological conditions may vary greatly, and natural and human influences may extend from beyond the watershed boundary. Examples of the latter include atmospheric deposition from pollution sources many kilometers away, ocean thermal conditions that reduce anadromous fish numbers that can spawn in fresh water, and national and international needs for wood products and trade that stimulate timber harvesting.

Perhaps the greatest opportunity for using watersheds in resource management is in providing a logical framework or checkpoint for planning efforts. As long as external influences and social dimensions are accounted for, watersheds remain uniquely suited for evaluating and planning management policies and practices, especially when soil and water resources are the focus. Because timber harvesting and related forest practices are of such interest and concern when managing watersheds in which forests are abundant, these will be examined in some detail in the remainder of this chapter. The discussion will emphasize forest practices and knowledge gained in western Oregon, but many of the general principles should be widely applicable.

Timber Harvesting

Timber harvesting can directly alter vegetation cover, soil conditions, and stream channels, which in turn may significantly influence water quantity and quality and related resources. Harvest systems, scheduling, layout, and operations represent the key areas in which management options are likely to be available for controlling the positive and negative effects of timber harvest (Murphy and Adams 2005). Technology and experience in timber harvesting have evolved greatly over the years, and a variety of important developments are highlighted below. These newer developments and their relative environmental performance should be carefully considered when reviewing research literature and other information on watershed effects of timber harvesting that may have occurred decades ago. It is also essential to clearly identify the specific practice in question, because the diverse practices associated with logging (e.g., road construction, felling, yarding, hauling, slash treatment) may individually yield widely different watershed effects in any given location (Adams and Andrus 1991).

Harvest Systems

There are many types of harvest systems and equipment options with which individual timber harvesting operations can be conducted. However, in a given geographical area, choices may be limited by local availability, trained personnel, or cost. Environmental performance among major systems (e.g., ground, cable, aerial) does vary somewhat, but generalizations are not particularly useful because each system can perform well when correctly matched to site conditions and when operations are carefully planned and conducted.

Ground-based harvest systems are most widely available and used, although individual equipment types and combinations vary considerably, especially given the trend toward increasingly mechanized systems. The most basic ground-based system is the draft animal, which is still in limited use throughout the United States. Unless soils are steep, wet, or weak, horses and other animals can perform well in pulling logs to a central landing location, but their low overall productivity and short-distance yarding capabilities are best suited to small operations with good road access. Note that timber harvest systems and forest roads are inextricably linked together; characteristics and decisions of one can greatly influence the other (see sections below).

Wheeled skidders, crawler tractors, and other tracked vehicles are often used to move (i.e., yard) logs or whole trees to the landing. Increasingly, powered vehicles also are used to fell, limb, and assemble trees or logs for yarding. The array of machines and mechanized logging procedures has become so complex that terminology has been defined for them (Kellogg et al. 1993). With any logging vehicle, there is potential for soil exposure, displacement, or compaction, which may lead to increased runoff and erosion, particularly in sloping terrain. Although some variation in soil effects from different vehicles has been observed (Cafferata 1992), these soil effects can occur with virtually any vehicle and are usually best managed through harvest scheduling, layout, supervision, or post-logging treatments (see sections below).

Cable harvest systems are widely used in areas of steep terrain in the Pacific Northwest. Often considerably more costly than ground-based systems, cable systems encompass a variety of equipment types and combinations. Historically, "highlead" cable systems with limited log lift capabilities were commonly used, in some cases resulting in considerable soil exposure and displacement along cable paths. More recent technical advancements have made "skyline" cable systems the norm, often with sophisticated carriages (Studier 1993) and intermediate supports (Mann 1984) that provide a high degree of log lift and control over soil or stream disturbance. Another trend has been the downsizing of cable equipment, which reduces costs and better matches the timber sizes now being cut (Kellogg 1981). Aerial harvest systems include balloons and helicopters, which also have seen some downsizing and improved cost competitiveness (still generally more costly than cable systems, however), and which offer another option for environmental control (Studier and Neal 1991).

Harvest Scheduling

Scheduling of timber harvest activities is important in both the short and long term. Seasonal conditions such as rain, snow, freezing, and thawing can greatly affect soil trafficability and other operational influences that may yield undesirable soil and water effects, especially when using ground-based systems. Soil disturbance generally is reduced when logging is conducted when soils are relatively dry, or deeply frozen or snow covered. Soil compaction (i.e., increased soil bulk density) from logging vehicles may occur even with dry soils (Cafferata 1992), however, and other management approaches such as harvest layout or post-harvest treatments (see sections below) may be needed if undesirable watershed effects are expected.

Harvest scheduling over the long term can be a consideration in larger watersheds where multiple operations are expected. For example, harvests may be scheduled to avoid or take advantage of the streamflow increases that can result from reductions in evapotranspiration following logging. Because such flow effects are proportional to the degree of harvest, and decline within a few decades as the new forest canopy develops (Harr 1983), within a watershed the sequence and nature of individual harvests over the years may merit evaluation. Other resource effects of logging can show similar patterns of response followed by a period of recovery (e.g., stream shade and temperature), which can also be considered in harvest scheduling within watersheds. Where there are sufficient data, the integration of multiple temporal and spatial factors to evaluate resource effects is well suited to computer analysis, which increasingly has been used to support harvest planning (Sessions 1992).

Harvest Layout and Operations

Once a general harvest area and system are identified, the specific layout and operational details help bring the plans to life. In and near any harvest area, key control points related to watershed resources should be identified. These include such features as steep slopes, streams, and perennially wet soils. With ground-based harvest systems, the layout and use of a planned skid trail network can be an effective means of controlling traffic patterns and soil impacts (Garland 1993). Most soil compaction occurs within the first few passes of logging machinery, so it is usually preferable to concentrate impacts within a relatively small area of defined trails rather than attempt to disperse vehicle traffic. The longevity of compaction makes planned trail networks especially attractive where intermediate stand entries are expected (i.e., cumulative effects are avoided). Trail layout and associated practices such as line pulling and log winching may add to logging costs, but these usually are not unreasonable and may even be offset by productivity gains from yarding over an efficiently designed trail system (Olsen et al. 1987).

When using cable systems, log suspension and control are greatly affected by harvest layout. Changes in landing location or orientation of the cable path can mean the difference between good suspension or heavy gouging and soil displacement, even on steep slopes. Uphill yarding generally is preferable to a downhill layout, which may concentrate soil exposure and surface runoff along the slope near the landing. Cable yarding layout has been aided in recent years by computer programs that can help define the operating limits of a given system and its most effective application for the local timber and terrain (e.g., Jarmer 1992).

Many watershed effects from logging can be avoided or minimized by harvest layouts that provide an area of little or no disturbance adjacent to streams (e.g., buffer strip). Appropriate widths and other characteristics of buffers have been the subject of considerable analysis and debate in the Pacific Northwest (Adams 2007; Belt et al. 1992). Design criteria can vary depending upon the specific site conditions and resource concerns, e.g., fish habitat or domestic water supply. As interest in increased protective measures has grown, so too has concern about costs and other operational constraints, which may now stem from not only stream buffers (Olsen et al. 1987), but also modified harvest practices upslope (Kellogg et al. 1991). In addition, the complexity of harvesting and resource interactions demands a thorough analysis of any environmental trade-offs, i.e., determining whether protection measures create new or more serious environmental problems such as buffer blowdown or more roads (Adams et al. 1988).

Harvest systems, scheduling, layout, and other plans are operationally implemented by forest workers. Despite their importance in accomplishing management objectives and avoiding resource problems, field personnel and their knowledge, skills, and abilities sometimes are overlooked during harvest planning. Timber harvesting is physically demanding and dangerous work that can result in relatively high turnover. Practices that protect or enhance watershed resources may be relatively challenging or require new equipment. Training and incentive programs may be desirable to maintain or improve job skills and environmental performance. Supervision and communication among forest workers also can be valuable tools, e.g., even the most well-intentioned worker may be unaware of a specific resource concern or a simple procedural means for avoiding a problem.

Forest Roads

Landslide surveys and other watershed studies in the Pacific Northwest have shown that forest roads sometimes can be key sites for runoff and erosion, particularly in steep terrain (Ice 1985; King and Tennyson 1984; Reid and Dunne 1984). Road construction typically results in exposed, excavated, and compacted soils, as well as new surface drainage features and stream crossings; each may contribute to watershed effects during or some time after construction. However, like timber harvesting, technology and procedures for road construction and use have advanced greatly in the past few decades, and these improvements should be considered when interpreting historical effects relative to expectations for both new and existing forest roads. In particular, careful consideration of road area, location, design, construction, and maintenance likely can avoid most watershed problems (Adams and Andrus 1990).

Road Area and Location

Some roads exist in nearly all forest watersheds, but they may be inadequate for current or projected needs for logging, fire control, recreation, and other purposes. Both environmental concerns and the high costs of road construction make minimal road construction desirable. When roads are built to support timber harvesting and log hauling, the expected harvest system and hauling routes can be important influences on road planning. Necessary road spacings and the resulting total area in logging roads, for example, are directly related to the maximum practical yarding distance for the logging system (Table 16.1). Newer computer programs for road network design can account for the needs of different logging systems as well as other factors such as terrain, transport routes, road standards, and cost components (Sessions 1992).

Historically, many forest roads were located near stream and river channels because of easy construction along moderate slopes and proximity to mills and other developed areas. Such locations increase the opportunities for sedimentation and other impacts, however, and new construction should favor upland road locations or undisturbed buffer strips between roads and water bodies. Stream crossings generally should be minimized, with right angle approaches to reduce soil disturbance near the stream. In steep terrain, upland road locations still present challenges because considerable soil excavation may be needed to create sufficient road width, and adequate road drainage becomes increasingly important to avoid erosion. Ridgetop roads can reduce excavation and drainage needs, but some side slope roads still will be necessary to access ridges and other areas.

Preliminary map and field surveys are essential for good road location. Topographic maps, aerial photos, soil surveys, geographic information systems, and other tools can be very useful in establishing general routes. Field reconnaissance invariably reveals other small-scale features (wet spots, rock outcrops, etc.) that modify road alignment. Additional adjustments may be necessary during the construction phase if excavation uncovers unfavorable localized soil or rock conditions.

Road Design

At one time, forest road construction consisted primarily of a bulldozer operator simply blading off the surface soil and compacting the underlying material to create a relatively hard and level track for vehicles. A few roads, particularly in gentle terrain, are still constructed this way, but forest road design generally has become a much more sophisticated process that may involve several steps

Table 16.1 General relations between timber harvesting systems, maximum yarding distance, and percent area in roads (compiled from various sources for Pacific Northwest conditions)

Harvest system	Maximum yarding distance (m)	Area in roads (%)
Tractor/skidder	450	4–15
Highlead/short skyline	450	3–10
Long/multispan skyline	1500	2–4
Helicopter	2300	1–3

and equipment types. In steep terrain, for example, slope failures can be markedly reduced through the use of a full-bench road design combined with trucking of the excavated material to a stable location (Sessions et al. 1987). Ridgetop roads provide similar benefits, but gradeability and surfacing of the steep roads used to access ridges become additional important design features (Anderson et al. 1987). In general, gravel surfacing reduces sediment losses from roads, although careful surfacing prescriptions (e.g., rock specifications) and subgrade preparation are needed to manage costs and enhance durability. Wet or weak soils, for example, may require the use of synthetic fabrics or other subgrade supplements to enhance surface life and trafficability.

Perhaps the most universally critical aspect of road design with respect to watershed concerns is the drainage system. Water should move efficiently from the road surface to stable areas where it will behave as if the road had not been there; i.e., it will infiltrate the soil. Road surfaces usually are designed with a crown, inslope, or outslope for immediate drainage, and where the road is cut into a slope, an inside ditch often is needed to collect and move water (Garland 1983). On long grades, ditches are supplemented with relief culverts or rolling dips to avoid gully formation. Spacing of ditch-relief culverts or road dips generally should decrease as the road grade or soil erodibility increases. A survey of ditch-relief culverts in the Oregon Coast Range, for example, showed that erosion both in ditches and at culvert outlets increased with long culvert spacings (Piehl et al. 1988a).

Design of stream crossings is another very important consideration, because these are sites of direct interaction with water resources. Several general design options usually are available, including fords, culverts, and bridges, in order of typically increasing cost. For smaller streams, simple fords are a relatively economical option that also may have low maintenance requirements (Warhol and Pyles 1989). Because vehicles drive directly through the stream, however, expected traffic frequency may be important where water quality is a concern. Culverts are the most commonly used stream crossing in the Pacific Northwest, although installation design varies widely. For larger streams and rivers, bridges may be the only viable option. Some of the key design considerations for stream crossings include local hydrology and peak flow events, fish migration, maintenance, management objectives, economics, and legal requirements (Pyles et al. 1989).

Because some of the most serious erosion problems at stream crossings occur during unusually large storms, state forest practice laws in the Pacific Northwest generally require that crossings be designed to withstand a 50-year frequency or larger flow. Although clear in intent, such directives have had variable implementation because data and methods for developing local estimates of peak flows often have been lacking. A survey of culverts in the central Oregon Coast Range, for example, showed over 40% would be unable to pass a 25 year peak flow without ponding above the top of the pipe inlet (Piehl et al. 1988b). Methods for predicting peak flows for stream crossing design have improved, however, including the refinement of predictive tools for areas of similar climate and other unifying hydrologic characteristics (Adams et al. 1986; Andrus et al. 1989).

Road Construction and Maintenance

Soils exposed during road construction are especially susceptible to erosion for a year or two following excavation, after which revegetation and other stabilization normally reduce erosion hazards. Construction timing relative to wet weather can thus be important, as can the use of plant seedings and physical barriers along road cut and fill slopes (Burroughs and King 1989). In steep terrain, control of road widths and excavated materials can be enhanced by the use of hydraulic excavators, which have become more cost-competitive and widely used in the Pacific Northwest (Balcom 1988).

The function of road drainage systems and other key construction features can only be insured through both routine and emergency maintenance procedures (Adams 1997). Surface grading, culvert and ditch cleaning, and supplemental gravel applications are common needs, especially as traffic levels increase or wet weather promotes rutting, slumping of cut banks, etc. Proximity to streams may be another important criterion for prioritizing road maintenance operations. Traffic control during wet weather can reduce sediment losses from roads, but even during dry weather overall sediment losses from forest roads can increase with heavier traffic loads (Bilby et al. 1989).

Permanent road closure may be a viable option where access no longer is a priority and where maintenance costs or risks of erosion are high. Although hydraulic excavators and other advanced equipment make possible the complete deconstruction of roads (i.e., fill material returned to approximate the original slope), such costly measures probably are unnecessary for the protection of watershed resources, as long as drainage is adequately provided for during closure (e.g., construction of water bars). Where older roads are still needed and potential watershed impacts remain a concern, upgrading of road standards (e.g., type or thickness of surfacing, culvert spacing or sizing, maintenance frequency) may be desirable.

Other Forest Practices

Forest management normally involves considerably more than just road construction and timber harvesting. Slash treatment, site preparation, reforestation, and stand treatments are common practices often stimulated by legal or economic concerns. In addition, insect and disease problems, wildfire, ice and wind damage, and other considerations can prompt some unique management responses. Because these practices may yield their own diverse watershed effects, management and policy decisions should proceed from a clear understanding of specific cause-and-effect relations among the many individual practices and resources that may be encountered.

Slash Treatment and Site Preparation

Slash that is left after logging may be treated in a variety of ways for one or more purposes, including wildfire hazard reduction and preparing the site for successful reforestation. Improved planting ease, control of competing vegetation, increased nutrient availability, and insect and disease control may be primary objectives. Because of the historical importance of fire in Pacific Northwest forests, prescribed burning treatments have been widely applied in the region with generally very positive results for reforestation, although questions are likely to persist about long-term effects and site-specific prescriptions (Walstad et al. 1990).

Both prescribed and natural fire, for example, may produce some undesirable effects on watershed resources, at least in the short term until revegetation occurs. If fires are of sufficient severity, soils and stream channels may be exposed to the erosive forces of water, wind, and gravity (Ice et al. 2004). Changes in physical properties that increase surface runoff (e.g., increased water repellency) may also occur (McNabb and Swanson 1990). Burning mineralizes some of the nutrients contained in slash, which may then leach from soils into water bodies. The most significant changes in water quantity and quality that have been noted in the region have occurred following wildfires or very hot slash burns; generally negligible changes are expected with low intensity prescribed burning (Beschta 1990). Control of burn intensity thus appears to be a primary means of avoiding undesirable impacts, although this approach needs to be weighed with such trade-offs as air quality (i.e., less intense burns produce more smoke) and reforestation benefits.

Slash treatment and site preparation often involve the use of heavy equipment that may produce its own effects on watershed resources. Piling of slash with crawler tractors, for example, may result in significant soil exposure and compaction. Site preparation with tractor-mounted brush rakes may produce similar effects, although compaction may primarily occur in the subsoil while the surface soil is somewhat loosened. These soil effects from post-logging treatments also may negate efforts made during logging (e.g., use of designated skid trails) to minimize compaction. Where soil compaction is an unavoidable result of logging or other practices, it may be alleviated through the use of soil tillage treatments (Cafferata 1992). The effectiveness of tillage implements varies widely, however, and other measures such as the construction of water bars may still be needed to protect watershed resources.

Herbicides are sometimes used as a site preparation measure to eliminate vegetation that would compete with desired species for moisture, light, nutrients, or space. Most herbicides act specifically to disrupt plant growth mechanisms and are of relatively low toxicity to humans and animals. They also are less likely than broadcast burning or mechanical methods to increase erosion (Table 16.2). Still, many people are concerned about such manufactured chemicals entering water supplies, in part due to the detection of more toxic contaminants found in earlier formulations of some herbicides (Walstad and Dost 1984). Toxicity is only part of what constitutes a chemical hazard, however. Equally important are the likelihood and degree of exposure, which are governed by the application methods and how the chemical interacts with the local environment (Norris et al. 1991). Contamination of water supplies often can be avoided by ensuring that herbicides are applied at minimum effective concentrations well away from water bodies. However, in situations where the risks from chemical use are determined to be high, alternative practices may be most appropriate.

Reforestation, Stand Management and Other Considerations

Reforestation after logging is a legal requirement in the Pacific Northwest. In some cases, the new forest may differ enough from the original forest in species or structure to affect water resources. For example, past logging practices in some locations apparently have encouraged the regrowth of riparian alder stands, which may affect water quantity (Hicks et al. 1991) and quality (Taylor and Adams 1986). Although land use laws now limit widespread changes in forest cover, some conversions to agriculture, residential, or industrial uses are possible, and these new uses may uniquely affect local watershed resources. Afforestation of agricultural or other open lands also may occur, and if a large enough part of the watershed is treated this way, streamflows may be reduced

Table 16.2 Total soil movement to hillslope erosion collection boxes 45 months after various site preparation and conifer release treatments in the Oregon Coast Range (Stednick, J.D, P.W. Adams, and W.R. Stack 1991, unpublished report to USDA Forest Service, Pacific Northwest Research Station, Portland, OR)

	Cumulative soil movement (kg)
SITE PREPARATION TREATMENTS	
Control – no site preparation treatment	3.8
Broadcast Burn – late summer, early fall	12.7
Aerial Spray – glyphosate, early fall	7.5
Manual Scalp – 1m radius, before planting	4.1
Spray and Burn – aerial picloram, June; fall burn	3.7
Slash and Burn – manual, June; burn late summer/early fall	9.3
CONIFER RELEASE TREATMENTS	
Control - no conifer release treatment	2.4
Manual Cut 1x – 1m radius, once, late spring	0.7
Manual Cut 2x – 1m radius, twice, late spring	1.5
Manual Cut 3x – 1m radius, three times, late spring	1.1
Aerial Spray – glyphosphate, early fall	2.1

because evapotranspiration loss from forest cover is higher than from most other types of vegetation.

Forest plantation and stand treatments may include brush release, fertilization, and insecticide and rodenticide applications. Brush release is usually performed with herbicides, but manual treatments are sometimes used. When carefully performed, both approaches generally have little or no effect on soil and water resources (Table 16.2), primarily because soil disturbance is avoided or of very limited areal extent. Guidelines for fertilizer, insecticide, and rodenticide applications generally follow those of herbicides, i.e., minimum effective amounts away from water bodies and runoff areas (e.g., ephemeral channels, road ditches). Insecticides and rodenticides may require a greater level of care, however, due to generally higher toxicities than herbicides and fertilizers.

In addition to silvicultural practices, other activities and influences on or near forest lands may be important to watershed resources. Livestock grazing may be a concern, particularly if animals are allowed to intensively graze riparian areas or to freely access streams for watering. Although most experience in grazing and watershed management in the Pacific Northwest has been from drier inland areas, the basic principles should be widely applicable: Carefully manage the level and season of grazing to maintain adequate plant cover and streambank and channel structure (Clary and Webster 1989). Concentrations of large, grazing wildlife (elk, deer, etc.) also may affect watershed resources, but management may be more difficult. Beaver can profoundly affect stream channels and water quality, in some cases causing problems (e.g., domestic water supplies) and in others providing apparent benefits (Leidholdt-Bruner et al. 1992).

Finally, although forest recreation is often dispersed, water bodies attract people and can create areas of concentrated use that may lead to erosion, biological contamination, or other problems (Clark et al. 1985; Cole 1989). In addition, use of 4-wheel drive vehicles, mountain bikes, or horses may lead to soil disturbance and erosion, particularly in wet or sloping ground. Controlling access to or intensity of use of riparian and other sensitive areas can be challenging, but is among the most effective approaches to limiting watershed impacts from recreation. Direct education and communication with recreation users can play a key role in raising awareness and cooperation in controlling local watershed impacts.

Conclusions

Watershed management is a growing, yet relatively complex and challenging approach to managing natural resources on forest lands. Although it lacks a fully comprehensive research base and widely accepted evaluation procedures, we appear to have both sufficient knowledge and experience to use improved practices and decisions to avoid most problems when managing forest resources in a watershed context. Education tapping this knowledge and experience can play a particularly important role in promoting positive practices and decisions, especially in light of prevailing watershed misconceptions, a lack of focus on cause-and-effect relations among watershed practices and resources, and the widely variable backgrounds of those involved with management and decision making in watersheds (Adams and Cleaves 1993). As we have learned throughout the realm of natural resource management, it is not only what we do, but how we choose to do it that makes the difference between positive or negative results.

Literature Cited

Adams, P.W. 2007. Policy and management for headwater streams in the Pacific Northwest: Synthesis and reflection. Forest Science 53(2): 104–118.

Adams, P.W. 1997. Maintaining woodland roads. Extension Circular 1139. Oregon State Univ. Cooperative Extension Service, Corvallis, OR. 8pp.

Adams, P.W., and Andrus, C.W. 1990. Planning secondary roads to reduce erosion and sedimentation in humid tropic steeplands, pp. 318–327. In: Research needs and applications to reduce erosion and sedimentation in tropical steeplands. Pub. No. 192. International Association of Hydrological Sciences, Washington, DC.

Adams, P.W., and Andrus, C.W. 1991. Planning timber harvesting operations to reduce soil and water problems in humid tropic steeplands, pp. 24–31. In: Proceedings of a symposium on forest harvesting in Southeast Asia. Forest Engineering Inc. and Oregon State Univ. College of Forestry, Corvallis, OR.

Adams, P.W., and Atkinson, W.A., editors. 1993. Watershed Resources: Balancing Environmental, Social, Political and Economic Factors in Large Basins. Conference proceedings, October 14–16, 1992, Red Lion Jantzen Beach, Portland, OR. College of Forestry, Oregon State Univ., Corvallis, OR. 162pp.

Adams, P.W., Beschta, R.L., and Froehlich, H.A. 1988. Mountain logging near streams: opportunities and challenges, pp. 153–162. In: Proceedings, International Mountain Logging and Pacific Northwest Skyline Symposium. College of Forestry, Oregon State Univ., Corvallis, OR.

Adams, P.W., Campbell, A.J., Sidle, R.C., Beschta, R.L., and coauthors. 1986. Estimating streamflows on small forested watersheds for culvert and bridge design in Oregon. Research Bull. 55. Forest Research Lab, Oregon State Univ., Corvallis, OR. 8pp.

Adams, P.W., and Cleaves, D.A. 1993. Approaches and structures for watershed education and research, pp. 110–118. In: P.W. Adams and W.A. Atkinson, editors. Watershed resources—Balancing environmental, social, political and economic factors in large basins. College of Forestry, Oregon State Univ., Corvallis, OR.

Agee, J.K. 1990. The historical role of fire in Pacific Northwest forest, pp. 25–39. In: J.D. Walstad, S.R. Radosevich, and D.V. Sandberg, editors. Natural and prescribed fire in Pacific Northwest forests. Oregon State Univ. Press, Corvallis, OR.

Anderson, P.T., Pyles, M.R., and Sessions, J. 1987. The operation of logging trucks on steep, low-volume roads, pp. 104–111. In: Fourth Intl. Conf. Low-Volume Roads, Vol. 2. Transportation Research Board, National Research Council, Washington, DC.

Andrus, C.W., Froehlich, H., and Pyles, M.R. 1989. Peak flow prediction for small forested watersheds along the southern Oregon and northern California coast. Water Note 1989-1. Water Resources Research Institute, Oregon State Univ., Corvallis, OR.

Balcom, J. 1988. Construction costs for forest roads. Research Bull. 64. Forest Research Lab, Oregon State Univ., Corvallis, OR. 21pp.

Belt, G.H., O'Loughlin, J., and Merrill, T. 1992. Design of forest riparian buffer strips for the protection of water quality: analysis of scientific literature. Report No. 8. Univ. of Idaho, Forest, Wildlife and Range Policy Group, Moscow, ID. 35pp.

Beschta, R.L. 1990. Effects of fire on water quantity and quality, pp. 219–232. In: J.D. Walstad, S.R. Radosevich, and D.V. Sandberg, editors. Natural and Prescribed Fire in Pacific Northwest Forests. Oregon State Univ. Press, Corvallis, OR.

Bilby, R.E., Sullivan, K., and Duncan, S.H. 1989. The generation and fate of road-surface sediment in forested watersheds in southwestern Washington. For. Sci. 35:453–468.

Burroughs Jr., E.R., and King, J.G. 1989. Reduction of soil erosion on forest roads. General Technical Report INT-264. USDA Forest Service, Intermountain Research Station, Ogden, UT. 34pp.

Cafferata, P. 1992. Soil compaction research, pp. 2–22. In: A. Skaugset, editor. Forest Soils and Riparian Zone Management: The Contributions of Dr. Henry A. Froehlich to Forestry. College of Forestry Conference Office, Oregon State Univ., Corvallis, OR.

Clark, R.N., Gibbons, D.R., and Pauley, G.B. 1985. Influence of forest and rangeland management on anadromous fish habitat in western North America: influences of recreation. General Technical Report PNW-178. USDA Forest Service, Forest and Range Experiment Station, Portland, OR. 38pp.

Clary, W.P., and Webster, B.F. 1989. Managing grazing of riparian areas in the Intermountain Region. General Technical Report INT-263. USDA Forest Service, Intermountain Research Station, Ogden, UT. 11pp.

Cole, D.N. 1989. Low-impact recreational practices for wilderness and backcountry. General Technical Report INT-265. USDA Forest Service, Intermountain Research Station, Ogden, UT. 131pp.

Garland, J.J. 1983. Designated skid trails minimize soil compaction. Extension Circular 1110. Oregon State Univ. Extension Service, Corvallis, OR. 6pp.

Garland, J.J. 1983. Designing woodland roads. Extension Circular 1137. Oregon State Univ. Cooperative Extension Service, Corvallis, OR. 28pp.

Harr, R.D. 1983. Potential for augmenting water yield through forest practices in western Washington and western Oregon. Water Resour. Bull. 19:383–393.

Hicks, B.J., Beschta, R.L., and Harr, R.D. 1991. Long-term changes in streamflow following logging in western Oregon and associated fisheries implications. Water Resour. Bull. 27(2):217–226.

Ice, G.G. 1985. Catalog of landslide inventories for the Northwest. Technical Bulletin 456. National Council of the Paper Industry for Air and Stream Improvement, Inc., New York. 78pp.

Ice, G.G., Neary, D.G., and Adams, P.W. 2004. Effects of wildfire on soils and watershed processes. J. For. 102(6):16–20.

Jarmer, C., and Sessions, J. 1992. Computer based skyline analysis: LOGGERPC 3.0 a new tool for logging planning, pp. 128–134. In: J. Sessions, editor. Proceedings of the Workshop on Computer Supported Planning of Roads and Harvesting. Oregon State Univ., College of Forestry, Corvallis, OR.

Kellogg, L.D. 1981. Machines and techniques for skyline yarding of small wood. Forest Research Lab Research Bull. 36. Oregon State Univ., Corvallis, OR. 15pp.

Kellogg, L.D., Bettinger, P., and Studier, D. 1993. Terminology of ground-based mechanized logging in the Pacific Northwest. Forest Research Lab Research Contribution 1. Oregon State Univ., Corvallis, OR. 12pp.

Kellogg, L.D., Pilkerton, S.J., and Edwards, R.M. 1991. Logging requirements to meet New Forestry prescriptions, pp. 43–49. In: J.F. McNeel and B. Andersson, editors. Proceedings of 14th Council on Forest Engineering Annual Meeting. Forest Engineering Research Institute of Canada, Vancouver, BC.

King, J.G., and Tennyson, L.C. 1984. Alteration of streamflow characteristics following road construction in north central Idaho. Water Resour. Res. 20:1159–1163.

Leidholdt-Bruner, K., Hibbs, D.E., and McComb, W.C. 1992. Beaver dam locations and their effects on distribution and abundance of coho salmon fry in two coastal Oregon streams. Northwest Sci. 66:218–223.

Mann, J.W. 1984. Designing double-tree intermediate supports for multispan skyline logging. Extension Circular 1165. Oregon State Univ. Cooperative Extension Service, Corvallis, OR. 8pp.

McNabb, D.H., and Swanson, F.J. 1990. Effects of fire on soil erosion, pp. 159–176. In: J.D. Walstad, S.R. Radosevich, and D.V. Sandberg, editors. Natural and Prescribed Fire in Pacific Northwest Forests. Oregon State Univ. Press, Corvallis, OR.

Murphy, G., and Adams, P.W. 2005. Harvest planning to sustain value along the forest-to-mill supply chain, pp. 17–23. In: C.A. Harrington and S.H. Schoenholtz, editors. Productivity of Western Forests: A Forest Products Focus. General Technical Report PNW-GTR-642. USDA Forest Service, Pacific Northwest Research Station, Portland, OR.

Norris, L.A., Lorz, H.W., and Gregory, S.V. 1991. Forest chemicals, pp. 207–296. In: W.R. Meehan, editor. Influences of Forest and Rangeland Management on Salmonid Fishes and Their Habitats. American Fisheries Society Special Publ. 19.

Olsen, E.D., Keough, D.S., and LaCourse, D.K. 1987. Economic impact of proposed Oregon Forest Practice Rules on industrial forest lands in the Oregon Coast Range: a case study. Res. Bull. 61. Forest Research Lab, College of Forestry. Oregon State Univ., Corvallis, OR. 15pp.

Piehl, B.T., Beschta, R.L., and Pyles, M.R. 1988a. Ditch-relief culverts and low-volume roads in the Oregon Coast Range. Northwest Sci. 62(3):91–98.

Piehl, B.T., Beschta, R.L., and Pyles, M.R. 1988b. Flow capacity of culverts on Oregon Coast Range roads. Water Resour. Bull. 24(3):631–637.

Pyles, M.R., Skaugset, A.E., and Warhol, T. 1989. Culvert design and performance on forest roads, pp. 82–87. In: Proceedings 12th Annual Council on Forest Engineering Meeting. Forest Engineering, Inc., Corvallis, OR.

Reid, L.M., and Dunne, T. 1984. Sediment production from forest road surfaces. Water Resour. Res. 20:1753–1761.

Satterlund, D.R., and Adams, P.W. 1992. Wildland Watershed Management. John Wiley & Sons, Inc., New York, NY. 448pp.

Sessions, J. 1992. Proceedings of the workshop on computer supported planning of roads and harvesting. Oregon State Univ. College of Forestry, Corvallis, OR.

Sessions, J., Balcom, J., and Boston, K. 1987. Road location and construction practices: Effects on landslide frequency and size in the Oregon Coast Range. West. J. Appl. For. 2(4):119–124.

Studier, D.D. 1993. Carriages for skylines. Forest Research Laboratory Research Contribution 3. Forest Research Lab, Oregon State Univ., Corvallis, OR. 14pp.

Studier, D.D., and Neal, J. 1991. Timber sale preparation: helicopter logging. USDA Forest Service, Pacific Northwest Region, Portland, OR.

Taylor, R.L., and Adams, P.W. 1986. Red alder leaf litter and streamwater quality in western Oregon. Water Resour. Bull. 22:629–635.

Walstad, J.D., and Dost, F.N. 1984. The health risks of herbicides in forestry: a review of the scientific record. Special Publication 10. Oregon State Univ., Forest Research Lab, Corvallis, OR. 60pp.

Walstad, J.D., Radosevich, S.R., and Sandberg, D.V. 1990. Natural and Prescribed Fire in Pacific Northwest Forests. Oregon State Univ. Press, Corvallis, OR. 332pp.

Warhol, T., and Pyles, M.R. 1989. Low water fords: An alternative to culverts on forest roads, pp. 77–81. In: Proceedings 12th Annual Council on Forest Engineering Meeting. Forest Engineering, Inc., Corvallis, OR.

Chapter 17
Research Opportunities in Hydrology and Biology in Future Watershed Studies

John D. Stednick

The effects of timber harvesting practices on water resources are mostly known from paired watershed studies. The first paired watershed study in the United States was in Colorado and designed to assess the effects of timber harvesting on water yield (Bates and Henry 1928). Many of these studies were designed to have demonstrable effects on water resources, specifically the timber harvesting was large in comparison to the watershed area (up to 100%), streamside vegetation buffers were not used, all timber including non-merchantable materials were removed, or the forest regeneration was suppressed by herbicide applications. Although these experiments helped identify the hydrologic processes affected by timber harvesting, they do not necessarily represent the effects of normal forest operations.

The scope of watershed studies was soon expanded beyond water quantity and the processes of the hydrologic cycle to include water quality. Measurements of inputs and outputs, as precipitation and streamflow, were used for chemical budgets. In the 1960s, 150 forested experimental watersheds were being studied; only 12 of these remain relatively active and half of those are long term ecological research sites (Ziemer and Ryan 2000). Many of the active research watersheds tend to focus on ecological processes in the watershed, rather than focus on hydrological processes as related to ecology. Nonetheless, several lessons learned from Alsea and other watershed studies (Stednick et al. 2004) are worth repeating.

The original Alsea Watershed Study and the ongoing research have provided some very useful lessons in forest hydrology and the effects of timber harvesting on hydrological responses. Timber harvesting of a large area in a watershed (Needle Branch) resulted in increased annual water yield. These water yield increases are higher in wetter years and lower in wetter years. As vegetation grows and the hydrological processes of interception, transpiration, and storage

John D. Stednick
Department of Forest, Rangeland, and Watershed Stewardship, Colorado State University, Fort Collins, CO 80523
jds@cnr.colostate.edu

J. D. Stednick (ed.), *Hydrological and Biological Responses to Forest Practices.*
© Springer 2008

increase, the annual water yield increase decreases. The hydrologic recovery to pretreatment conditions or expected annual water yields was estimated to take 31 years. The original streamflow response in Needle Branch showed increased peak flows for some years, but the mean was not significantly different from the pretreatment period. Peak flows tend to be higher for lower recurrence intervals, the higher recurrence intervals tend to not be significantly different. Continued streamflow monitoring suggested that the peak flow response in Needle Branch now approximates pretreatment conditions. Low flows were temporarily increased after harvesting, but mean low flow and low flow days were not significantly different immediately after harvesting. Additional streamflow monitoring suggests that as the streamside or riparian vegetation matures large daily variations in streamflow, especially in summer months are observed and low flows are often lower than expected. The Alsea was and continues to be operated as a cause and effect study site. The casual mechanisms need to be further addressed. A process oriented approach can provide predictive relationships that can be applied elsewhere.

Both the original harvest units in Deer Creek and the additional harvesting did not show any significant change in the hydrological metrics used: annual water yield, peak flow, low flows, or low flow days. With 39% of the watershed now harvested, and with a 31 year recovery period, the equivalent clearcut area is small (5.4%) and water resources responses are not detectable, thus the management schedule and application of BMPs in Deer Creek have been successful in maintaining water quantity and quality.

The water quality responses to timber harvesting in Needle Branch were dramatic, especially stream temperature and dissolved oxygen. These responses do not represent standard operating procedures, but do illustrate the importance of streamside vegetative buffers. The timber harvesting operations in Deer Creek, where harvest units had streamside vegetation left in place, showed no water quality changes. The results from the Alsea Watershed Study were the basis for development of forestry practices or best management practices (BMPs) designed to minimize the effects of forestry practices on water quality. Additional water resources monitoring is helping to identify the natural range of variability in water quantity and quality and may be used to further refine BMP definitions. A long-term water resource monitoring program is needed to better characterize this variability.

The effects of timber harvesting practices on water resources are variable (Stednick 1996). Study results may be a function of climate, road layout, harvesting methods, location of harvest unit in the watershed, and the post-treatment monitoring period climate; i.e. unusual storm events or drought conditions not observed in the calibration period. A summary of the effects of timber harvesting on annual water yield was compiled in 1967 (Hibbert 1967); and then later updated (Bosch and Hewlett 1982; Stednick 1996). Some of the original data used in the 1967 and 1982 efforts were unrecoverable for the latest study. It is critical to store all data in a retrievable format and allow others access to these data. Meta-data are needed to describe sampling, measurement, validation and/or analytical procedures, something more common in geographical information coverages. Data access can be improved, but it is important to recognize that the accessible data may not answer the management or research question posed. In one particular case in Utah, the effects of timber harvesting on water yield and peak flows were answered with adjusted streamflow records, ignoring data stationarity and the hydrologic appropriateness of the adjustment (Troendle and Stednick 1999). The long-term record may not be as useful as expected, especially if care is not taken to ensure continuity, such as when the gauging station has been relocated (Troendle et al. 2003). Data records need to be evaluated for data worth including a description of data quality, and standardization of archiving techniques. Many datasets from long term monitoring efforts are not available to outside parties. Outside perspectives may be fruitful in data reevaluation. Reanalysis of existing streamflow data on a more holistic basis is needed. Forest hydrology research from the past needs to be revisited (DeWalle 2003).

Frequent streamflow measurements with current meters have been replaced with electronic loggers of stage at artificial control sections, flumes and weirs. Weir and flume calibration maintenance needs to be an integral part of any hydrometerological monitoring program. Streamflow metrics originally included annual water yield and sometimes peak flows. The more time integrative metrics, annual water yield, have less uncertainty for comparative purposes. Many of the management questions today relate to sustainability of forest resources, and other metrics need be examined. Low flows have been recognized as important for maintenance of connectivity of fish habitat and water resources management. Additional metrics may include persistence of low flows. Less time integrative metrics could be flashiness of peak flows or time-to-peak. What streamflow metric is best to identify watershed health?

There is a continuing need for well-supported programs of field data collection and field process studies (Sidle 2006). Streamflow records can be complemented with precipitation records. Aside from research level studies, streamflow records and precipitation (or other hydrometeorological data) are often collected by different agencies. To illustrate this disconnect, the streamflow water year is October through September, while meteorological years are January through December. Other types of hydrological data may not be collected at all, making characterization of internal watershed process (e.g. change in storage) difficult.

Watershed studies can be hampered by inadequate or unrealistic goals, inadequate planning and funding, and lagging interest of sponsors during necessarily long-term studies. Maintaining long-term streamflow gauging stations is expensive, with little short term return, and often the value of long-term records is not appreciated until those data are needed to respond to a management question (Stednick et al. 2004). An excellent guideline for watershed studies has been developed (Callaham 1990).

The difficulty of justifying long-term monitoring is evident in the loss of gauging stations across the nation. The period of record for streamflow monitoring is surprisingly short. Some of the longest streamflow records for experimental watersheds include Coweeta (73 years), Fraser (65 years), Andrews (59 years) and Fernow (52 years) Experimental Forests. These few watersheds do not represent all the hydrologic regions in the United States, nor may they represent the variability in climate that has come to be better appreciated. Long-term records are required for trend detection. With continued human population growth, additional hydrologic modification (water diversions and transfers) is inevitable, and changes in air temperature and precipitation may further compound the signal from the treatment.

Past work in forestry (and agriculture) has resulted in important insights into how managed ecosystems respond to changes in soil moisture and runoff, but less is known about unmanaged ecosystems. Research needs to determine how variation in the global water cycle affects hydrologic processes at the watershed scale, and how changes in hydrologic processes in watersheds influence ecosystem processes (Lucier et al. 2006). The hydrologic cycle has critical roles in providing feedbacks between organisms and the physical environment.

Some people question whether watershed studies warrant the large capital investment. The capital investment includes equipment and maintenance, personnel to maintain equipment and process data, and the time and effort to effect technology transfer. There is also the loss of undisturbed or control watersheds as development or other land uses encroach on these watersheds. Few areas are not subject to these increasing pressures. Efforts should be made to continue monitoring on undisturbed or control watersheds, with clear study objectives. Similarly, the treatment watershed is changing over time with site revegetation, nutrient cycling changes, and forest stand development influencing hydrologic processes and rates.

Some may argue that the hydrological processes in wildland watersheds are completely understood, and now computer simulations can replace physical studies (DeWalle 2003). Yet others would argue that the fundamental mechanisms of streamflow generation are still not understood completely, and thus the ability to simulate hydrological processes in complex and heterogeneous terrain requires additional data collection. Process representations, while not completely understood or accompanied by complete datasets, are slow to make their way into watershed models and are often abbreviated. In particular, the subsurface is often generalized, and the larger scale modeling efforts have less certainty.

How individual land use practices aggregate to a watershed scale response remains unanswered (Thomas 1990). The cumulative effects of timber harvesting may be assessed by synergistically supplementing experimental results into numerical simulations (Ziemer et al. 1991; Alila and Beckers 2001; Dunne 2001). Hydrologic models can filter out the effects of climate variability (Bowling et al. 2000) and link land use practices to physical processes (Beckers and Alila 2004) if those processes are suitably represented. But usually measurements directly characterizing land uses are unavailable e.g. maps of land use vegetation, soil type, soil moisture, and groundwater. Recent efforts have been geared to link specific field studies with specific watershed models. A rather simple approach was used to model the effects of forest roads on watershed response based on flow interception and channel network routing (Bowling and Lettenmaier 2001; Wemple and Jones 2003).

Earlier watershed studies tended not to look specifically at large flood events, due in part to the low frequency of occurrence of such events and possibly to avoid issues that were politically sensitive (DeWalle 2003). Not unexpectedly, forest industry would be sensitive to research results that might limit management activities. Recent analyses may have placed too great a reliance on percentage changes in flows without consideration of the downstream impact for extreme events (Thomas and Megahan 1998; Beschta et al. 2000; Jones 2000). Given the complex interaction between streamflow and the type of timber harvesting, watershed area harvested, location of area harvested, road design, construction and maintenance, site preparation and revegetation, precipitation events before and after treatment especially the low recurrence interval events an accurate assessment may be difficult (or impossible) to attain from reanalysis of existing data alone (DeWalle 2003). Aside from watershed studies, monitoring programs on forest lands are typically too brief to sample the variability of natural and disturbed hydrologic regimes (Dunne 2001). Purely empirical resolutions to questions such as whether timber harvest increases the magnitude of peakflows will always suffer from a small sample size. Current forest hydrology models are attempting to address the influence of low recurrence interval events on watershed responses.

The Alsea Watershed Study contributed substantially to knowledge of the life history and biology of coho salmon and cutthroat trout. Later work in the New Alsea Watershed Study further identified the variability in salmonid populations over time, and identified habitat characteristics associated with these populations. Given the variability of salmonid populations in control and managed watersheds, it is recommended that fish populations and biomass be monitored for at least 20 years. Such an opportunity also exists in Canada. A single-watershed, intensive case study of Carnation Creek on Vancouver Island, British Columbia has generated the longest series of continuous data on fish-forestry interactions. The study incorporates preharvest (1970–1975), during-harvest (1976–1981 when 41% of the basin was logged) and post-harvest (1982-present) observations. The project was designed initially to examine the effects of progressive clearcutting and three different types of streamside forest harvest treatments on stream channels and fish populations (Chamberlin 1988; Hogan et al. 1998). Current research is focused on the mechanisms and rates of natural resources recovery by quantifying long term changes in biological and physical watershed processes as the second forest grows. Monitoring efforts include attributes of the hydrologic regime, hillslopes, channel stream networks, aquatic habitats, and fish populations (Tschaplinksi 2006).

Other than aquatic habitat suitability indices, there are few models available to predict or estimate fish populations or biomass. There are no models that link land use activities to fish population dynamics. Although Flynn Creek is undisturbed, the productivity for anadromous salmonids of this and other coastal watersheds may not represent their historic potential. Ongoing salmon harvest, commercial and sport, has reduced the number of spawning fish returning to the watersheds. And the historic decline of beaver because of widespread trapping has decreased salmonid habitat in the form of pools. Beaver populations have been seen to vary even over this study period. The concrete fish traps in each of the study watersheds have only been maintained to be kept free of large woody material; the traps themselves have resulted in channel downcutting and may interfere with passage of spawning salmon. To address this concern, all the fish trap structures were removed in the summer of 2005. A biological response would not be unexpected, but care must be exercised to not confuse this with treatment effects. The biological processes regulating salmonid productivity are known, but coupling these to hydrological processes, at the channel, hillslope and watershed scale is lacking. Better understanding of the hydrological processes of how water travels from the hillslope to the channel will improve modeling of water quality and quantity; thus land use planning decisions can be assessed for resource sustainability.

Hydrologic processes create the template for biologic processes. Aquatic habitats are inventoried and departures from the undisturbed condition documented. The natural or historical range of variability in water resources and fisheries needs to be better understood to assess land use activity effects on these resources. Habitat heterogeneity may be more important in providing aquatic habitat, but appropriate metrics have yet to be developed. Understanding these linkages should have the highest research priority. The economic, recreational, cultural and even social importance of salmon in the Pacific Northwest has become even more acute. A better understanding of salmonid habitat requirements is needed to better identify management opportunities from habitat restoration to land use activities, and more informed management decisions made. Aquatic habitat conservation plans or strategies strive to maintain and restore spatial and temporal connectivity within and between watersheds. However, the cadastral system often does not follow watershed boundaries, and management activities are done on an ownership basis instead. This approach is not sustainable.

Improved understanding of the linkage between land use practices and hydrologic responses, both short- and long-term are needed for sustainable management. Natural resource management often tends to be resource specific and does not address the interconnectedness of resource management. Long-term monitoring is needed for assessment of hydrologic response. The effects of land use practices can only be evaluated in the context of the monitoring period.

Silvicultural prescriptions that include timber harvesting continue to evolve, but few studies have assessed the effects of these new practices on natural resources. Specifically most of today's timber harvest comes from intensively managed, regenerated second-growth forests on private lands. The New Alsea Watershed Study provides the opportunity to assess the effects of new timber harvesting practices on water and fishery resources. Needle Branch will be harvested as two separate units over the next 5 years. The first unit will harvest the middle-third of the watershed, from the private land up to the fish/no fish delineation. The second unit will be harvesting the upper portion of the watershed.

A nested-watershed approach will be used on Needle Branch. A second stream gauging station has been established in the upper watershed at the point where anadromous fish are not expected to be present due to channel gradient and discontinuous habitat during low flow conditions. The two gauging stations will be operated simultaneously for 2 years to develop a pretreatment relationship. The Needle Branch watershed will be harvested as two units. The first unit between the gauging stations will be harvested in 2 years, and the upper watershed unit will be harvested in 5 years. This will be an opportunity to assess the effects of contemporary forest management practices in regenerated forest on water and fishery resources. Hopefully, this new study will be as successful as the original Alsea Watershed Study.

Literature Cited

Alila, Y., and Beckers, J. 2001. Using numerical modeling to address hydrologic forest management issues in British Columbia. Hydrol. Process. 15:3371–3387.

Bates, C.G., and Henry, A.J. 1928. Forest and streamflow experiment at Wagon Wheel Gap, Colorado. U.S. Weather Bureau Monthly Weather Review. 79pp.

Beckers, J., and Alila, Y. 2004. A model of rapid preferential hillslope runoff contributions to peak flow generation in a temperate rain forest watershed. Water Resour. Res. W03501 10. 1029/2003WR002582.

Beschta, R.L., Pyles, M.R., Skaugset, A.E., and Surfleet, C.G. 2000. Peakflow responses to forest practices in the western Cascades of Oregon, USA. J. Hydrol. 233:102–120.

Bosch, J.M., and Hewlett, J.D. 1982. A review of catchment experiments to determine the effect of vegetation changes on water yield and evapotranspiration. J. Hydrol. 55:3–23.

Bowling, L.C., and Lettenmaier, D.P. 2001. The effect of forest roads and harvest on catchment hydrology in a mountainous maritime environment, pp. 145–164. In: M. Wigmosta and S. Burges, editors. Land Use and Watersheds: Human Influence on Hydrology and Geomorphology in Urban and Forest Areas. Water and Science Application 2. American Geophysical Union, Washington, DC.

Bowling, L.C., Storck, P., and Lettenmaier, D.P. 2000. Hydrologic effects of logging in Western Washington, United States. Water Resour. Res. 36:3223–3240.

Callaham, R.Z. 1990. Guidelines for management of wildland watershed projects. Wildland Resources Center, Univ. of California, Berkeley, CA. 36pp.

Chamberlin, T.W., editor. 1988. Proceedings of the Workshop: Applying 15 years of Carnation Creek Results. Workshop held January 13–15, 1987. Pacific Biological Station, Nanaimo, BC. 239pp.

DeWalle, D.R. 2003. Forest hydrology revisited. Hydrologic Processes 17:1255–1256.

Dunne, T. 2001. Introduction: problems in measuring and modeling the influences of forest management on hydrologic and geomorphic processes, pp. 77–86. In: M. Wigmosta and S. Burges, editors. Land Use and Watersheds: Human Influence on Hydrology and

Geomorphology in Urban and Forest Areas. Water and Science Application 2. American Geophysical Union, Washington, DC.

Hibbert, A.R. 1967. Forest treatment effects on water yields. In: W.E. Sooper and H.W. Lull, editors. Interational Symposium of Forest Hydrology. Pergamon Press, Oxford.

Hogan, D.L., Tschaplinski, P.J., and Chatwin, S. 1998. Carnation Creek and Queen Charlotte Islands Fish/Forestry Workshop: Applying 20 Years of Coastal Research to Management Solutions. Ministry of Forests Research Program, Victoria, BC. 41pp.

Jones, J.A. 2000. Hydrologic processes and peak discharge response to forest removal, regrowth, and roads in 10 small experimental basins, western Cascades, Oregon. Water Resour. Res. 36:2621–2642.

Lucier, A., Palmer, M., Mooney, H., Nadelhoffer, K., and coauthors. 2006. Ecosystems and Climate Change: Research Priorities for the U.S. Climate Change Science Program. Recommendations from the Scientific Community. Report on an Ecosystem Workshop, prepared for the Ecosystems Interagency Working Group. Special Series No. SS-92-06. Univ. of Maryland Center for Environmental Science, Chesapeake Biological Laboratory, Solomons, MD. 50pp.

Sidle, R.C. 2006. Field observations and process understanding in hydrology: essential components in scaling. Hydrol. Process. 20:1439–1445.

Stednick, J.D. 1996. Monitoring the effects of timber harvest on annual water yields. J. Hydrol. 176:79–95.

Stednick, J.D., Troendle, C.A., and Ice, G.G. 2004. Chapter 13. Lessons for watershed research in the future, pp. 277–287. In: G.G. Ice, and J.D. Stednick, editors. A Century of Forested Wildland Watershed Lessons. Soc. Amer. Foresters, Bethesda, MD.

Thomas, R.B. 1990. Problems in determining the return of a watershed to pretreatment conditions: techniques applied to a study at Caspar Creek, California. Water Resour. Res. 26:2079–2087.

Thomas, R.B., and Megahan, W.F. 1998. Peak flow responses to clearcutting and roads in small and large basins, western Cascades, Oregon: a second opinion. Water Resour. Res. 34:3393–3403.

Troendle, C.A., Nankervis, J.M., and Porth, L.S. 2003. The impact of Forest Service activities on the stream flow regime in the South Platte River. Report submitted to USDA Forest Service, Rocky Mountain Region, Lakewood, CO. 50pp.

Troendle, C.A., and Stednick, J.D. 1999. Discussion of "Effects of basin scale timber harvest on water yield and peak streamflow". J. Amer. Water Resour. Assoc 35:177–181.

Tschaplinksi, P.J. 2006. Project results span 35 years. For. Res. Ext. Partnership 8(2):12–13.

Wemple, B.C., and Jones, J.A. 2003. Runoff production on forest roads in a steep, mountain catchment. Water Resour. Res. 39:1220.

Ziemer, R.R., Lewis, J., Rice, R.M., and Lisle, T.M. 1991. Modeling the cumulative watershed effects of forest management strategies. J. Environ. Qual. 20:36–42.

Ziemer, R.R., and Ryan, D.F. 2000. Current status of experimental paired-watershed research in the USDA Forest Service. EOS, Transactions, American Geophysical Union 81(48).

Index

Ecological Studies

Volumes published since 2001

Printed in the United States of America